Catalysts for Alcohol-Fuelled Direct Oxidation Fuel Cells

RSC Energy and Environment Series

Editor-in-Chief:
Professor Laurence Peter, University of Bath, UK

Series Editors:
Professor Heinz Frei, Lawrence Berkeley National Laboratory, USA
Professor Ferdi Schüth, Max Planck Institute for Coal Research, Germany
Professor Tim S. Zhao, The Hong Kong University of Science and Technology, Hong Kong

Titles in the Series:
1: Thermochemical Conversion of Biomass to Liquid Fuels and Chemicals
2: Innovations in Fuel Cell Technologies
3: Energy Crops
4: Chemical and Biochemical Catalysis for Next Generation Biofuels
5: Molecular Solar Fuels
6: Catalysts for Alcohol-Fuelled Direct Oxidation Fuel Cells

How to obtain future titles on publication:
A standing order plan is available for this series. A standing order will bring delivery of each new volume immediately on publication.

For further information please contact:
Book Sales Department, Royal Society of Chemistry, Thomas Graham House, Science Park, Milton Road, Cambridge, CB4 0WF, UK
Telephone: +44 (0)1223 420066, Fax: +44 (0)1223 420247
Email: booksales@rsc.org
Visit our website at www.rsc.org/books

Catalysts for Alcohol-Fuelled Direct Oxidation Fuel Cells

Edited by

Zhen-Xing Liang
South China University of Technology, Guangzhou, China

Tim S. Zhao
The Hong Kong University of Science and Technology, Hong Kong

RSC Publishing

RSC Energy and Environment Series No. 6

ISBN: 978-1-84973-405-9
ISSN: 2044-0774

A catalogue record for this book is available from the British Library

Published by The Royal Society of Chemistry,
Thomas Graham House, Science Park, Milton Road,
Cambridge CB4 0WF, UK

Registered Charity Number 207890

For further information see our web site at www.rsc.org

Printed in the United Kingdom by Henry Ling Limited, Dorchester, DT1 1HD, UK

Preface

Energy and environment issues are of paramount importance to achieve the sustainable development of our society. Alcohol-fuelled direct oxidation fuel cells (DOFCs), as a clean and highly efficient energy harvesting engine, have attracted intensive research activities over recent decades. Catalysts are the heart material that determines the performance of DOFCs. The rapid advances in electrocatalysts, particularly nano-sized ones, have left current information available only in scattered journals. To be truly useful to both present and future researchers in the field, this book is intended to devote an insightful review of the reaction nature, systematically summarize the recent advances in nanocatalysts, and convey a more global perspective. Contributions by leading experts will serve as a central source of reference for the fundamentals and applications of the electrocatalysts, establishing the state of the art, disseminating the latest research discoveries, and providing a potential textbook to senior undergraduate and graduate students.

Chapter 1 provides an informative summary on the electrocatalysis of the alcohol oxidation reactions and the platinum-based electrocatalysts. Chapters 2–6 deal with the recent advances in both the fundamental understanding and the material science in DOFCs. Special emphasis is placed on the newly emerging nanocatalysts developed over the past several years, including the novel nanostructured electrocatalysts, the gold-leaf-based nanocatalysts, the palladium-based nanocatalysts, bioelectrocatalysts, and correlation of the "structure–activity" relationship as well. Based on the discussion, the challenges and perspectives of the nanocatalysts in alcohol-fuelled DOFCs are extensively discussed in Chapter 7. In addition to typical alcohol oxidation reactions, recent developments of nanocatalysts for other fuel (formic acid, borohydride, sugars, *etc.*) oxidation reactions are also included in this book.

RSC Energy and Environment Series No. 6
Catalysts for Alcohol-Fuelled Direct Oxidation Fuel Cells
Edited by Zhen-Xing Liang and Tim S. Zhao

Published by the Royal Society of Chemistry, www.rsc.org

The editorial board expresses their appreciation to the contributing authors, who have set the high standards for *Catalysts for Alcohol-Fuelled Direct Oxidation Fuel Cells*. Last, but not least, the editors acknowledge the efforts of the professional staff at the Royal Society of Chemistry for providing invaluable editorial assistance.

T. S. Zhao
The Hong Kong University of Science & Technology, Hong Kong, China

Contents

RSC Energy and Environment Series No. 6
Catalysts for Alcohol-Fuelled Direct Oxidation Fuel Cells
Edited by Zhen-Xing Liang and Tim S. Zhao

Published by the Royal Society of Chemistry, www.rsc.org

CHAPTER 1

Electrocatalysis of Alcohol Oxidation Reactions at Platinum Group Metals

CLAUDE LAMY*[a] AND CHRISTOPHE COUTANCEAU[a,b]

[a] Groupe de Recherches du CNRS no. 3339 (PACS), University of Montpellier, 2, place Eugène Bataillon, 34095 Montpellier, France;
[b] Université de Poitiers, IC2MP, UMR CNRS 7285, 4 rue Michel Brunet, 86022, Poitiers cedex, France
*E-mail: Claude.Lamy@iemm.univ-montp2.fr

1.1 Introduction

Electrocatalysis, *i.e.* the heterogeneous catalysis of electrochemical reactions occurring at the electrode/electrolyte interface, has mainly concerned technological investigations related either to energy storage (*e.g.* water electrolysis) or to energy conversion (*e.g.* fuel cells).[1,2] However, except for the hydrogen electrode, which is now well known, and the oxygen electrode, which has been extensively studied, other electrodes of practical interest, such as soluble fuel electrodes, need much more investigation. Among them, alcohol electrodes are particularly suitable for use in a direct oxidation fuel cell (DOFC) because of several favourable features, such as a high theoretical energy density (4–9 kWh kg^{-1} compared to 33 kWh kg^{-1} for molecular hydrogen) and a great facility of handling.

Moreover, alcohols, which may be produced from the biomass, are very interesting fuels due to a lot of advantages: high solubility in aqueous

RSC Energy and Environment Series No. 6
Catalysts for Alcohol-Fuelled Direct Oxidation Fuel Cells
Edited by Zhen-Xing Liang and Tim S. Zhao

Published by the Royal Society of Chemistry, www.rsc.org

electrolytes, relatively high reactivity, ease of storage and supply, low toxicity (except for methanol). They can be directly electro-oxidized in a direct alcohol fuel cell (DAFC).This explains why many fundamental investigations were undertaken in the last three decades on the electrochemical oxidation of several alcohols: methanol,[3–5] ethanol,[6–8] ethylene glycol,[6,9] glycerol,[10] propanol[11] and butanol,[12,13] and also on related compounds: formic acid,[14–16] formaldehyde,[17] carbon monoxide,[18] *etc.* Until now, the most promising and most studied fuels for application in a DOFC, with the direct oxidation of the organic molecule, are alcohols such as methanol and ethanol.

However, a lot of electrocatalytic problems still arise due to the relative complexity of the reaction mechanisms. These include the effect of the nature of the reaction products, the structure of the adsorbed intermediates, the nature and structure of the electrode material, the molecular structure of the organic compounds, the pH and the anions of the supporting electrolyte, and the role of the water adsorption residues.

Furthermore, the catalyst structure, such as the particle size, the composition and the degree of alloying, are also of great importance, since most of the alcohol oxidation reactions are structure sensitive and because it is of great interest to reduce the platinum group metal loading (either by metal dispersion or by synthesis of multimetallic catalysts) in order to reduce the cost of the DAFC system. These topics will be illustrated mainly with results obtained in this laboratory.

1.2 Thermodynamics and Kinetics of Alcohol Oxidation Reactions

In a DOFC the total electro-oxidation to CO_2 of an aliphatic oxygenated compound C_xH_yO containing one oxygen atom (mono-alcohols, aldehydes, ketones, ethers, *etc.*) involves the participation of water (H_2O) or of its adsorbed residue (OH_{ads}) provided by the cathodic reaction (electro-reduction of dioxygen).

The overall electro-oxidation reaction in acid medium to reject the carbon dioxide produced can thus be written as follows:

$$C_xH_yO + (2x-1)H_2O \rightarrow xCO_2 + nH^+ + ne^- \tag{1.1}$$

with $n = 4x + y - 2$. Such an anodic reaction is very complicated from a kinetics point of view since it involves multielectron transfers and the presence of different adsorbed intermediates and several reaction products and by-products. However, from thermodynamic data it is easy to calculate the reversible anode potential, the cell voltage under standard conditions, the theoretical energy efficiency and the energy density.

1.2.1 Thermodynamic Data

According to reaction (1.1) the standard Gibbs energy change $-\Delta G_1^{\circ}$, allowing calculation of the standard anode potential, $E_1^{\circ} = \frac{-\Delta G_1^{\circ}}{nF}$, can be evaluated from the standard energy of formation ΔG_i^{f} of reactant i:

$$-\Delta G_1^\circ = x\Delta G^f_{CO_2} - \Delta G^f_{C_xH_yO} - (2x-1)\Delta G^f_{H_2O} \quad (1.2)$$

In the cathodic compartment the electro-reduction of oxygen occurs, as follows:

$$1/2 O_2 + 2H^+ + 2e^- \rightarrow H_2O \quad (1.3)$$

with $\Delta G_2^\circ = \Delta G^f_{H_2O} = -237.1$ kJ mol^{-1}, leading to a standard cathodic potential, E_2°:

$$E_2^\circ = E^\circ_{O_2} = -\frac{\Delta G_2^\circ}{2F} = \frac{237.1\times 10^3}{2\times 96485} = 1.229 \text{ V } vs. \text{ SHE} \quad (1.4)$$

where SHE is the standard hydrogen electrode, acting as a reference electrode.
In the fuel cell the electrical balance corresponds to the complete combustion of the organic compound in the presence of oxygen, as follows:

$$C_xH_yO + (x + \frac{y}{4} - \frac{1}{2})O_2 \rightarrow xCO_2 + \frac{y}{2}H_2O \quad (1.5)$$

with $\Delta G_r^\circ = (2x + \frac{y}{2} - 1)\Delta G_2^\circ - \Delta G_1^\circ = x\Delta G^f_{CO_2} + \frac{y}{2}\Delta G^f_{H_2O} - \Delta G^f_{C_xH_yO}$, leading to the equilibrium standard cell voltage:

$$E_{eq}^\circ = -\frac{\Delta G_r^\circ}{nF} = -\frac{\Delta G_2^\circ}{2F} + \frac{\Delta G_1^\circ}{nF} = E_2^\circ - E_1^\circ = E^\circ_{O_2} - E^\circ_{alcohol} \quad (1.6)$$

Then it is possible to evaluate the specific energy W_e in kWh kg^{-1}:

$$W_e = \frac{(-\Delta G_r^\circ)}{3600\,M} \quad (1.7)$$

with M the molecular mass of the compound. Knowing the enthalpy change ΔH° from thermodynamic data:

$$\Delta H_r^\circ = (2x + \frac{y}{2} - 1)\Delta H_2^\circ - \Delta H_1^\circ = x\Delta H^f_{CO_2} + \frac{y}{2}\Delta H^f_{H_2O} - \Delta H^f_{C_xH_yO} \quad (1.8)$$

one may calculate the reversible energy efficiency under standard conditions:

$$\varepsilon_{rev} = \frac{\Delta G_r^\circ}{\Delta H_r^\circ} \quad (1.9)$$

For the oxidation of methanol and ethanol, the corresponding equations are:

$$CH_3OH + H_2O \rightarrow CO_2 + 6H^+ + 6e^- \qquad E^\circ_{MeOH} = 0.016 \text{ V } vs. \text{ SHE} \tag{1.10}$$

$$CH_3CH_2OH + 3H_2O \rightarrow 2CO_2 + 12H^+ + 12e^- \qquad E^\circ_{EtOH} = 0.085 \text{ V } vs. \text{ SHE} \tag{1.11}$$

whereas the electro-reduction reaction of molecular oxygen occurs at the cathode:

$$O_2 + 4H^+ + 4e^- \rightarrow 2H_2O \qquad E^\circ_{O_2} = 1.229 \text{ V } vs. \text{ SHE} \tag{1.12}$$

where $E^\circ{}_i$ are the standard electrode potentials *vs.* SHE.

This corresponds to the overall combustion reaction of alcohols in oxygen:

$$CH_3OH + 3/2O_2 \rightarrow CO_2 + 2H_2O \qquad E^\circ_{O_2/MeOH} = 1.213 \text{ V} \tag{1.13}$$

$$CH_3CH_2OH + 3O_2 \rightarrow 2CO_2 + 3H_2O \qquad E^\circ_{O_2/EtOH} = 1.145 \text{ V} \tag{1.14}$$

where the cell voltages are calculated under standard conditions.

For higher alcohols, such as *n*-propanol, taken as an example, the following calculations can be made:

$$C_3H_7OH + 5H_2O \rightarrow 3CO_2 + 18H^+ + 18e^- \tag{1.15}$$

$$\begin{aligned} -\Delta G^\circ_1 &= 3\Delta G^f_{CO_2} - \Delta G^f_{C_3H_7OH} - 5\Delta G^f_{H_2O} \\ &= -3 \times 394.4 + 168.4 + 5 \times 237.1 = 171 \text{ kJ mol}^{-1} \end{aligned} \tag{1.16}$$

$$E^\circ_1 = -\frac{\Delta G^\circ_1}{18F} = \frac{171 \times 10^3}{18 \times 96485} = 0.098 \text{ V } vs. \text{ SHE} \tag{1.17}$$

$$C_3H_7OH + 9/2O_2 \rightarrow 3CO_2 + 4H_2O \tag{1.18}$$

$$\begin{aligned} \Delta G^\circ_r &= 9\Delta G^\circ_2 - \Delta G^\circ_1 = 3\Delta G^f_{CO_2} + 4\Delta G^f_{H_2O} - \Delta G^f_{C_3H_7OH} \\ &= -3 \times 394.4 - 4 \times 237.1 + 168.4 = -1963 \text{ kJ mol}^{-1} \end{aligned} \tag{1.19}$$

The standard cell voltage is thus:

$$\begin{aligned} E^\circ_{eq} &= -\frac{\Delta G^\circ_r}{18F} = \frac{1963 \times 10^3}{18 \times 96485} = \frac{237.1 \times 10^3}{2 \times 96485} - \frac{171 \times 10^3}{18 \times 96485} \\ &= 1.229 - 0.098 = 1.131 \text{ V} \end{aligned} \tag{1.20}$$

and the specific energy is:

$$W_e = \frac{1963}{3600 \times 60} = 9.09 \text{ kWh kg}^{-1} \tag{1.20a}$$

The enthalpy change of reaction (1.18) is:

$$\Delta H_r^\circ = -3 \times 395.5 - 4 \times 285.8 + 302.6 = -2027 \text{ kJ mol}^{-1} \tag{1.21}$$

so that the reversible energy efficiency is :

$$\varepsilon_{rev} = \frac{\Delta G_r^\circ}{\Delta H_r^\circ} = \frac{1963}{2027} = 0.968 \tag{1.22}$$

Similar calculations can be made with many oxygenated fuels, including polyols (ethylene glycol, glycerol), propargyl alcohol, ethers and polyethers [dimethyl ether, CH_3OCH_3, ethyl methyl ether, $CH_3OC_2H_5$, diethyl ether, $C_2H_5OC_2H_5$, dimethoxymethane, $(CH_3O)_2CH_2$, trimethoxymethane, $(CH_3O)_3CH$, trioxane, $(CH_2O)_3$].

The energy density of the fuel, W_e, the cell voltage of the cell at equilibrium, E°_{eq}, and the reversible energy efficiency of the cell, ε_{rev}, for several alcohols can be calculated under standard conditions (25 °C, liquid phase). The results are summarized in Table 1.1.

For all oxygenated compounds listed in Table 1.1, the cell voltage varies from 1.2 to 1.0 V, which is very similar to that of a hydrogen/oxygen fuel cell (E°_{eq} = 1.23 V). The energy density varies between half to one of that of gasoline (10–11 kWh kg^{-1}), so these compounds are good alternative fuels to hydrocarbons. Furthermore, the reversible energy efficiency ε_{rev} is close to 1, while that of the H_2/O_2 fuel cell is 0.83 at 25 °C (standard conditions). From these data, it appears that amongst the mono-alcohols, methanol and ethanol lead to higher cell voltages and reversible energy efficiency under standard conditions.

Table 1.1 Thermodynamic data associated with the electrochemical oxidation of some alcohols (under standard conditions).

Fuel	$\Delta G_1^\circ/kJ\ mol^{-1}$	E_1°/V vs. SHE	$\Delta G_r^\circ/kJ\ mol^{-1}$	E°_{cell}/V	$W_e/kWh\ kg^{-1}$	$\Delta H_r^\circ/kJ\ mol^{-1}$	ε_{rev}
CH_3OH	−9.3	0.016	−702	1.213	6.09	−726	0.967
C_2H_5OH	−97.3	0.084	−1325	1.145	8.00	−1367	0.969
C_3H_7OH	−171	0.098	−1963	1.131	9.09	−2027	0.968
n-C_4H_9OH	−409	0.177	−2436	1.052	9.14	−2676	0.910
CH_2OH–CH_2OH	−25.5	0.026	−1160.8	1.203	5.20	−1189	0.976
CH_2OH–$CHOH$–CH_2OH	1	−0.01	−1661.6	1.230	5.02	−1650	0.993

1.2.2 Kinetics Problems

The electro-oxidation of aliphatic oxygenated compounds, even the simplest one, *i.e.* methanol, involves the transfer of many electrons ($n = 6$ for methanol). The reaction mechanism is thus complex, the oxidation reaction occurring through many successive and parallel paths involving many adsorbed intermediates and by-products. The oxidation reaction needs a convenient electrocatalyst to increase the reaction rate and to modify the reaction pathway in order to reach more rapidly the final step, *i.e.* the production of carbon dioxide. The relative slowness of the reaction, and the difficulty to break the C–C bond at low temperatures (25–80 °C), lead to high anodic overvoltages η_a, which will greatly reduce the operating cell voltage (Figure 1.1).

Thus, the practical electrical efficiency of a fuel cell is dependent on the current that is delivered by the cell and is lower than that of the reversible efficiency due to the irreversibility of the electrochemical reactions involved at the electrodes. The practical efficiency of a fuel cell can be expressed as follows:

$$\varepsilon_{cell} = \frac{n_{exp} \times F \times E(j)}{-\Delta H_r^\circ} = \frac{nFE_{eq}^\circ}{(-\Delta H_r^\circ)} \times \frac{E(j)}{E_{eq}^\circ} \times \frac{n_{exp}}{n} = \varepsilon_{rev} \times \varepsilon_E \times \varepsilon_F \quad (1.23)$$

with the cell voltage $E(j) = E^\circ_{eq} - (|\eta_a| + |\eta_c| + R_e j)$ at a current density j and R_e the electrolyte and interfacial specific resistances.

From eqn (1.23) it follows that the increase of the practical fuel cell efficiency can be achieved by increasing the voltage efficiency ($\varepsilon_E = E(j)/E^\circ_{eq}$) and the faradic efficiency ($\varepsilon_F = n_{exp}/n$), the reversible yield being fixed by the thermodynamic data.

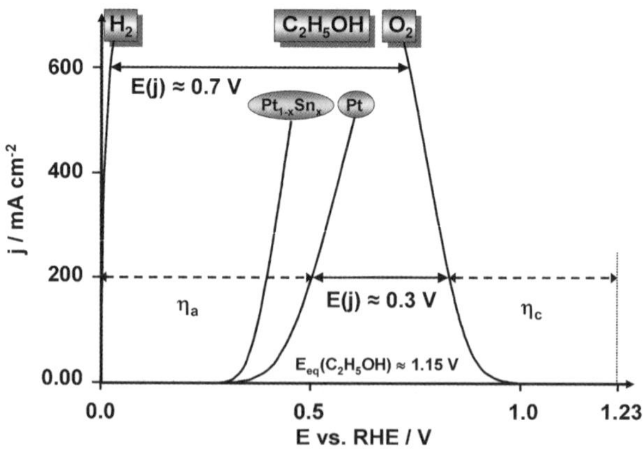

Figure 1.1 Current density (j) *vs.* electrode potential (E) curves for H_2 and EtOH electro-oxidation at different Pt-based catalytic anodes, and oxygen electro-reduction at a Pt cathode.

For a given electrochemical system the increase of the voltage efficiency is directly related to the decrease of the overpotentials of the oxygen reduction reaction, η_c, and alcohol oxidation reaction, η_a, which needs to enhance the activity of the catalysts at low overpotentials and low temperature, whereas the increase of the faradic efficiency is related to the ability of the catalyst to oxidize completely (or not) the fuel into carbon dioxide, *i.e.* it is related to the selectivity of the catalyst. Indeed, in the case of ethanol, taken as an example, acetic acid and acetaldehyde are formed at the anode,[7] which corresponds to the number of electrons involved as 4 and 2, respectively, against 12 for the complete oxidation of ethanol to carbon dioxide. The enhancement of both these efficiencies is a challenge in electrocatalysis.

In most cases, η_a is at least 0.5 V for a reasonable current density (100 mA cm^{-2}) so that the cell voltage, including an overvoltage $\eta_c = 0.3$ V for the cathodic reaction, will be of the order of 0.4 V, and the voltage efficiency will be $\varepsilon_E = 0.4/1.2 = 0.33$ under the operating conditions. Such a drawback of the direct alcohol fuel cell can only be removed by improving the kinetics of the electro-oxidation of the fuel. This needs a relatively good knowledge of the reaction mechanisms, particularly of the rate-determining step, and to search for electrode materials (Pt–X binary and Pt–X–Y ternary electrocatalysts) with improved catalytic properties.

1.3 Preparation and Physicochemical Characterization of Platinum-Based Nanocatalysts

Several methods, developed to synthesize platinum-based nanocatalysts, with control of the particle size, atomic composition and metal loading, are shown

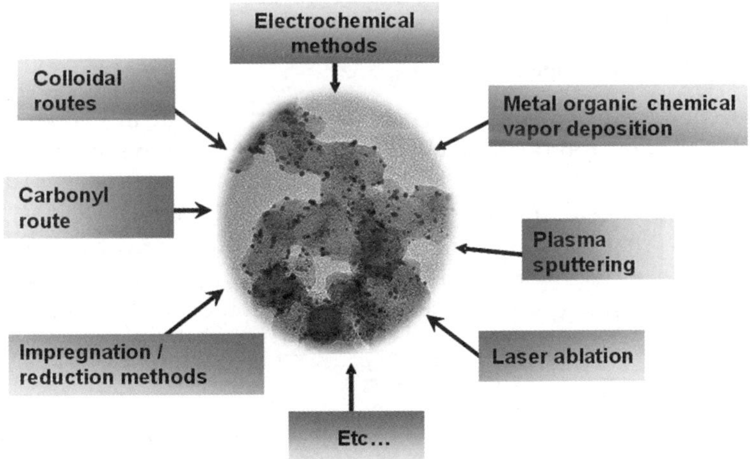

Figure 1.2 Different methods to synthesize platinum-based nanocatalysts with control of the particle size, atomic composition and metal loading.

in Figure 1.2: chemical methods (impregnation–reduction, colloidal, carbonyl complex route), physicochemical methods [electrochemical pulse deposition, metal organic chemical vapour deposition (MOCVD)] and physical methods [plasma enhanced physical vapour deposition (PVD), laser ablation], *etc*. Some of these methods will be developed further.

1.3.1 Synthesis by Chemical Methods

1.3.1.1 Impregnation–Reduction Methods

The co-impregnation–reduction method is extensively used in the domain of heterogeneous catalysis for the synthesis of supported metallic nanoparticles.[19] In the case of fuel cell anodes, a carbon support (Vulcan XC-72 previously heated at 400 °C for 4 h under a pure nitrogen atmosphere) is used. This support is first oxidized by contact with *aqua regia* (solution of $^{1}/_{3}HNO_3$ + $^{1}/_{3}HCl$ + $^{1}/_{3}H_2O$) for 5 min in order to form surface functional groups (hydroxyl and carboxylic acid groups) which will be further used to disperse the metal by exchanging protons with the metal salts. The carbon powder is then put in an aqueous solution (ultrapure water from Millipore, MilliQ 18 MΩ cm) containing chloroplatinic acid (H_2PtCl_6 from Alfa Aesar) and other metal salts ($SnCl_4$ and $ReCl_3$ from Sigma, as examples), and the pH is adjusted to about 1. After 24 h, the metallic catalysts are deposited on the carbon substrate, and the mixture is dried overnight at 70 °C in an oven.

The resulting powder is then calcined in air for 4 h at 300 °C and reduced in a pure hydrogen atmosphere for 4 h at 300 °C. PtSn/C and PtRe/C catalytic powders with different atomic ratios can be prepared in this way with a metal loading close to 30 wt% on carbon.[20,21]

1.3.1.2 Colloidal Methods

1.3.1.2.1 Bönnemann method

The procedure described by Bönnemann *et al.*[22,23] was slightly modified and adapted to the preparation of mono- and multimetallic catalyst precursors.[24,25] The synthesis is carried out under a controlled atmosphere (argon) free of oxygen and water, with non-hydrated metal salts (99.9% $PtCl_2$, $RuCl_3$, $SnCl_2$, *etc.*). The first step consists of the preparation of a tetraalkyl triethylborohydride reducing agent $(Nalk_4)^+(BEt_3H)^-$, which will also act as a surfactant after metal reduction, preventing any agglomeration of the metallic particles:

$$MCl_n + n\left[(Nalk_4)^+(BEt_3H)^-\right] \rightarrow M\left[(Nalk_4)^+Cl^-\right]_n + nBEt_3 + {}^{n}/_{2}H_2 \qquad (1.24)$$

The colloidal precursors are then dispersed on a carbon support (*e.g.* Vulcan XC-72) and calcined at 300 °C for 1 h under an air atmosphere to remove the organic surfactant. The Bönnemann method allows different possibilities for the synthesis of multimetallic supported catalysts:

- Synthesis of multimetallic catalysts with a controlled atomic ratio by mixing different metal salts before the reduction step and formation of the precursor colloid (co-reduction method):

$$\begin{aligned} & x\mathrm{PtCl_2} + y\mathrm{MCl}_n{}^{+}(2x+ny)\left[(\mathrm{Nalk_4})^{+}(\mathrm{BEt_3H})^{-}\right] \rightarrow \\ & \mathrm{Pt}_x\mathrm{M}_y\left[(\mathrm{Nalk_4})^{+}\mathrm{Cl}^{-}\right]_{(2x+ny)} + (2x+ny)\mathrm{BEt_3} + (x+ny/2)\mathrm{H_2} \end{aligned} \quad (1.25)$$

- Synthesis of catalysts with controlled atomic ratios by codeposition of different metal colloids before the calcination step and formation of the catalytic powder (codeposition method):

$$x\mathrm{Pt}[(\mathrm{Nalk_4})+\mathrm{Cl}^{-}]_2 + y\mathrm{M}\left[(\mathrm{Nalk_4})^{+}\mathrm{Cl}^{-}\right]_n \xrightarrow{\mathrm{THF,300\,^{\circ}C,air,C}} (\mathrm{Pt}_x + \mathrm{M}_y)/\mathrm{C} \quad (1.26)$$

- Synthesis by simply mixing the different catalytic powders (ball milling method): xPt/C + yM/C.

1.3.1.2.2 Water-in-oil microemulsion method

The "water in oil" microemulsion method[26] is derived from that developed by Boutonnet *et al.*[27] The catalysts are prepared by mixing $NaBH_4$ (99% from Acros Organics), as reducing agent, with a microemulsion carrying the specific reactants dissolved in an aqueous phase (MilliQ Millipore ultrapure water, 18.2 MΩ cm). Poly(ethylene glycol dodecyl ether) (Brij30 from Fluka) was chosen as surfactant and the organic phase was *n*-heptane (99% from Acros). Desired amounts of the metal salts are dissolved in ultrapure water in order to obtain metallic nanoparticles with controlled size and compositions after the reduction process with $NaBH_4$. The following concentrations of reactants are used: [water]/[Brij30] molar ratio = 5, Brij30 = 16.5 wt%, [metal salts] = 0.205 mol L^{-1}, $NaBH_4/Pt^{IV}$ molar ratio = 15.

Carbon (Vulcan XC-72), previously treated under N_2 at 400 °C for 4 h, is added directly to the colloidal solution to obtain the desired metal loading and the mixture is kept under stirring for 2 h. In the present work, all the catalysts were synthesized in order to obtain 40 wt% metal loading. The mixture was filtered on a 0.22 μm Durapore membrane filter (Millipore). The resulting powder was abundantly rinsed with ethanol, acetone and ultrapure water. The carbon-supported catalysts were dried overnight in an oven at 75 °C. This method leads to small particles (2–4 nm diameter) with a narrow size distribution.

1.3.1.2.3 Polyol method

This method, well described by Fievet *et al.*,[28] allows obtaining metal nanoparticles by reduction of metallic salts in ethylene glycol acting as a reducing agent and can be performed without any surfactant. Anode (PtRu/

C)[29] as well cathode (PtCo/C)[30] bimetallic nanostructured catalysts can be prepared by this method according to the following procedure.

$H_2PtCl_6{\cdot}6H_2O$ and MCl_n (99.9% purity, Alfa Aesar) are dissolved in ethylene glycol (puriss. p.a., ≥99.5%, Fluka) with nominal concentrations leading to the desired Pt/M atomic ratio. Then the pH is adjusted to 11 with a 1 M NaOH solution in ethylene glycol. Vulcan XC-72 thermally treated for 4 h at 400 °C under nitrogen (U Quality from l'Air Liquide) is added to the solution and the mixture is ultrasonically homogenized for 5 min. The mixture is then deoxygenated by bubbling pure nitrogen for 15 min and heated at reflux (197.3 °C) for 2 h, maintaining nitrogen bubbling, with the rate of temperature ramp being set at 9.7 °C min^{-1}. After cooling, the particle cleaning step consists of decreasing the pH of the mixture down to pH = 1 by addition of a 0.5 M $HClO_4$ solution, then filtering the reaction mixture and rinsing the catalytic powder with ultrapure water (MilliQ, Millipore, 18.2 MΩ cm). Finally, thermal treatment of the PtM/C powder is performed at 200 °C under air for 1 h to remove the remaining ethylene glycol.

1.3.1.3 Carbonyl Complex Route

Pt/Ni[31] and Pt/Cr[32] alloy nanoparticle catalysts with high metal loadings (20–40 wt%) can be prepared *via* a carbonyl complex route,[33] followed by H_2 reduction at high temperatures. Pt and Ni or Cr carbonyl complexes are synthesized simultaneously using methanol as the solvent through the reaction of Pt and Ni or Cr salts with CO at about 50 °C for 24 h with constant mechanical stirring. The resulting carbonyl complexes are a mixture of Pt and Ni or Cr carbonyl complexes or a Pt–X molecular complex, identified by their infrared spectra.[31] After the synthesis of Pt–X carbonyl complexes, Vulcan XC-72 carbon is added to the mixture under a N_2 gas flow and stirred at about 55 °C for more than 6 h. Subsequently, the solvent is removed and the catalyst powder is subjected to heat treatment at different temperatures under nitrogen and hydrogen, respectively. In the case of Pt/Ni, an alloying temperature of about 300 °C, significantly lower than that given by other authors,[34] is found suitable; it is low enough to be beneficial to the formation of small alloy particles with a narrow particle size distribution and high enough to avoid the formation of a mixture of a pure Pt phase and an alloy phase, as shown by XRD measurements for T = 200 °C. After heat treatment, the sample is washed with ultrapure water until chloride ions are no longer detected and then dried under nitrogen at about 130 °C.

1.3.2 Synthesis by Electrochemical Deposition

PtRu/C catalysts with different particle sizes and compositions can be prepared by galvanostatic pulse electrodeposition from a mixture of metal salts. Given amounts of K_2PtCl_6 and K_2RuCl_5 (99.9% from Alfa Aesar) are dissolved in 1.0 M H_2SO_4 (Suprapur from Merck) solution in ultrapure water (MilliQ

Millipore 18 MΩ cm). The metallic catalysts are deposited on a homemade carbon gas diffusion electrode: 4 mg cm^{-2} of a mixture (Vulcan XC-72 carbon + 15 wt% PTFE) deposited on a carbon cloth (B-1 designation A, 116 g m^{-2}, 0.35 mm thickness from E-TEK division). The galvanostatic pulse deposition of the metals is performed in a two-electrode cell using a high-power potentiostat (Wenking model HP 88) and a waveform generator (Hewlett Packard 33120A), processed by homemade software. More details are given elsewhere.[35] The square wave signal of the current density applied to the electrodes for the reduction and deposition of the metals is the following: 0.1 s (t_{on}) at $j = -20$ mA cm^{-2}, and 2.5 s or 0.3 s (t_{off}) at $j = 0$.

1.3.3 Synthesis by Plasma-Enhanced PVD

Anodic catalysts can also be deposited by plasma sputtering on a diffusion layer.[36,37] To carry out the sputtering, the platinum (and/or ruthenium) target(s) face(s) the diffusion layer in a vacuum reactor. The biased target is sputtered by energetic argon ions created by inductive plasma excited by a planar coil through a dielectric window [transformer coupled plasma (TCP) antenna]. The antenna is powered by a rf generator (13.56 MHz, 1000 W). The Ar working pressure is kept at 0.5 Pa and the target–substrate distance is set at 5.5 cm. Sputtered electrodes of different structures can be obtained either by simultaneous deposition of Ru and Pt or by deposition of Ru on a Pt-coated diffusion layer or alternative depositions of Pt and Ru on the diffusion layer.[38] For Pt and Ru co-deposition the power supply for the Pt target and Ru target are simultaneously and independently adjusted between 100 and 300 W in order to obtain the desired atomic composition of the catalysts. For double-layer and multi-layer PtRu catalysts, a power supply of 200 W is used, leading to a deposition rate of 14 μg cm^{-2} min^{-1} for platinum and 5.5 μg cm^{-2} min^{-1} for ruthenium. The deposition time is modulated in order to obtain the desired atomic compositions.

1.3.4 Physicochemical Characterizations

The catalysts so-prepared were characterized by transmission electron microscopy and energy-dispersive X-ray spectroscopy (EDX) using a JEOL 2100 UHR microscope/EDX analyzer (200 kV) equipped with a LaB_6 filament. Images were taken with a Gatan Ultrascan 2k × 2k camera. The mean particle size and size distributions were determined by measuring the diameter of isolated particles using ImageJ free software,[39] although particle agglomeration is present in all catalysts. Between 200 and 300 particles were considered for each catalyst in order to have an acceptable statistical sample. The microstructure of the catalytic powders was evaluated by X-ray diffraction (XRD). The powder diffraction patterns were recorded on a Bruker D5005 Bragg–Brentano (θ–θ) diffractometer operated with a copper tube powered at 40 kV and 40 mA (unfiltered Cu$K\alpha 1$ = 1.542 Å). Measurements were recorded from $2\theta = 15°$ to $2\theta = 90°$ in step mode, with steps of 0.06° and a fixed acquisition time of 10 s/step. The catalyst composition was determined using an inductive coupling plasma–optical emission spectrometer (ICP-OES, Perkin–Elmer Optima 2000 DV) or EDX.

1.4 Experimental Methods for the Elucidation of Reaction Mechanisms

1.4.1 Cyclic Voltammetry and CO Stripping

The electrochemical measurements were carried out in a standard three-electrode electrochemical cell. The supporting electrolyte was either 0.1 M $HClO_4$ or 0.5 M H_2SO_4 (prepared with concentrated ultrapure acids from Merck Suprapur diluted in ultrapure water). In acidic medium, the reference electrode was a mercury/mercury(I) sulfate electrode in contact with a saturated K_2SO_4 solution, connected through a Luggin capillary to the working electrode compartment, or a reversible hydrogen electrode (RHE). In alkaline medium, the reference electrode was a mercury/mercury(I) oxide electrode in contact with a 1 M NaOH solution, connected through a Luggin capillary to the working electrode compartment, or a RHE. All potentials are referred to that of the RHE.

The activity towards 0.1 M alcohol oxidation of supported Pt-based catalysts is evaluated by cyclic voltammetry at a low sweep rate (5 or 10 mV s^{-1}). For CO stripping, CO is adsorbed at 0.1 V *vs.* RHE for 3 min, and the two first cycles are recorded at 10 or 20 mV s^{-1}. For each experiment the electrolyte is purged and maintained under a pure N_2 atmosphere (U quality from L'Air Liquide).

1.4.2 Infrared Reflectance Spectroscopy

The first *in situ* IR reflectance spectroscopy under electrochemical control of the working electrode in a three-electrode cell was realized by Beden *et al.*[40] using the so-called electrochemically modulated IR reflectance spectroscopy (EMIRS). This technique is based on the electro-modulation of the electrode potential at a low frequency (typically 8 Hz) and the detection of the modulated reflectance signal by a lock-in amplifier. The technique is such that it gives the pseudo first derivative of the IR absorbance signal of the adsorbed species only, since the solution species are not submitted to the modulation (at the same low frequency) of the electrical field in the double layer. Experimental details of this external reflection technique are fully described in textbooks.[41]

Later on, owing to the development of Fourier transform infrared (FTIR) spectroscopy, this technique was adapted to the *in situ* observation of adsorbed species and reaction products at the electrode/electrolyte interface.[42,43] Some important details of these techniques are described below.

The IR cell is equipped with a CaF_2 window and a working electrode connected to a glass rod fitted through the top of the cell (allowing it to be pressed down against the CaF_2 window).[42] A Pt wire is used as counter electrode, and a RHE is used as reference electrode. The cell potential is controlled by using a potentiostat and a waveform generator, connected to an XY recorder, to obtain the current–potential curves. A Bruker IFS 66v FTIR spectrometer was modified for beam reflection on the electrode surface at a 65° incident angle. To remove interference from atmospheric water and CO_2, the beam path is vacuum evacuated. An Infrared Associates liquid N_2-cooled HgCdTe detector is used. The spectral resolution is 4 cm^{-1}. The reproducibility of the spectral acquisition is controlled by repeating each experiment three times, with freshly prepared electrodes.

The working electrode consists of a mirror-polished platinum disc, or a gold disc or a glassy carbon (GC) disc (8 mm diameter), glued to a glass tube as the holder. For nanoparticle powders the catalyst suspension is deposited on the GC disk and the solvent is evaporated under a pure N_2 stream.

In situ IR spectra are recorded using either single potential alteration IR reflectance spectroscopy (SPAIRS),[43] also called linear potential sweep Fourier transform IR reflectance spectroscopy (LPS-FTIRS)[44] or subtractively normalized interfacial Fourier transform IR reflectance spectroscopy (SNIFTIRS).[42]

With the SPAIRS technique,[43] the electrode reflectivity is recorded each 100 mV during the first slow voltammetric scan at 1 mV s^{-1}. Each spectrum results in the co-addition of 128 interferograms. The data acquisition requires 20 s, *i.e.* over ~20 mV. Final spectra are normalized as a change in absorbance, ΔA_n :

$$\Delta A_{\mathrm{n}} = \frac{\Delta A}{N} = -\frac{1}{N}\left(\frac{\Delta R}{R}\right) = -\frac{1}{N}\left(\frac{R_E - R_{E\mathrm{ref}}}{R_{E\mathrm{ref}}}\right) \tag{1.27}$$

where $N = (E - E_{\mathrm{ref}})/100$ mV, R_E is the reflectivity taken during the voltammetric scan, $R_{E\mathrm{ref}}$ that recorded at the most negative potential before cycling, and E and E_{ref} are the two corresponding potentials in mV; N is a number without units. For $R_E > E_{\mathrm{ref}}$, a negative peak means the production of species at the vicinity of the electrode surface.

With the SNIFTIRS technique,[42] the working electrode potential is modulated between two potential values (E_{i} and E_{f}) according to a square wave signal. The reflectivity is obtained at these two electrode potentials (frequency of potential modulation typically 0.025 Hz) and results in the co-addition of 128 interferograms 30 times at each potential. The final spectra are calculated as:

$$\left(\frac{\Delta R}{R}\right) = \left(\frac{R_{E\mathrm{f}} - R_{E\mathrm{i}}}{R_{E\mathrm{i}}}\right) \tag{1.28}$$

where E_{i} is the initial and E_{f} the final potential of the modulation and $\Delta E = E_{\mathrm{f}} - E_{\mathrm{i}}$ = constant. In this case, a negative peak means the production of species and a positive band indicates the consumption of species at the electrode surface. For rough electrodes, no polarization of the light beam is used to carry out the FTIRS experiments.

1.4.3 Differential Electrochemical Mass Spectrometry

The differential electrochemical mass spectrometry (DEMS) setup is described elsewhere.[45] Briefly, it consists of a differentially pumped two-chamber system with a Balzers QMS 112 quadrupole mass spectrometer, a Pine Instruments potentiostat and a computerized data acquisition system. The potential scan rate is 10 mV s^{-1}. All experiments are carried out at room temperature. A thin film of the metal catalyst is deposited on a mirror-finish polished glassy carbon disk as substrate (*ca.* 6 mm diameter), leading to a catalyst film with a low platinum loading (28 μg_{metal} cm^{-2}). Then, an aqueous Nafion solution is successively pipetted on the catalyst film and dried in an argon stream to fix the particles to the substrate. The electrode is then mounted in the thin-layer flow through the DEMS cell so that the catalyst film is exposed to the solution through a centred circular tightener (6 mm inner diameter, *ca.* 100 μm thick), forming a thin electrolyte layer of *ca.* 5 μL volume at the electrode. The electrolyte flow rate is about 5–6 μL s^{-1}. The gaseous products are evaporated into the mass spectrometer through a porous membrane (*e.g.* Scimat, 60 μm thick, 50% porosity, 0.2 μm pore diameter). Two Pt wires are used as counter electrodes in the thin layer cell. The reference electrode is a saturated calomel electrode (SCE) connected to the outlet of the DEMS cell through a Teflon capillary, but the electrode potentials are referred to the RHE. CO or an alcohol is preadsorbed at a constant electrode potential for 10 min from a CO-saturated or an alcoholic acidic solution. The mass spectrometer is calibrated by CO stripping on a polycrystalline Pt electrode,[45] which leads to a sensitivity factor of the DEMS: $K^* = 2Q_{MS}/Q_F = 3.4 \times 10^{-5}$, where Q_{MS} is the mass spectrometric charge, Q_F is the faradic charge from the CO stripping cyclic voltammogram corrected from the double-layer charge, and 2 is the number of electrons per CO for oxidation to CO_2.

1.4.4 Chromatographic Techniques

The quantitative analysis of reactants and reaction products can be performed by chromatographic techniques (gas chromatography, liquid chromatography). On-line chromatographic techniques, particularly high-performance liquid chromatography, were developed to analyze quantitatively the reaction products of alcohol electro-oxidation.[46]

Experiments are carried out either in a two-compartment three-electrode glass-cell[46] or in a direct alcohol fuel cell.[7] In the first case, the working compartment (receiving the working electrode and the reference electrode) and the counter-electrode compartment are separated by an ion exchange membrane (Nafion 423). All the electrolyses are performed at a controlled potential and the experimental set-up (cyclic voltammetry, electrolyses, analyses, *etc.*) have been previously described.[46] In the second case, to analyze quantitatively the reaction products at the outlet of the anode side of the DAFC, fuel cell tests are carried out at 80 °C in a single cell with a 25 cm^2 geometric surface area using a Globe Tech test bench. All experiments are carried out at a controlled current density j using a high-power potentiostat (Wenking model HP 88) interfaced with a PC to apply constant current sequences and to store the data, and with a variable resistance in order to fix the current applied to the cell. The membrane electrode assembly, the fuel cell experimental conditions and the analytical set-up have been previously fully described.[7]

1.5 Main Parameters of the Electrode Material Controlling the Electro-reactivity of Alcohols

The nature and the structure of the electrode material play a major role in electrocatalysis, particularly for structure-sensitive reactions such as the electro-oxidation of alcohols.

1.5.1 Chemical Nature of the Electrode Material

The effect of the chemical nature of the electrode is illustrated by the oxidation of *n*-butanol in alkaline medium on different noble metal electrodes: Pt, Rh, Pd and Au.[12] The results obtained (Figure 1.3) show that the highest electrocatalytic activity in alkaline solution is reached with gold, although the potential range where the oxidation takes place is much too positive for practical applications. Palladium is much less active for alcohol oxidation, whereas rhodium is usually inactive. In acid solutions, platinum appears the only noble metal catalyst available.

1.5.2 Effect of Crystallographic Structure

The electronic structure of the electrode greatly influences its electrocatalytic activity.[47] It has been proved that the electro-oxidation of methanol is a surface-sensitive reaction, since its electro-oxidation on platinum single crystals involves different adsorption steps, resulting from interactions of the adsorbed species with the oriented metallic surface.[48,49] The initial activity of the Pt(110) plane is much higher than that of the other low-index planes, but the poisoning phenomenon is so rapid that it causes a fast decrease of the current densities after a few sweeps. The Pt(111) plane appears to be less sensitive to poisoning phenomena, although the current densities are smaller. Finally, the Pt(100) plane is totally blocked over a large range of potentials (until 0.7 V *vs.* RHE), but the current increases sharply once the adsorbed blocking species are removed at higher potentials (Figure 1.4). These structural

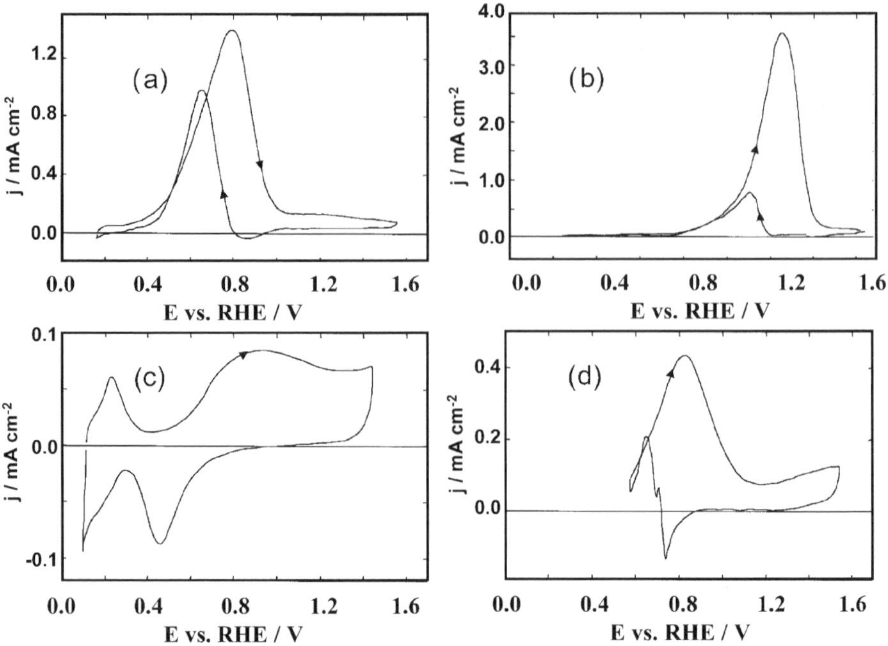

Figure 1.3 Voltammograms showing the oxidation of 0.1 M *n*-butanol in 0.1 M NaOH on different noble metal electrodes (25 °C, sweep rate 50 mV s^{-1}): (a) platinum, (b) gold, (c) rhodium and (d) palladium.

effects were confirmed by *in situ* IR spectroscopy, which showed the presence of a high coverage of linearly bonded carbon monoxide (CO_L) at a Pt(110) plane, but different adsorbed CO species (CO_L, CO_B, CO_m, where CO_B and CO_m = bridge-bonded and multi-bonded CO, respectively) are present on the other crystal planes (Figure 1.4).[50–52] Further, strong lateral interactions between these different species on Pt(100) lead to a very low activity of this electrode at low potentials.[50]

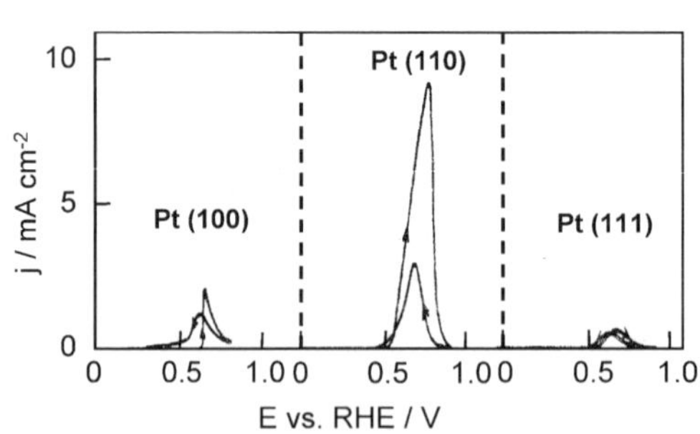

Figure 1.4 Voltammograms of the oxidation of 0.1 M CH_3OH in 0.5 M $HClO_4$ on platinum low-index single-crystal planes (room temperature, 50 mV s^{-1} sweep rate, first sweep).

1.5.3 Modification of the Electrode Properties by Metal Adatoms

The modification of the platinum surface by underpotential deposition (upd) or spontaneous deposition of foreign metal adatoms is a convenient method to enhance the electrocatalytic properties of platinum.[53] However, the effect of adatoms has been shown to be not as effective for the electro-oxidation of methanol as for the oxidation of other alcohols such as ethylene glycol.

For the oxidation of ethylene glycol in alkaline medium, chosen as an example, platinum electrodes were modified by upd of seven metal adatoms: Bi, Cd, Cu, Pb, Re, Ru and Tl.[54] The different adatoms used may be classified into three groups, according to their effect on the electrocatalytic activity of platinum. The first group (Pb, Bi, Tl) has a pronounced enhancement effect on the rate of oxidation, multiplying the current density by a factor from 6 to 15. The second one (Cd) does not increase the current density but shifts the oxidation curve towards more negative potentials (by 200 mV), which is also a positive catalytic effect, as far as the overvoltage is reduced. The third group (Cu, Re, Ru) practically does not change the catalytic activity of the electrode surface. However, in acid medium, the effect of the same series of adatoms is quite different, particularly for Bi and Tl, which greatly poison the electrode surface.[9]

1.5.4 Binary and Multimetallic Electrodes

The electrocatalytic activity may also be modified by metal alloying, *i.e.* adding to platinum a second metallic component (a "promoter") in order to decrease the formation of poisoning species or to promote their oxidation at lower potentials.[55,56] Numerous metals have been explored, to generate O or OH species at lower potentials on the platinum surface. Historically, systematic screening of the possible metals showed that only a few metals led to encouraging results.[57] Ruthenium, tin and molybdenum were the most promising ones. In a more recent systematic study, theoretical calculations have been carried out to determine the effects of alloying platinum with transition metals of Groups 4 to 6 on the formation of adsorbed species.[58] Such calculations can only provide an insight on this effect but not on the stability of the selected alloys. According to the authors of this study, Ru and Cr are potentially interesting alloying components, but in a proton exchange membrane fuel cell (PEMFC) the electrolyte is strongly acidic and thus a corrosive medium environment prevails, so that a more noble metal (*e.g.* Pt with Ru) is necessary.

In the case of methanol electro-oxidation, the reason for the enhancement of activity is generally related to the higher CO tolerance of PtRu. This is the bifunctional mechanism according to which adsorbed CO species are oxidized by OH species generated on the Ru surface atoms.[59,60] Another explanation for the enhanced CO tolerance is the electronic effect,[61] according to which the presence of ruthenium brings about a change in the electronic state of platinum, leading to a change in CO binding. A mechanism combining both effects is also possible. Thus a bifunctional catalyst should be able to catalyze two different reaction steps (adsorption and surface reaction), so that active sites with different properties are necessary.

Another way to increase both the active surface area and the number of oxygenated species at the electrode surface is to prepare alloy particles or deposits and then to dissolve the non-noble metal component.[62] This technique, which is similar to that used to prepare Raney-type catalysts, yields very high surface area electrodes, and hence some improvements in the electrocatalytic activities are achieved compared to that of pure platinum. However, it is always difficult to be sure whether the mechanism enhancing the activity is due to this effect or the possible presence of remaining traces of the dissolved metal. Results with Pt_3Cr and Pt_3Fe were shown to be encouraging.[62] Concerning the behaviour of PtFe particles for methanol electro-oxidation, it was concluded that the catalytic effect is due to Fe alloyed to platinum.[63]

Finally, trimetallic catalysts have been developed to enhance the electroactivity of platinum-based catalysts, either for methanol or ethanol electro-oxidation. When added to PtRu catalysts, molybdenum lead to higher activity towards methanol electro-oxidation at low overpotentials.[64] Such anodes (prepared by electrodeposition of metals on a carbon gas diffusion layer[35]) lead to an improvement of the electrical performance of a single DMFC at 90 °C. The trimetallic $Pt_{0.7}Ru_{0.26}Mo_{0.04}/C$ catalyst gives a higher cell performance than a $Pt_{0.8}Ru_{0.2}/C$ anode prepared in the same way. For ethanol electro-oxidation, a trimetallic $Pt_{0.86}Sn_{0.1}Ru_{0.04}/C$ catalyst made by a colloidal route leads to a cell performance twice as high as that obtained with a $Pt_{0.9}Sn_{0.1}/C$ catalyst, under the same experimental conditions.[7]

The higher capacity of the third metal to form oxy-hydroxyl species at low potentials is often proposed as an explanation for the activity enhancement of methanol oxidation. In the case of ethanol electro-oxidation at PtSnRu and PtSn, the open circuit voltage (ocv) is similar for both catalysts and it was shown that the presence of ruthenium does not lead to a noticeable change in the product distribution in the anode outlet.[7] Here, the role of ruthenium seems more difficult to explain. Its main role may be to limit strong adsorption of the surface poisons by diluting adsorption sites,[65] or to provide adsorbed OH species as soon as tin starts to form higher oxide species that are not catalytically active.[66]

However, in the case of multimetallic catalysts, the problem of the stability of the surface layer is crucial. Preferential dissolution may occur, leading to a modification of the nature, and therefore of the properties, of the electrocatalyst. A modification of the size of the nanoparticles, and of their crystallographic structure, is also possible and should be verified. All these problems of ageing are crucial for applications to fuel cells.

1.5.5 Effect of the Particle Size and Carbon Support

The effects of the dispersion of the electrocatalyst and of particle size on the electro-oxidation of alcohols have been the subject of numerous studies. The main objective is to determine the optimum size of the platinum particles in order to increase the effectiveness factor of platinum. Such a size effect, which is widely accepted in the case of the reduction of oxygen,[67,68] is still a subject of discussions for the oxidation of methanol. According to some authors,[69] an optimum of 2 nm for the platinum particle size exists, but on studying nanoparticles of size down to 1.4 nm, other authors observed no size effects.[70] According to another study,[71] the rate of oxidation of methanol remains constant for particles greater than 4.5 nm, but decreases with the size for smaller particles (until 2.2 nm).

Such kinds of information can be obtained from voltammetric measurements:[72] it is possible to determine the quantity of electricity involved in the adsorption of hydrogen, or in the electro-oxidation of previously adsorbed CO (CO stripping), and then to estimate the real surface area and the roughness factor (γ) of a Pt/C catalyst. From this real surface area and the Pt loading, it is possible to estimate the specific surface area, S (in $m^2\ g^{-1}$), as follows:

$$S = 100\gamma / W \tag{1.29}$$

with W (in $\mu g\ cm^{-2}$) the Pt loading of the carbon powder.

This specific surface area is useful to characterize easily the supported catalyst. Using the simple idea that all the particles can be assumed spherical, the particle diameters d in nm can be obtained from the value of S by:

$$d = 6000 / S\rho \tag{1.30}$$

with ρ the density of the platinum (21.4 $g\ cm^{-3}$). The estimated value of d can be then compared to the size observed by microscopic techniques.

It is also convenient to compare the behaviour of different dispersed catalysts with the help of the two following quantities: MA, the mass activity in $A\ g^{-1}$, and SA, the specific activity in $A\ cm^{-2}$.[72] Both MA and SA increase with the amount of surface oxides on the carbon powder. MA does not depend on the platinum specific surface area, but depends on the carbon structure only. SA decreases with increasing the platinum specific surface area.[72]

In the latter study, it was reported that the oxidation state of the carbon powder support has a greater effect than that of the particle size on the kinetics of the electro-oxidation of methanol at platinum particles.[72] This analysis could explain the discrepancies encountered in the literature. When the high surface area carbon powder (which is the support commonly used for the electrocatalyst) is partially covered by oxidized groups, the strength of the Pt–OH bonds at the particle surface can be modified, thus altering the rate of surface reactions between adsorbed OH and adsorbed species resulting from the alcohol chemisorption.

1.5.6 Platinum-Based Nanoparticles and Nanocrystals

A crucial point is the method used to prepare nanostructured catalysts, which may lead to different catalyst structures and compositions, in a controlled way, by varying only slightly the experimental conditions. The colloidal route for catalyst preparation derived from the method developed by Bönnemann *et al.*[22] is a very convenient technique for synthesizing PtRu compounds with a controlled size, composition and structure.[24,25] Catalytic powders could be obtained in the following ways: (i) Pt/C and Ru/C, from deposition of a Pt colloid or a Ru colloid solution on carbon; (ii) PtRu/C, from deposition of alloyed PtRu colloid particles (obtained by the co-reduction of platinum and ruthenium salts) on carbon; (iii) Pt+Ru/C, from the deposition of a mixture of Pt and Ru colloid solutions on carbon. Amounts of each reagent can be adjusted to obtain different Pt/Ru atomic ratios and various metal loadings on carbon. Analysis of the XRD patterns led to the conclusion that catalyst (ii) is a true alloy of PtRu and that preparation method (iii) leads to Pt particles decorated by Ru particles.[24]

An important conclusion is that the best catalyst for CO and methanol electro-oxidation reactions is not the alloyed one as expected, nor the mixture of the Pt/XC-72 and Ru/XC-72 powders, but the one consisting of the dispersion of Pt colloid + Ru colloid on the same carbon support, the Pt+Ru/XC-72 catalyst. This catalyst leads to higher current densities for the electro-oxidation of methanol than the other catalysts with the same atomic ratio for potentials lower than 0.5 V *vs.* RHE.

1.6 Survey of the Results on the Electro-oxidation of Several Alcohols

The electro-oxidation of alcohols, even the simplest one, *i.e.* methanol, is a complex reaction, involving multi-electron transfers and several intermediate steps. The complete oxidation to CO_2, except for methanol, is never observed at room temperature. This is due to the difficulty of breaking the C–C bond and of finding multifunctional electrocatalysts, in order to simultaneously activate one of the following main reaction steps: cleavage of C–H bonds with hydrogen adsorption and oxidation, breaking of the C–C bond, removal from the electrode active surface sites of strongly adsorbed intermediates (C_1 or C_n poisoning species) at low overpotentials, activation of the water molecule at low potentials to further oxidize the alcohol adsorbed residues, completion of the oxidation reaction to CO_2 by providing extra oxygen atoms, *etc.*

In order to improve the fuel utilization in a DAFC it is important to investigate the reaction mechanism and to develop active electrocatalysts able to activate each reaction path. The elucidation of the reaction mechanism, thus, needs to combine pure electrochemical methods (cyclic voltammetry, CO "stripping", rotating disc electrodes, *etc.*) with other physicochemical methods, such as *in situ* spectroscopic methods (IR[41] and UV-VIS[73] reflectance spectroscopy, or mass spectrometry such as EQCM[74] and DEMS[45]), or radiochemical methods[75] to monitor the adsorbed intermediates, and on-line chromatographic techniques[46] to analyze quantitatively the reaction products and by-products.

In the following sections we will illustrate, with the help of several examples dealing with the electro-oxidation of alcohols, how the use of coupled experimental techniques will be able to elucidate most of the reaction mechanisms.

1.6.1 The Electro-oxidation of Methanol

The oxidation of methanol has been thoroughly studied for many years, so that the reaction mechanism is now well established.[48] The overall oxidation reaction involves six electrons, and one water molecule, as follows:

$$CH_3OH + H_2O \rightarrow CO_2 + 6H^+ + 6e^- \quad (1.10)$$

1.6.1.1 IR Studies of Methanol Adsorption and Oxidation

It was first shown by EMIRS that the main poisoning species formed during the chemisorption and oxidation of methanol on a platinum electrode is carbon monoxide, either linearly bonded (CO_L) or bridge bonded (CO_B) to the surface.[40] The coverage degree of the electrode surface by CO_L can reach 90% on a pure platinum electrode, so that most of the active sites are blocked.

Besides $(CO)_{ads}$, some other adsorbed species, such as $(-CHO)_{ads}$ or $(-COOH)_{ads}$, were identified, on a platinum electrode, by IR reflectance spectroscopy, *e.g.* EMIRS[50,76,77] and Fourier transform IR reflectance spectroscopy.[78]

Intermediate products, like formaldehyde (HCHO) and formic acid (HCOOH), and the final product (CO_2) are formed on smooth polycrystalline platinum electrodes and were analyzed quantitatively by liquid or gas chromatography.[79] It is now well established that the electro-oxidation of methanol occurs through different steps, involving several adsorbed intermediate species, and in addition to the main reaction product (carbon dioxide), some amounts of side products, such as HCHO or HCOOH, are detected.

The reaction mechanism of methanol oxidation (n_{th} = 6 for complete oxidation to CO_2) on a smooth Pt electrode is, therefore, relatively complex (Figure 1.5).[48]

In this mechanism, the adsorbed formyl-like species $(-CHO)_{ads}$ plays a key role and the strategy to develop efficient electrocatalysts will be to favour this adsorbed species, and to avoid the subsequent formation of adsorbed CO, either to prevent its formation by blocking neighbouring Pt sites or to oxidize it at lower potentials. To reach this goal the activity of the electrocatalyst can be significantly increased by using platinum-modified electrodes, including alloys and nanoparticles. Unfortunately, for the electro-oxidation of methanol, only platinum–ruthenium[80–82] and platinum–tin[83] alloys showed a better electrocatalytic activity than platinum itself. In order to understand the mechanism of

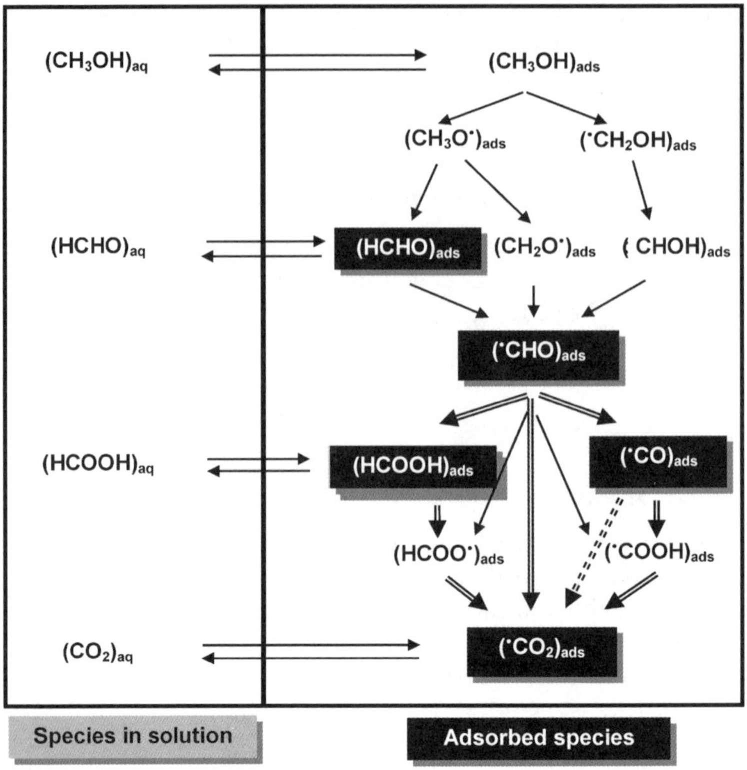

Figure 1.5 Detailed reaction mechanism of the oxidation of methanol on a platinum electrode.

methanol oxidation at platinum-based electrodes, several *in situ* spectroscopic techniques, such as IR reflectance spectroscopy, have been used.[48,84–86]

The following results concern the investigation by FTIR reflectance spectroscopy (SPAIRS and SNIFTIRS) of the mechanism of methanol oxidation at smooth Pt, $Pt_{0.9}Ru_{0.1}$ and $Pt_{0.5}Ru_{0.5}$ bulk alloys, since these catalysts display the greatest electrocatalytic activities towards the oxidation of CO and of CH_3OH.[87] Figure 1.6 shows the SPAIR spectra recorded for these three catalysts in a 0.5 M $HClO_4$ + 0.1 M CH_3OH solution during the first voltammetric cycle at 1 mV s^{-1}.

For Pt (Figure 1.6a) the first spectra corresponding to low potentials do not show any particular bands apart from those at 1640 and 1100 cm^{-1}, assigned to adsorbed water and perchlorate ions, respectively. For potentials greater than 0.37 V *vs.* RHE, an absorption band is seen at 2050 cm^{-1} which becomes more intense as the electrode potential increased up to 0.57 V and then disappears. This band is assigned to linearly adsorbed carbon monoxide (CO_L) coming from the dissociation of methanol. From 0.42 V *vs.* RHE a new band at around 2345 cm^{-1}, corresponding to carbon dioxide, becomes visible and more intense at 0.67 V. The fact that this band appears at rather low potentials

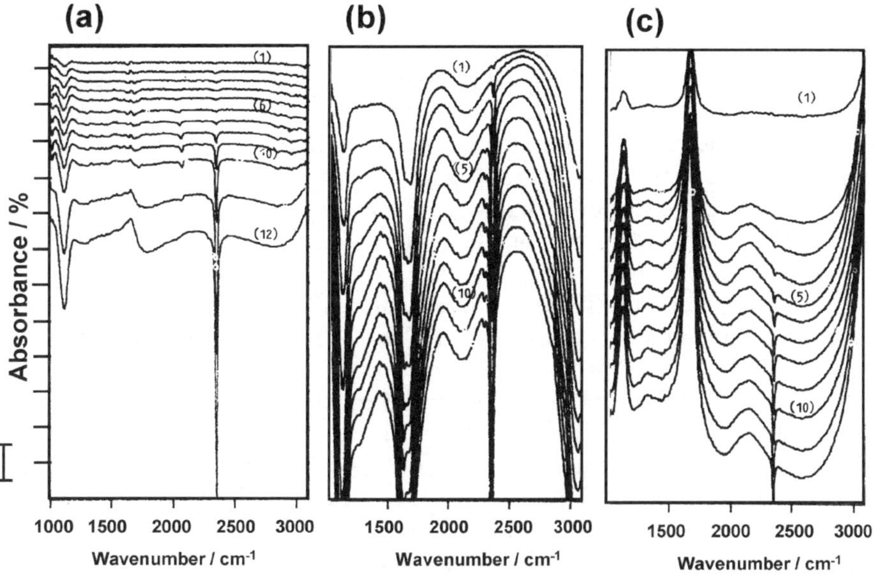

Figure 1.6 SPAIR spectra of methanol oxidation in a 0.5 M $HClO_4$ + 0.1 M CH_3OH solution on a smooth metal electrode: (a) pure Pt, (b) $Pt_{0.9}Ru_{0.1}$, (c) $Pt_{0.5}Ru_{0.5}$. Reference spectrum taken at +70 mV *vs.* RHE, sweep rate 1 mV s^{-1}, forward scan. Electrode potential in mV *vs.* RHE: (1) 120; (2) 170; (3) 220; (4) 270; (5) 320; (6) 370; (7) 420; (8) 470; (9) 520; (10) 570; (11) 620; (12) 670.

and is relatively intense suggests that at least part of the methanol can be directly oxidized to carbon dioxide. At around 1040 cm^{-1}, and for low potentials, a band can be seen and is assigned to the C–O bonding in the methanol molecule. The C–H bond stretching mode of the methyl group was highlighted through the band at 2900 cm^{-1} (the opposite sign of this band compared to the CO_2 band indicates the consumption of methanol).

In the case of the $Pt_{0.9}Ru_{0.1}$ alloy, the SPAIR spectra are quite different (Figure 1.6b). The first remark concerns the appearance, at low potentials, of a weak band for the ν(C=O) stretching mode, at 2050 cm^{-1}, corresponding to CO_L, with a maximum of intensity at 0.32 V *vs.* RHE, instead of 0.52 V for pure Pt. However, the main feature of these spectra is the band at 2345 cm^{-1}, corresponding to the formation of CO_2, which appears at potentials as low as 0.17 V *vs.* RHE (instead of 0.42 V for pure platinum). The very large intensity of this band confirms that CO_2 can be easily formed in large amounts on this alloy. Owing to this fact, a small band at 2280 cm^{-1} is clearly seen and can be attributed unambiguously to $^{13}CO_2$, with an intensity 100 times less than that of the main band of $^{12}CO_2$, as expected from the isotopic composition of natural carbon. In the range 1600–1750 cm^{-1}, a double absorption band (at 1690 and 1630 cm^{-1}, respectively) is observed, corresponding to a carbonyl group, probably of formaldehyde. This assumption is confirmed by the presence of a band at 1450 cm^{-1}, typical of formaldehyde.

Another set of SPAIR spectra was recorded for the alloy containing 50 at% Ru (Figure 1.6c) under the same conditions as for pure platinum and $Pt_{0.9}Ru_{0.1}$. An obvious and important remark is the quasi-absence of the band corresponding to CO_L. The band at 2345 cm^{-1}, assigned to the formation of CO_2, can be seen at lower potentials (from 0.22 V) in comparison to that obtained on pure platinum, but at higher potentials than for $Pt_{0.9}Ru_{0.1}$. In the same spectra series, two other bands, at 1200 and 1250 cm^{-1}, are observed and attributed to the formation of small amounts of formic acid and formaldehyde, respectively.

All these observations are in agreement with the voltammetric results, showing a maximum of activity of the $Pt_{0.9}Ru_{0.1}$ alloy for the electro-oxidation of methanol.

SNIFTIRS experiments were also carried out with the same electrodes. This mode of recording IR spectra allows us to obtain information coming practically only from the adsorbed species, opposite to the SPAIRS technique, which gives information both on adsorbed species and on products situated near the electrode surface. Figure 1.7 shows SNIFTIR spectra corresponding to pure platinum and to the $Pt_{0.9}Ru_{0.1}$ alloy. For pure Pt (Figure 1.7a), the bands corresponding to CO_L and CO_B can be clearly seen (at 2050 and 1850 cm^{-1}, respectively). The band at 2345 cm^{-1}, corresponding to adsorbed CO_2 produced from methanol oxidation, is visible at mean potentials greater than $\bar{E}$ = 0.42 V *vs*. RHE.

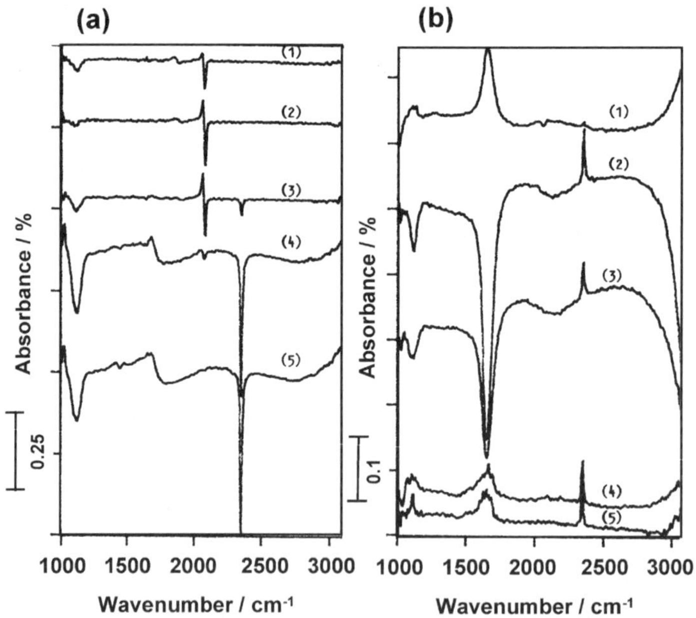

Figure 1.7 SNIFTIR spectra of methanol oxidation at (a) Pt and (b) $Pt_{0.9}Ru_{0.1}$; 0.5 M $HClO_4$ + 0.1 M CH_3OH; potential windows (in mV *vs*. RHE): (1) (70–370); (2) (170–470); (3) (270–570); (4) (370–670); (5) (470–770).

For $Pt_{0.9}Ru_{0.1}$ (Figure 1.7b), the CO_2 band is observed at lower potentials ($\bar{E} = 0.22$ V). The CO_L band is practically absent with this alloy. However, the band corresponding to adsorbed water is very intense. This could suggest that the surface is not totally covered with adsorbed species coming from methanol dissociation, leaving free sites to adsorb water weakly, particularly at the Ru atoms, since at pure Pt the band for adsorbed water was comparatively small.

1.6.1.2 DEMS Study of Methanol Adsorption and Oxidation

DEMS was also used for methanol stripping experiments, which can give some information on the electrode coverage by species coming from the adsorption and oxidation of methanol.[88] First, it can be seen from the cyclic voltammograms (CVs) and the mass spectrometric cyclic voltammograms (MSCVs) recorded on a co-reduced $Pt_{0.8}Ru_{0.2}/C$ catalyst as an example (Figure 1.8), that the coverage of the electrode from methanol adsorption (curves 2) is much lower than that from CO adsorption (curves 1).

Secondly, the coverage by the species coming from methanol adsorption on each catalyst can be evaluated (Figure 1.9). It appears that this coverage is

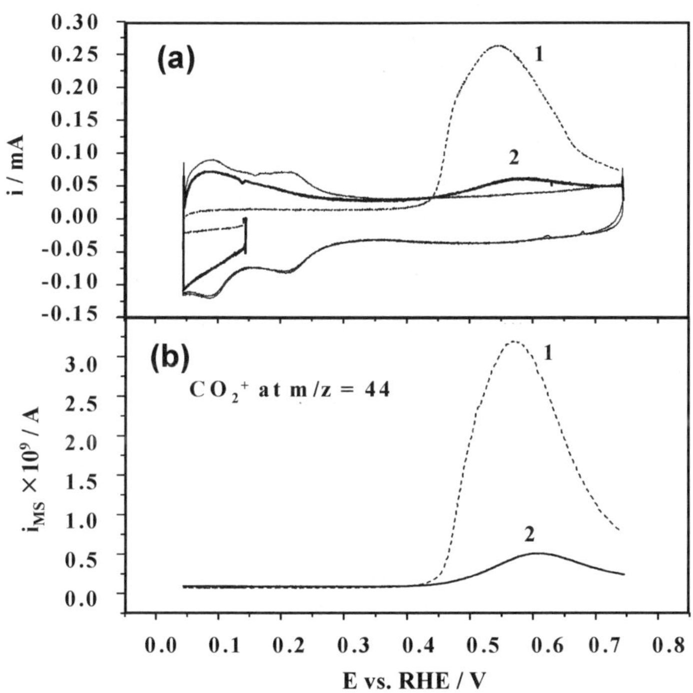

Figure 1.8 (a) CVs and (b) MSCVs of CO_2^+ recorded in a DEMS cell during (1) a CO stripping experiment and (2) a methanol stripping experiment on a co-reduced $Pt_{0.8}Ru_{0.2}/C$ catalyst (0.5 M H_2SO_4, $v = 10$ mV s^{-1}, $T = 25$ °C, electrolyte flow = 5 μL s^{-1}; metal loading = 28 μg_{metal} cm^{-2}).

always higher at a co-deposited (non-alloyed Pt + Ru electrode) than at a co-reduced catalyst, whatever the methanol adsorption potential is, which can be an explanation for the higher activity towards methanol oxidation of the co-deposited catalyst compared with the co-reduced one.

Thirdly, the increase in the ruthenium content leads to a decrease in the amount of adsorbed species from methanol. Moreover, from the determination of the faradic and spectroscopic charges, the number of electrons involved in the oxidation of species coming from methanol adsorption to produce CO_2 can be evaluated according to eqn (1.31):

$$n_{DEMS} = K^* \times \frac{Q_F}{Q_{MS}} \tag{1.31}$$

where $K^* = 3.4 \times 10^{-5}$ is the sensitivity factor of the mass spectrometer as determined from CO stripping.[88]

A number equal to 2 indicates that only CO is adsorbed at the surface, whereas a number higher that 2 indicates that both CHO and CO species are adsorbed at the surface. From the results given in Table 1.2, the calculated number of electrons is always close to 2, which indicates that the main adsorbed species is CO. However, this number is higher for the co-reduced than for the co-deposited catalyst, and becomes greater than 2 with an increase in the ruthenium amount, which likely indicates that more CHO_{ads} is formed at co-reduced $Pt_{0.8}Ru_{0.2}/C$ and $Pt_{0.5}Ru_{0.5}/C$ catalysts.

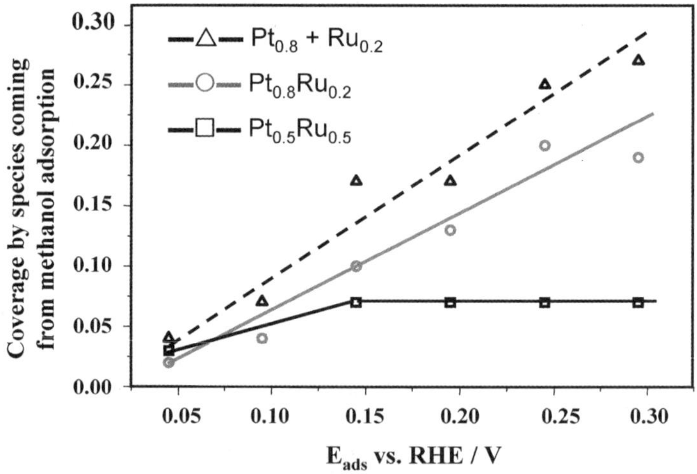

Figure 1.9 Comparison of the coverage in species coming from the adsorption of methanol for different PtRu/C catalysts. The experiments were realized under the same experimental conditions as in Figure 1.8, with co-reduced $Pt_{0.8}Ru_{0.2}/C$ and $Pt_{0.5}Ru_{0.5}/C$ catalysts, and a codeposited $Pt_{0.8}+Ru_{0.2}/C$ catalyst.

Table 1.2 Faradic (Q_F) and spectrometric (Q_{MS}) charges recorded during the electro-oxidation of species coming from preadsorbed methanol at 0.3 V *vs.* RHE, and the number of electrons involved in the oxidation of the adsorbed species.[a]

Catalyst	Faradic charge, Q_F/mC	MSCV charge, Q_{MS}/nC	Electron number, n
$Pt_{0.8}$+$Ru_{0.2}$/C co-deposited	1.3	23.2	1.9
$Pt_{0.8}Ru_{0.2}$/C co-reduced	1	16.4	2.1
$Pt_{0.5}Ru_{0.5}$/C co-reduced	0.4	6.1	2.2

[a]Conditions: 0.5 M H_2SO_4 + 0.1 M CH_3OH, $T = 25\ °C$, $v = 10\ mV\ s^{-1}$, electrolyte flow = 5 μL s^{-1}, metal loading = 28 μg cm^{-2}.

1.6.1.3 Mechanism of the Electro-oxidation of Methanol

The mechanism of the electro-oxidation of methanol on platinum has been previously discussed, mainly after the identification of the adsorbed species, both reactive intermediates and poisoning species,[87] and the analysis of reaction products and by-products.[79] Methanol is firstly dissociatively adsorbed at Pt-based electrodes, leading to the cleavage of several C–H bonds, and gives finally the so-called formyl-like species −CHO. From this species, different steps can occur, but with platinum the dissociation of this species gives rapidly adsorbed CO, which is responsible for the electrode poisoning. This is the explanation of the rather poor performance of Pt catalysts, due to the relatively high potential necessary to oxidize such CO species into CO_2. These different steps can be summarized as follows:

$$Pt + CH_3OH \rightarrow Pt - CH_3OH_{ads} \rightarrow Pt - CHO_{ads} + 3H^+ + 3e^- \qquad (1.32)$$

$$Pt - CHO_{ads} \rightarrow Pt - CO_{ads} \rightarrow CO_2 \qquad (1.33)$$

The key step in the reaction mechanism appears to be the formation of the intermediate species $-(CHO)_{ads}$, which facilitates the overall reaction. The kinetics of its further desorption and/or oxidation into reaction products controls the reaction mechanism, leading to complete oxidation. The usual path is the formation of CO_{ads}:

$$Pt - (CHO)_{ads} \rightarrow Pt - (CO)_{ads} + H^+ + e^- \qquad (1.34)$$

followed by its oxidation to CO_2.

An alternative path to the spontaneous formation of the poisoning species (eqn 1.34) is the oxidation of the $-(CHO)_{ads}$ species, with adsorbed $-(OH)_{ads}$ species coming from the dissociation of water according to the following reactions:

$$Pt + H_2O \rightarrow Pt-(OH)_{ads} + H^+ + e^- \quad (1.35)$$

$$Pt-(CHO)_{ads} + Pt-(OH)_{ads} \rightarrow 2Pt + CO_2 + 2H^+ + 2e^- \quad (1.36)$$

One parallel reaction, leading to adsorbed formate, has also been observed:

$$Pt-(CHO)_{ads} + Pt-(OH)_{ads} \rightarrow Pt + Pt-(COOH)_{ads} + H^+ + e^- \quad (1.37)$$

then leading by further oxidation to the formation of carbon dioxide:

$$Pt-(COOH)_{ads} \rightarrow Pt + CO_2 + H^+ + e^- \quad (1.38)$$

On the other hand, adsorbed CO can be oxidized through the reaction:

$$Pt-(CO)_{ads} + Pt-(OH)_{ads} \leftrightarrow Pt + Pt-(COOH)_{ads} \quad (1.39)$$

followed by reaction (1.38), or reaction:

$$Pt-(CO)_{ads} + Pt-(OH)_{ads} \rightarrow 2Pt + CO_2 + H^+ + e^- \quad (1.40)$$

This mechanism takes into account the formation of all the products as detected by IR reflectance spectroscopy (SNIFTIRS or SPAIRS)[89] and liquid or gas chromatography:[79] CO_2 from steps (1.36), (1.38) or (1.40), formation of formaldehyde according to step (1.32), or formation of formic acid after steps (1.37) or (1.39).

Thus, the crucial aspect is to determine the fate of the $-(CHO)_{ads}$ species. Possible mechanisms for its oxidative removal are schematically shown in Scheme 1.1.

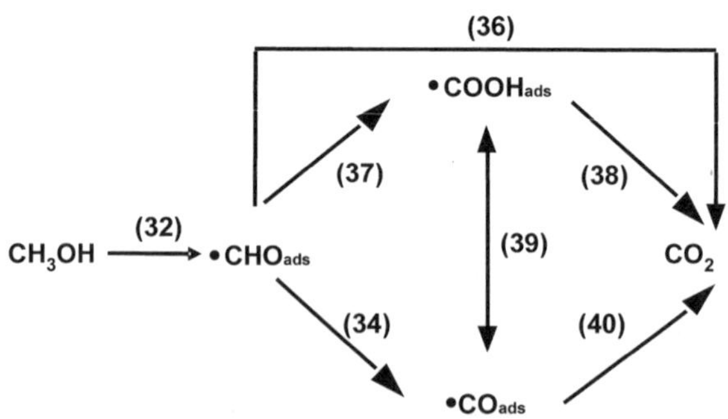

Scheme 1.1 Schematic representation of the mechanism of oxidation of $-(CHO)_{ads}$.

From this scheme, it appears that desorption and oxidation of the formyl species can follow different pathways through competitive reactions. This scheme summarizes the main problems and challenges to improve the kinetics of the electro-oxidation of methanol.

On a pure platinum surface, step (1.34) is spontaneously favoured, and the formation of adsorbed CO is a fast process, even at low potentials. Thus, the coverage of adsorbed CO is high and explains the poisoning phenomena encountered at a platinum electrode. This poisoning species can be removed, by oxidation through step (1.40) into CO_2, only at potentials at which oxygenated species are present at the electrode surface. For platinum, such oxygenated species, arising from the dissociation of water, step (1.35), appears only for potentials greater than 0.5–0.6 V.[90,91]

An alternative route is the direct oxidation of $-(CHO)_{ads}$ into CO_2 by step (1.36) or through step (1.37) followed by step (1.38), with the intermediate formation of $-(COOH)_{ads}$ species. In both cases, the reaction needs again the presence of an extra oxygen atom which can be provided only by the dissociation of water at the catalytic surface, so that several PtM catalysts (with M = Ru, Mo, Sn, Fe, Ni, *etc.*), where the added metal M is able to dissociate water at lower potentials than Pt, have been thoroughly investigated.

With PtRu catalysts, which are the best bimetallic catalysts for methanol oxidation, it appears clearly from the literature, and this is fully confirmed by IR reflectance spectroscopic studies, that the presence of adsorbed $-(OH)_{ads}$ on ruthenium sites at low potentials leads to the oxidation of adsorbed CO at potentials much lower than those on pure platinum. It is also probable that $-(CHO)_{ads}$ can be oxidized directly to carbon dioxide, without the formation of adsorbed CO poisoning species.[3]

The following steps for methanol oxidation at $Pt_{1-x}Ru_x$ electrodes can then be postulated (bifunctional mechanism[80]), as follows:

- Activation of the methanol molecule by Pt:

$$Pt + CH_3OH \rightarrow Pt - CH_3OH_{ads} \tag{1.41}$$

$$Pt - CH_3OH_{ads} \rightarrow Pt - CHO_{ads} + 3H^+ + 3e^- \tag{1.42}$$

$$Pt - CHO_{ads} \rightarrow Pt - CO_{ads} + H^+ + e^- \tag{1.34}$$

- Activation of the water molecule, either by Pt or by Ru:

$$Pt + H_2O \rightarrow Pt - OH_{ads} + H^+ + e^- \tag{1.35}$$

$$Ru + H_2O \rightarrow Ru - OH_{ads} + H^+ + e^- \tag{1.43}$$

- Surface reactions between the adsorbed species:

$$Pt-CHO_{ads}+Pt-OH_{ads}\rightarrow CO_2+2Pt+2H^++2e^- \quad (1.36)$$

$$Pt-CHO_{ads}+Ru-OH_{ads}\rightarrow CO_2+Pt+Ru+2H^++2e^- \quad (1.44)$$

$$Pt-CO_{ads}+Pt-OH_{ads}\rightarrow CO_2+2Pt+H^++e^- \quad (1.40)$$

$$Pt-CO_{ads}+Ru-OH_{ads}\rightarrow CO_2+Ru+Pt+H^++e^- \quad (1.45)$$

The last topic to comment on is the difference between the optimum compositions in Ru of the PtRu alloys for the oxidation of CO (50 at% Ru) and methanol (10–20 at% Ru). Gasteiger *et al.*[81] proposed an explanation based on a statistical modelling of the surface of an alloy containing x at% Ru. The structural atomic model assumed by these authors needs a maximum number of threefold platinum sites for the adsorption of the CH_3OH molecule, and adjacent to one Ru atom to adsorb the OH species. This would correspond to a surface with 8–14 at% Ru, depending on the orientation of the low-index single-crystal planes: 8% for (111), 10% for (100) and 14% for (110) crystallographic orientations, respectively. Our FTIR measurements showed unambiguously that the amount of CO_L adsorbed species, coming from methanol adsorption at $Pt_{0.9}Ru_{0.1}$, is very small: there is no accumulation of such species on the electrode surface, since they are oxidized as soon as they are formed. In the case of CO oxidation, the main adsorbed species is obviously CO, principally linearly bonded (CO_L), but the maximum coverage is reached at lower potentials for the optimum composition of PtRu alloys (50 at% Ru). In the case of methanol, the dissociation of the molecule gives different adsorbed species, such as $-CH_3OH_{ads}$, $-CHO_{ads}$ and $-CO_{ads}$. Their distribution depends on the surface composition, and mainly on the OH coverage, which is higher in the presence of ruthenium. From this remark, it is obvious that the optimum composition of the $Pt_{1-x}Ru_x$ alloys for the oxidation of methanol and of CO should be different.

Previously reported results[82,87,92] provided evidence that the best activity of the PtRu electrodes for methanol oxidation is obtained with about 20% of Ru surface atoms. This can be roughly explained by the bifunctional mechanism,[80] taking into account the number of surface sites necessary to accommodate all the intermediates adsorbed, as observed by IR reflectance spectroscopy.[87]

A detailed analysis of the number of surface sites necessary to accommodate all the adsorbed intermediates (see Scheme 1.2) involved, either in the oxidation of CO or in that of methanol, should describe correctly the optimum composition found for the $Pt_{1-x}Ru_x$ alloys. In the case of CO oxidation, one Pt atom adsorbs CO, whereas one Ru atom adsorbs OH, which leads to the

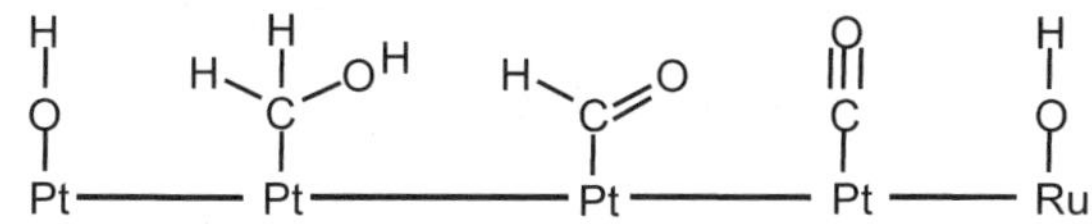

Scheme 1.2 Structure of the different adsorbed intermediates involved in methanol electro-oxidation on $Pt_{1-x}Ru_x$ electrodes.

optimum 50/50 atomic composition. In the case of methanol oxidation, the activation reactions [(1.34), (1.35), (1.41)–(1.43)] need four Pt surface atoms (three for the CH_3OH adsorption residues and one for H_2O) and one Ru surface atom for H_2O activation, leading to the formation of OH_{ads} (Ru does not adsorb CH_3OH, nor dissociate the CH_3OH molecule), so that the optimum surface atomic composition of the PtRu electrodes is four Pt for one Ru, *i.e.* $x = 0.2$. If we take into account the presence of other adsorbed species on Pt (*e.g.* $-COOH_{ads}$) or multi-site adsorption (*e.g.* CO_B), the number of Pt atoms could vary from 5 (corresponding to an optimum composition of 16.7 at% Ru) to 7 (*i.e.* 12.5 at% Ru). This crude model will give an optimum surface composition of 12.5–20 at% Ru. This value should be compared with that proposed by Gasteiger *et al.*,[81] which is around 10 at% at room temperature.

1.6.2 The Electro-oxidation of Ethanol

The reaction mechanisms for the anodic oxidation of alcohols other than methanol are more difficult to elucidate since the number of electrons exchanged greatly increases and the activation of the C–C bond breaking is relatively difficult at moderate temperatures (50–90 °C). The complete oxidation of ethanol involves 12 electrons per molecule, *i.e.*:

$$C_2H_5OH + 3H_2O \rightarrow 2CO_2 + 12H^+ + 12e^- \qquad (1.11)$$

whereas that of *n*-propanol involves 18 electrons:

$$C_3H_7OH + 5H_2O \rightarrow 3CO_2 + 18H^+ + 18e^- \qquad (\%)$$

and more generally for an aliphatic mono-alcohol:

$$C_nH_{2n+1}OH + (2n-1)H_2O \rightarrow nCO_2 + 6nH^+ + 6ne^- \qquad (1.46)$$

so that the reaction mechanism involves many adsorbed intermediates and by-products occurring in several steps.

The electrocatalytic oxidation of ethanol has been investigated for many years on several platinum-based electrodes, including Pt/X alloys (with X = Ru, Sn, Mo, *etc.*), and dispersed nanocatalysts.

Smooth platinum electrodes are rapidly poisoned by some strongly adsorbed intermediates, such as CO, resulting from the dissociative chemisorption of the

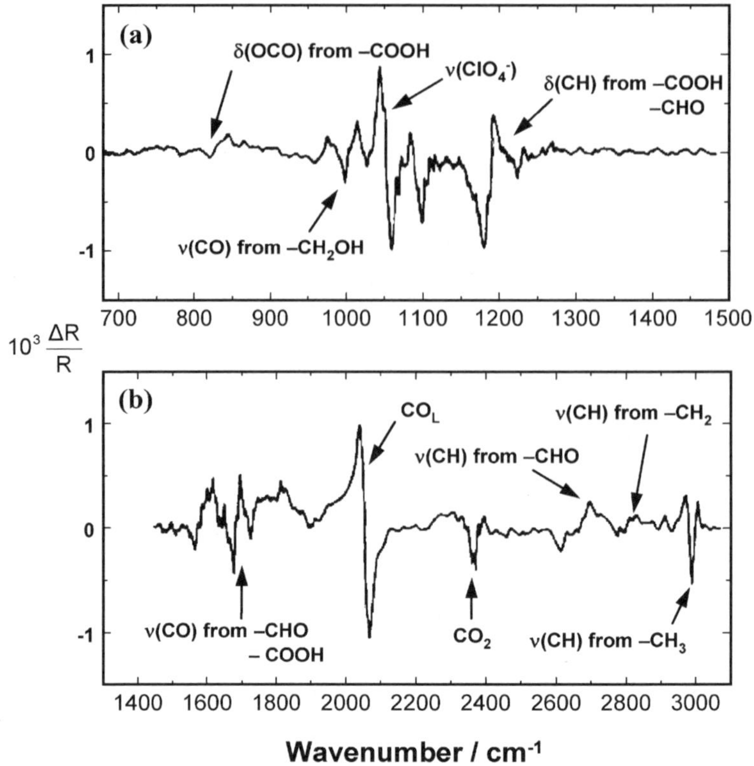

Figure 1.10 IR reflectance spectra (EMIRS) of the adsorbates resulting from the chemisorption of ethanol on a platinum polycrystalline electrode in acid medium (0.5 M $HClO_4$, room temperature): (a) 10^{-3} M C_2H_5OH; (b) 10^{-1} M C_2H_5OH.

molecule, as shown by the first experiments in IR reflectance spectroscopy (EMIRS).[93] Both kinds of adsorbed CO, either linearly bonded or bridge bonded to the platinum surface, are observed. Besides, other adsorbed species have been identified by IR reflectance spectroscopy, including reaction intermediates such as acetaldehyde and acetic acid, and other by-products (Figure 1.10).[94–96]

These first results on smooth platinum obtained by EMIRS were confirmed by more recent investigations using SNIFTIRS[42] and SPAIRS,[43] able to observe intermediate species produced on carbon-supported (*e.g.* Vulcan XC-72) Pt-based nanoparticles.

1.6.2.1 IR Study of the Adsorption and Oxidation of Ethanol

Among several binary Pt-based electrodes, PtSn nanocatalysts lead to the highest activity for ethanol oxidation, so a detailed analysis of the reaction intermediates was carried out by *in situ* IR reflectance spectroscopy.[66]

The SNIFTIR spectra (Figure 1.11a) and SPAIR spectra (Figure 1.11b), obtained on a Pt/C electrode during the adsorption and oxidation of ethanol, are shown at several potentials. The SNIFTIRS technique, which is convenient for detecting adsorbed species,[42] yields a band at about 2050 cm^{-1}, which is due to CO_L for mean potentials between 0.3 and 0.7 V *vs.* RHE.[44,50] The SPAIRS technique allows observation of several absorption bands. At 0.6 V *vs.* RHE, a negative band at 2345 cm^{-1} is observed, which is attributed to the formation of CO_2. For higher potentials, negative absorption bands are observed close to 1720, 1400 and 1280 cm^{-1}. These bands are respectively attributed to the carbonyl C=O stretching mode (ν_{co}) from –COOH or –CHO, to C–O symmetric stretching of adsorbed CH_3COO^-, and to the coupled C–O stretching and OH deformation from –COOH (1370 and 1280 cm^{-1}).[97] The bands located between 1280 and 1400 cm^{-1} are probably due to the symmetric CH_3 bending mode of acetic acid (AA).[98] These absorption bands originate from AA and/or acetaldehyde.[95,96] Actually, it is very difficult to differentiate a carbonyl group of an acid group from that of an aldehyde group[93] because the

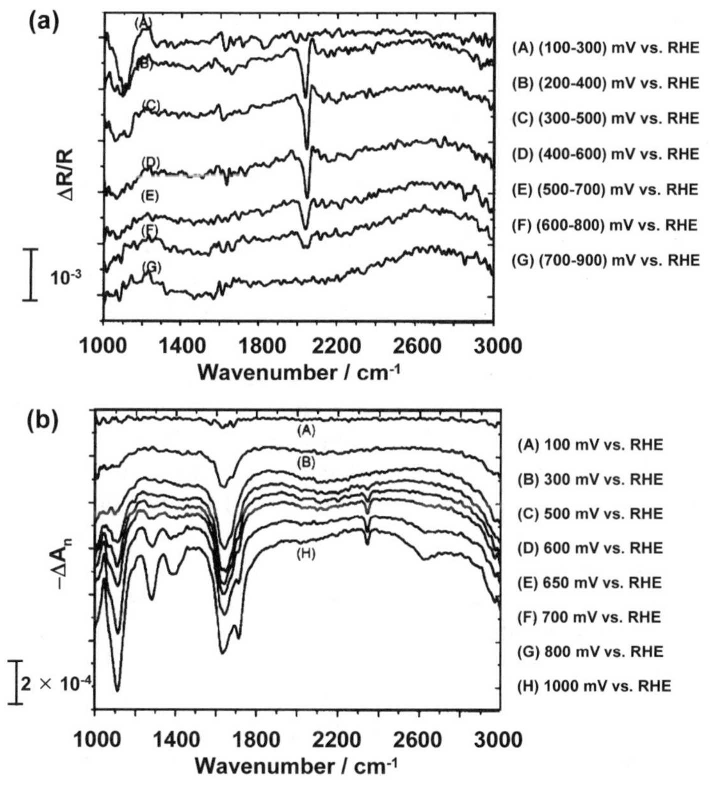

Figure 1.11 IR reflectance spectra recorded in 0.1 M $HClO_4$ + 0.1 M ethanol on a Pt/C electrode at several potentials: (a) SNIFTIR spectra from 100 to 900 mV *vs.* RHE at constant modulation amplitude of 200 mV; (b) SPAIR spectra from 100 to 1000 mV *vs.* RHE.

position of the corresponding bands are only separated by around 5 cm^{-1} and the IR set-up resolution is 4 cm^{-1}.

The wavenumber of the absorption band at about 1280 cm^{-1} remains unchanged whatever the potential. Normally, a shift in the frequency of an absorption band is characteristic of an adsorbed species and can be explained either by the Stark effect, which is observed only for adsorbed species for which the location of the absorption band is potential dependent,[40] or by a change in species coverage, or by a difference in the Pt–adsorbate bond strength, as suggested by Rice *et al.* for adsorbed CO.[99] Because these effects are related to adsorbed species, and do not occur for the species responsible for the absorption band located close to 1280 cm^{-1}, it is likely that this absorption band corresponds to non-adsorbed species like AA close to the electrode surface.

On platinum, the poisoning by CO species occurs at rather low potentials, since the CO absorption band close to 2050 cm^{-1} appears during the second modulation potential between 0.2 and 0.4 V *vs.* RHE, in agreement with previous studies by DEMS, IR spectroscopy and gas chromatography.[93–95] On the other hand, Iwasita and Pastor[95] have detected some traces of methane at low potentials. From prolonged electrolysis experiments and *in situ* IR reflectance spectroscopy, the main reaction products and intermediates involved in ethanol electro-oxidation on a platinum catalyst have been determined as CO and CH_3CO species as adsorbed intermediates, CH_3CHO as intermediate and final product, and CH_3COOH, CO_2 and CH_4 as final products.

Modification of platinum by tin greatly enhances the electro-oxidation rate of ethanol[7,100] and may change the reaction mechanism, so that the adsorbed species and reaction intermediates were also investigated by IR reflectance spectroscopy. Figure 1.12 shows the SNIFTIR (Figure 1.12a) and SPAIR (Figure 1.12b) spectra of the species obtained on a $Pt_{0.83}Sn_{0.17}/C$ catalyst under the same experimental conditions as in Figure 1.11 for a Pt catalyst.

It appears that on PtSn, CO is formed at lower potentials than on Pt. Indeed, on PtSn the absorption band located close to 2050 cm^{-1} is observed during the first modulation potential between 100 and 300 mV *vs.* RHE, whereas this absorption band appears only during the second modulation on a Pt catalyst. On Pt the CO absorption band intensity increases from the second modulation potential (200–400 mV *vs.* RHE), then decreases from the fourth modulation potential (400–600 mV *vs.* RHE) and disappears at higher potentials. In the case of the PtSn catalyst, the intensity of the CO absorption band increases for the two first modulation potentials (100–300 and 200–400 mV *vs.* RHE) and then decreases for higher potentials (disappearing at the last modulation potential between 700 and 900 mV *vs.* RHE). PtSn activates the dissociative adsorption of ethanol at potentials lower than Pt does. The bipolar shape of the SNIFTIR band suggests that more CO is present at the initial potential than at the final modulation potential; then, considering the two first modulations of the potential for each catalyst, the oxidation of CO starts from

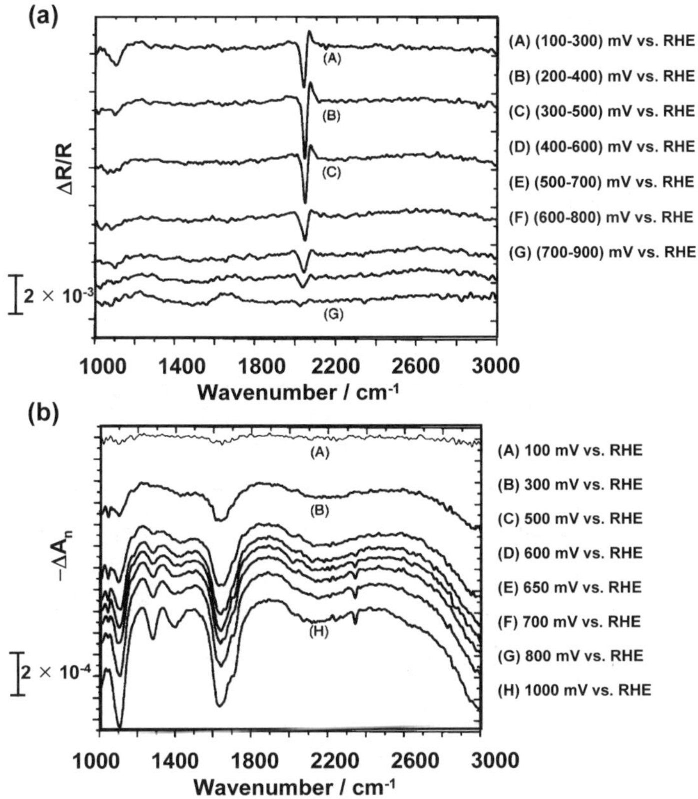

Figure 1.12 IR reflectance spectra recorded in 0.1 M $HClO_4$ + 0.1 M ethanol on a $Pt_{0.83}Sn_{0.17}$/C electrode at several potentials: (a) SNIFTIR spectra from 100 to 900 mV *vs.* RHE at a constant modulation amplitude of 200 mV; (b) SPAIR spectra from 100 to 1000 mV *vs.* RHE.

potentials at least as low as 0.2 V *vs.* RHE at a PtSn catalyst and at potentials higher than 0.3 V at a Pt catalyst.

SPAIR spectra obtained on Pt and PtSn catalysts during ethanol electro-oxidation (Figure 1.11b and Figure 1.12b) show that CO_2 (wavenumber of the absorption band located at 2345 cm^{-1}) is only detected at 0.6 V *vs.* RHE on both Pt and PtSn catalysts. However, as shown in Figure 1.13a, the intensity of the CO_2 IR band is higher with PtSn than for Pt at potentials lower than 0.65 V *vs.* RHE.

On the other hand, the absorption band due to a carbonyl group (at about 1725 cm^{-1}) displays, at high potentials, a lower intensity on PtSn (the absorption band located close to 1725 cm^{-1} remains as a shoulder even at high potentials) than on Pt (Figure 1.13b), which indicates that the formation of C_2 species (presumably acetaldehyde) resulting from a non-dissociative adsorption of ethanol is lower on a PtSn catalyst.

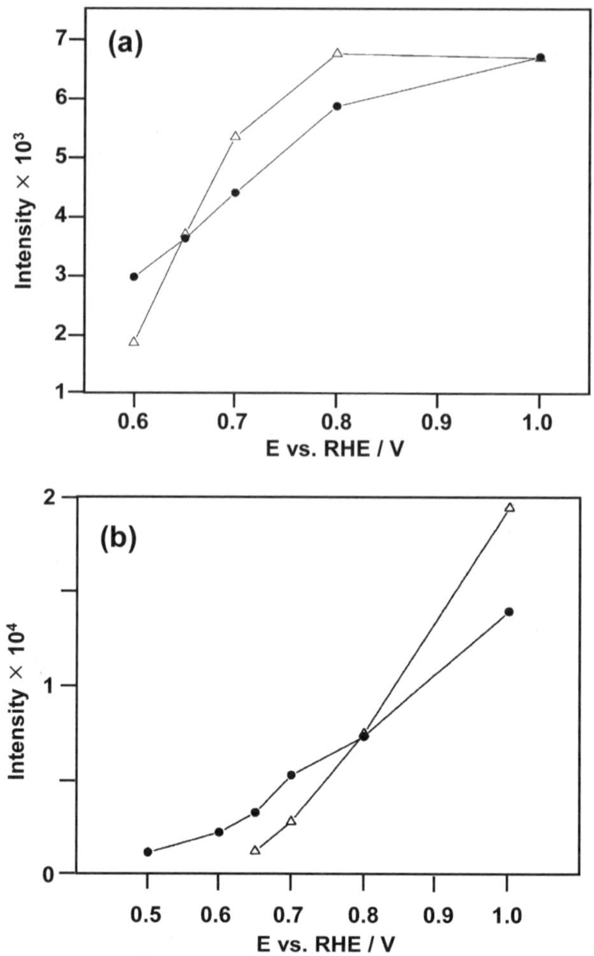

Figure 1.13 (a) Intensity of the CO_2 absorption band located close to 2345 cm^{-1} drawn from Figure 1.11b for a Pt catalyst (△) and Figure 1.12b for a PtSn catalyst (●), (b) intensity of the absorption band due to the ν(C–OH) mode of acetic acid located close to 1280 cm^{-1} drawn from Figure 1.11b for a Pt catalyst (△) and from Figure 1.12b for a PtSn catalyst (●).

In situ IR reflectance spectroscopy investigation of the oxidation of ethanol appears thus as an efficient method to elucidate some mechanistic aspects of the reaction. However, the quantitative analysis of the reaction products remains difficult due to different parameters: the characteristic absorption band may not be monopolar (this is the case for CO, for example) and the difficulty in obtaining a quantitative relationship between IR extinction coefficients and concentration for adsorbed intermediates and reaction products or by-products.

1.6.2.2 DEMS Study of Ethanol Electro-oxidation

A quantitative study of the electro-oxidation of ethanol on carbon-supported Pt, PtRu and Pt_3Sn catalysts was made by Behm and co-workers[101] using combined voltammetric and on-line DEMS measurements.

The three reaction products considered in the ethanol oxidation reaction are acetaldehyde, acetic acid and carbon dioxide. In order to obtain quantitative results on these three products, the authors made the choice to record the following MS ionic currents: $m/z = 22$ for CO_2 (corresponding to the CO_2^{2+} ion) and $m/z = 29$ for CH_3CHO (corresponding to the COH^+ main fragment). Indeed, both CO_2^+ and CH_3CHO^+ ions present a signal for $m/z = 44$ and it is not possible to distinguish them by recording this ionic current. Acetic acid could not be analyzed quantitatively by DEMS measurement owing to its low vapour pressure and its quantification was made indirectly by calculating the difference between the faradic current measured during ethanol oxidation and the partial current of ethanol oxidation to CO_2 and acetaldehyde.

The average current efficiency for complete oxidation of ethanol to CO_2 can be calculated using the following equation:

$$A_q(CO_2) = \frac{6\,Q_i}{K^*_{22}\,Q_f} \text{ or } A_i(CO_2) = \frac{6I_i}{K^*_{22}I_f} \qquad (1.47)$$

where Q_f and I_f are the faradic charge and faradic current during ethanol oxidation, respectively, and Q_i and I_i are the corresponding mass spectrometric charge and current for $m/z = 22$. K^*_{22} is the calibration constant for $m/z = 22$ determined from CO oxidation on a platinum catalyst.[88] The same approach is applied for the current efficiency of acetaldehyde formation, which is determined by:

$$A_q(CH_3CHO) = \frac{2Q_i}{K^*_{29}\,Q_f} \text{ or } A_i(CH_3CHO) = \frac{2I_i}{K^*_{29}I_f} \qquad (1.48)$$

K^*_{29} is the calibration constant for $m/z = 29$ determined from ethanol oxidation on a gold electrode, which presents a current efficiency close to 90% for the formation of acetaldehyde.[102] The authors also monitored the mass spectrometric signal for $m/z = 15$ and $m/z = 30$, corresponding to fragments of methane (CH_3^+) and ethane ($C_2H_6^+$), respectively.

Figure 1.14 shows a typical set of results obtained during the electro-oxidation of ethanol on a PtRu/C catalyst with the corresponding voltammogram (Figure 1.14a) and three mass spectrometric cyclic voltammograms (MSCVs) (Figures 1.14b–d).

The $m/z = 15$ ionic current (CH_4) follows exactly the $m/z = 29$ signal (CH_3CHO) at potentials greater than 0.3 V *vs.* RHE, indicating that CH_3^+ originates from the acetaldehyde fragment. A low signal appears for $m/z = 15$ (corresponding to methane) at potentials below 0.3 V *vs.* RHE, and the ethane ionic current is not measurable. These observations are also valid for a Pt_3Sn/C

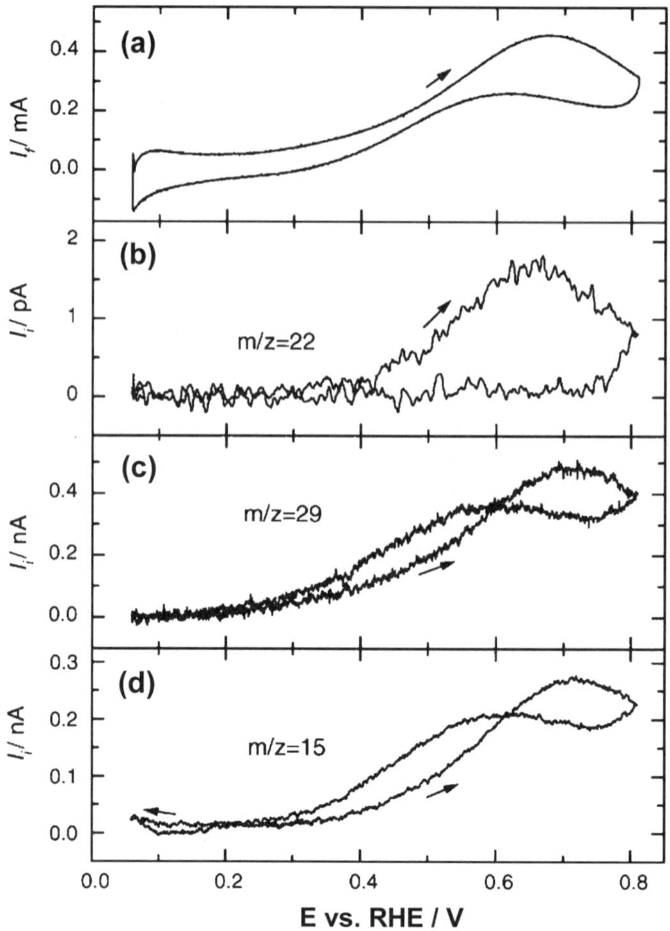

Figure 1.14 Simultaneously recorded (a) CV and (b–d) MSCVs for (b) m/z = 22, (c) m/z = 29 and (d) m/z = 15 for the oxidation of ethanol on a PtRu/C catalyst in a 0.5 M H_2SO_4 + 0.1 M ethanol solution; scan rate 10 mV s^{-1}. The arrows indicate the direction of the potential scan.

catalyst but are at variance with the result obtained on a Pt/C catalyst for which methane and ethane desorption are clearly noticeable at low potentials (below 0.3V *vs.* RHE). These species were also reported as being desorbed from Pt and PtRu catalysts containing 8 or 15 at% of Ru.[103] Both signals recorded for m/z = 22 and m/z = 29 give an interesting indication of the mechanism of ethanol oxidation. Considering the difference between the positive-going scan and the negative-going scan for the m/z = 22 ionic current, the incomplete oxidation of ethanol prevails in the negative-going scan. Furthermore, the formation of acetaldehyde, characterized by the m/z = 29 ionic current, begins at *ca.* 0.2 V *vs.* RHE on a PtRu/C catalyst, whereas it appears at higher potentials on a Pt/C catalyst. Over this potential, a strong increase in the formation of acetaldehyde on a Pt/C catalyst was observed, whereas the formation rate of acetaldehyde is lower on PtRu/C than on Pt/C.

From the mass spectrometric data and eqns (1.47) and (1.48), it is possible to calculate the relative current due to the formation of carbon dioxide, acetaldehyde and finally acetic acid from the ethanol oxidation reaction. These partial currents are shown for Pt/C, PtRu/C and Pt_3Sn/C catalysts in Figure 1.15.[101] This figure clearly presents the efficiency of the three different catalysts towards the formation of reaction products resulting from the electro-oxidation of ethanol. This quantitative analysis allows us to evaluate the total number of exchanged electrons during the oxidation reaction and the global current efficiency (A_q) and product yield (W_q) of the reaction calculated from the total charge involved for each partial current (Table 1.3).

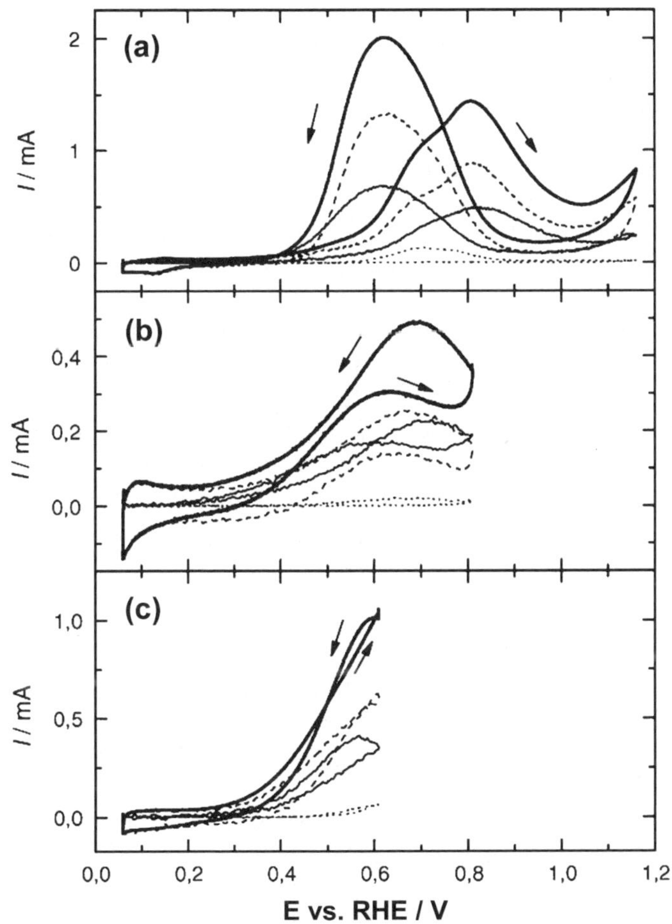

Figure 1.15 Faradic current signal and partial reaction currents during ethanol oxidation to CO_2 (*dotted line*), acetaldehyde (*thin solid line*) and acetic acid (*dashed line*) on (a) Pt/C, (b) PtRu/C and (c) Pt_3Sn/C catalyst electrodes.

Table 1.3 Average current efficiency (A_q) and product yields (W_q) for CO_2, CH_3CHO and CH_3COOH formation during the oxidation of ethanol in a 0.5 M H_2SO_4 + 0.1 M ethanol solution integrated over a complete signal cycle.

	Pt/C	PtRu/C	Pt_3Sn/C
$A_q(CO_2)$ (%)	2.7	2.6	2.6
$A_q(CH_3CHO)$ (%)	37	53	41
$A_q(CH_3COOH)$ (%)	60	44	56
$W_q(CO_2)$ (%)	1.3	1	1.2
$W_q(CH_3CHO)$ (%)	55	69	59
$W_q(CH_3COOH)$ (%)	44	30	40

This quantitative analysis allows a comparison in the product yield, W_q, for the ethanol oxidation reaction between different catalysts. In the discussed example, the three catalysts considered present close yields, with a low CO_2 production (for PtRu/C and Pt_3Sn/C catalysts, CO_2 is only produced during the positive-going scan), whereas acetaldehyde and acetic acid present a product yield of 55–70% and 45–30%, respectively. A slight increase in the acetaldehyde yield can be observed for the PtRu catalyst, leading to a lower faradic efficiency for ethanol oxidation, compared to that obtained on Pt/C and Pt_3Sn/C catalysts.

On the other hand, experiments carried out at high ethanol concentration (0.5 mol L^{-1}) showed the presence of the ethyl acetate ester characterized by the presence of fragments at m/z = 61, 73 and 88 at ratios typical of the ethyl acetate mass spectrum.[104] The ethyl acetate is formed by the following chemical reaction between the electrochemically formed acetic acid and ethanol (eqn 1.49) and confirms the formation of acetic acid.

$$CH_3COOH + CH_3CH_2OH \rightleftarrows CH_3COOCH_2CH_3 + H_2O \qquad (1.49)$$

1.6.2.3 HPLC Investigation of Ethanol Electro-oxidation

The adsorption and electro-oxidation of ethanol on Pt electrodes was also studied by combining cyclic voltammetry and long-term electrolysis with the on-line analysis by chromatographic techniques of the reaction products and by-products.[94]

The cyclic voltammograms of platinum, recorded in a three-electrode cell containing 0.1 M perchloric acid solutions with various concentrations of ethanol (from 5×10^{-3} to 0.2 M), are shown in Figure 1.16.

The voltammetric profiles of all the investigated solutions have the same shape and the current densities, j, increase with increasing the ethanol concentration, c. A logarithmic analysis of the curves of $\log j$ *vs.* $\log c$, for the two potentials 0.8 and 1.2 V *vs.* RHE, corresponding to peaks (A) and (B), respectively, gives a linear relation shift leading to an overall reaction order, m,

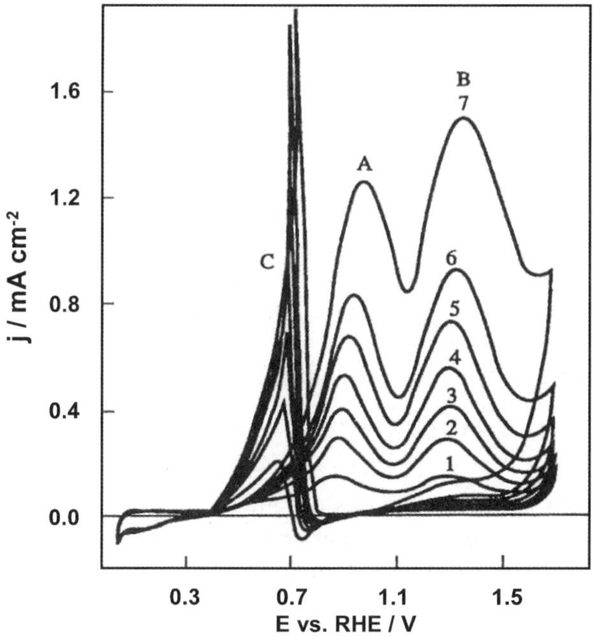

Figure 1.16 Voltammograms of a platinum electrode recorded in a 0.1 M perchloric acid solution at 0.05 V s^{-1} and 10 °C with various concentrations of ethanol: (1) 5×10^{-3} M; (2) 10^{-2} M; (3) 1.5×10^{-2} M; (4) 2.5×10^{-2} M; (5) 5×10^{-2} M; (6) 0.1 M; (7) 0.2 M.

close to 1. In a first approximation, the reaction is controlled by a first-order process in the two oxidation peaks (A) and (B).

Long-term electrolyses were carried out over a range of ethanol concentrations from 3.6×10^{-3} to 3.6×10^{-2} M with a three-plateau program sequence.[94] A first potential plateau (E_{el}), set in the potential region of peak (A) at 0.8 V *vs.* RHE for 10 s, is used to oxidize ethanol. This oxidation plateau is followed by a short potential pulse (E_{ox} = 1.6 V *vs.* RHE for 0.1 s), which allows cleaning of the electrode surface from the poisoning species (CO and C_2 residues) formed during the ethanol adsorption. A last potential plateau, E_{ads} = 0.2 V for 0.2 s, is added to reduce the surface oxygenated species and to adsorb ethanol, before its oxidation.

The mean number of electrons, $\tilde{n}$, associated with the overall oxidation process, for each electrolysis, is calculated from the analytical results and from the experimental value of the quantity of electricity, Q_{ex}, read on a coulometer during the experiment:

$$\tilde{n} = Q_{ex}/(F \times V_o \times \delta c) \qquad (1.50)$$

where δc is the concentration variation of the electrolyzed solution at a given electrolysis time and V_o is the volume of the electrolyte in the working compartment.

At low concentrations, the mean number of electrons $\tilde{n}$ is close to 4, which corresponds to the formation of acetic acid:

$$CH_3-CH_2OH+H_2O \rightarrow CH_3-COOH+4H^++4e^- \quad (1.51)$$

The experimental quantity of electricity, Q_{ex}, increases with the initial concentration, while the mean number of electrons, $\tilde{n}$, decreases and reaches a constant value ($\tilde{n} \approx 2$) for concentrations greater than 0.1 M. These results are in agreement with SNIFTIRS results (see Section 1.6.2.1.), which showed that the amount of acetic acid is greater at low ethanol concentrations, while at higher ethanol concentrations the IR spectrum indicates a higher amount of acetaldehyde. In the latter case, the main reaction would be:

$$CH_3-CH_2OH \rightarrow CH_3-CHO+2H^++2e^- \quad (1.52)$$

involving the transfer of 2 electrons per molecule.

Only weak traces of carbon dioxide (10^{-4} M) were detected by gas chromatography. The variation of the concentrations of ethanol and of the main reaction products (acetaldehyde and acetic acid), as analyzed by HPLC, are given in Figure 1.17.

The analysis of these concentration profiles suggests that the reaction mechanism does not only involve successive reactions, such as:

$$\text{Ethanol} \rightarrow \text{Acetaldehyde} \rightarrow \text{Acetic acid}$$

but also the direct oxidation of ethanol to acetic acid (eqn 1.51) is observed for potentials greater than 0.8 V *vs.* RHE.

Another observation is that the mean number of electrons may reach 4 at low ethanol concentrations. This can only be consistent with the above

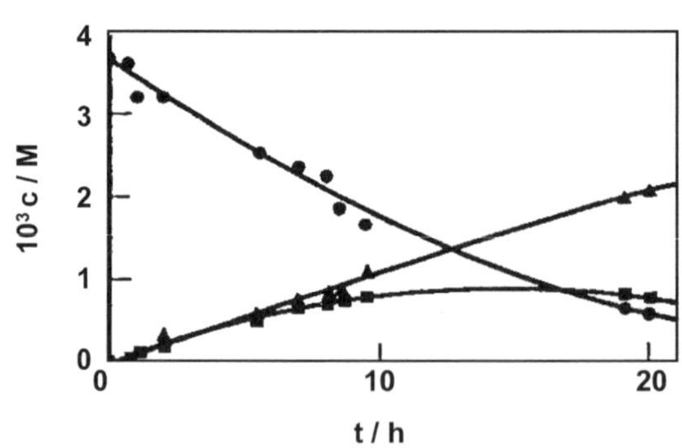

Figure 1.17 Concentration *vs.* time curves of the different products involved in the prolonged electrolysis of 3.6 mM ethanol at 0.8 V *vs.* RHE on a platinum electrode in 0.1 M $HClO_4$ and at 10 °C: (●) ethanol consumption; (■) acetaldehyde and (▲) acetic acid production.

explanation if it is assumed that ethanol oxidation involves two parallel reactions:

$$CH_3-CH_2OH \rightarrow CH_3-CHO + 2H^+ + 2e^- \quad (v_1) \qquad (1.52)$$

$$CH_3-CH_2OH + H_2O \rightarrow CH_3-COOH + 4H^+ + 4e^- \quad (v_2) \qquad (1.51)$$

where v_1 and v_2 are the reaction rates, with $v_1 > v_2$ for ethanol concentrations greater than 5×10^{-2} M and $v_1 < v_2$ for ethanol concentration lower than 5×10^{-2} M.

Several added metals were considered to improve the kinetics of ethanol oxidation at platinum-based electrodes, including ruthenium,[105,106] lead or tin.[100,107,108] Amongst them, tin appeared to be very promising. Figure 1.18 shows the polarization curves of ethanol electro-oxidation recorded at a slow sweep rate (5 mV s^{-1}) on different platinum-based electrodes. The $Pt_{0.9}Sn_{0.1}$/C catalyst displays the best activity in the potential range from 0.15 to 0.5 V *vs.* RHE, as it gives higher oxidation current densities than the other catalysts.

The role of tin can also be examined through the identification by *in situ* IR reflectance spectroscopy of the intermediate species resulting from ethanol adsorption and through the analysis by HPLC of reaction products and by-products in the anodic compartment of an electrochemical cell with two electrodes ($Pt_{0.83}Sn_{0.17}$/GC anode and glassy carbon cathode) separated by a Nafion 420 membrane.[21]

Long-term experiments on the electro-oxidation of ethanol were also carried out in a direct ethanol fuel cell (DEFC) of 25 cm^2 electrode surface area with a membrane-electrode assembly, the anode of which contained dispersed Pt-

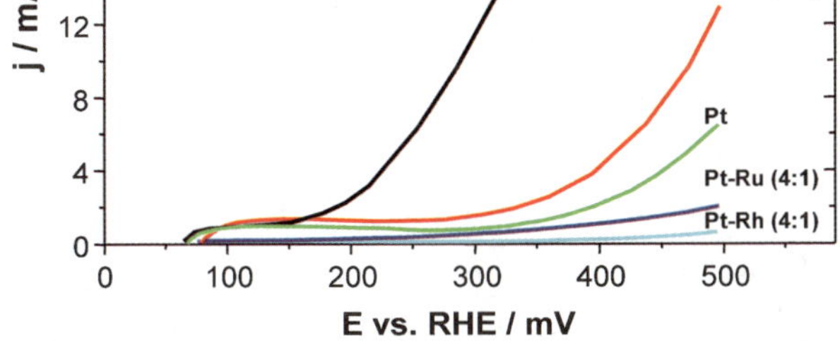

Figure 1.18 Electro-oxidation of ethanol on Pt/C and on different Pt-based bimetallic catalysts with 0.1 mg_{Pt} cm^{-2} loading; 0.1 M $HClO_4$ + 1 M C_2H_5OH; 5 mV s^{-1}; 3000 rpm; 25 °C.

based catalysts. The reaction products at the outlet of the anodic compartment are analyzed quantitatively by on-line HPLC analysis coupled with the fuel cell test bench.[7]

In order to determine the reaction products, the current density is kept constant and the cell voltage is simultaneously measured as a function of time. The ethanol flow rate was chosen close to 2 mL min^{-1} in order to perform the experiments for at least 4 h. The experiments were carried out using Pt/C, PtSn/C and PtSnRu/C as anode catalysts, and in each case no other reaction products than acetic acid (AA), acetaldehyde (AAL) and carbon dioxide (CO_2) could be detected by HPLC.[7] Depending on the electrode potential, AAL, AA, CO_2 and traces of CH_4 are obtained in different amounts, the main products being AAL and AA (Table 1.4), with the latter being considered as a final product because it is not oxidized under smooth conditions. Long-term electrolysis experiments on a Pt catalyst show that AAL is detected at potentials as low as 0.35 V *vs.* RHE, whereas no acetic acid is detected in this potential range.

From the results given in Table 1.4 it appears that the addition of tin to platinum greatly favours the formation of AA compared to AAL. This can be explained by the bifunctional mechanism,[80] according to which ethanol is adsorbed dissociatively at platinum sites, either *via* an O-adsorption or a C-adsorption process,[95] followed by the oxidation of these adsorbed species by oxygenated species formed on Sn at lower potentials, then leading to AA.

Table 1.4 indicates that alloying platinum with tin leads to an important change in the product distribution: increase of the AA chemical yield and decrease of the AAL and CO_2 chemical yields. The presence of tin seems to allow, at lower potentials, the activation of water molecules and the oxidation of AAL to AA. In the same manner, the amount of CO_2 decreased, which can be explained by the need of several adjacent platinum atoms (3 or 4) to realize

Table 1.4 Distribution of the reaction products resulting from the oxidation of 2 M ethanol on Pt/C, $Pt_{0.9}Sn_{0.1}/C$ and $Pt_{0.86}Sn_{0.1}Ru_{0.04}/C$ anode catalysts under DEFC working conditions at 80 °C for 4 h.

Anode electrocatalyst		*Pt/C*	*$Pt_{0.9}Sn_{0.1}/C$*	*$Pt_{0.86}Sn_{0.1}Ru_{0.04}/C$*
Metal loading/wt%		*60*	*60*	*60*
Current density/mA cm^{-2}		*8*	*32*	*32*
Cell voltage/V		*0.3–0.35*	*0.45–0.49*	*0.5–0.55*
Acetaldehyde	Concentration/mM	19	10	9
	Chemical yield/%	48	15.4	15.2
Acetic acid	Concentration/mM	13	50	44
	Chemical yield/%	32	77	75
CO_2	Concentration/mM	8	5	6
	Chemical yield/%	20	7.6	9.8

the dissociative adsorption of ethanol into CO, *via* breaking the C–C bond. In the presence of tin, "dilution" of platinum atoms can limit this reaction. The effect of tin, in addition to the activation of water molecules, may be also related to some electronic effects (ligand effects) on the CO oxidation reaction.[109]

1.6.2.4 Reaction Mechanisms of Ethanol Electro-oxidation

Voltammetric results, completed by the different spectroscopic and chromatographic analyses, allow the proposal of a detailed reaction mechanism of ethanol oxidation, on Pt-based electrodes, involving parallel and consecutive oxidation reactions, where the adsorption steps play a key role.

The first step is the dissociative adsorption of ethanol leading to acetaldehyde according to the following reactions:

$$Pt + CH_3-CH_2OH \rightarrow Pt-(CHOH-CH_3)_{ads} + H^+ + e^- \qquad (1.53)$$

$$Pt-(CHOH-CH_3)_{ads} \rightarrow Pt + CHO-CH_3 + H^+ + e^- \qquad (1.54)$$

Acetaldehyde has to be adsorbed to be further oxidized either into AA or CO_2. To complete the oxidation reaction leading to both these species, extra oxygen atoms are needed, which have to be brought by activated (adsorbed) water molecules at the platinum surface, giving adsorbed hydroxyl $-(OH)_{ads}$:

$$Pt + H_2O \rightarrow Pt-(OH)_{ads} + H^+ + e^- \quad \text{at } E > 0.6\text{ V } \textit{vs.}\text{ RHE} \qquad (1.35)$$

Thus, as soon as AAL is formed, it can adsorb on platinum sites, leading to a $-(CO-CH_3)_{ads}$ species (adsorbed acetyl):

$$Pt + CHO-CH_3 \rightarrow Pt-(CO-CH_3)_{ads} + H^+ + e^- \qquad (1.55)$$

Further oxidation, without breaking of the C–C bond, may occur at potentials > 0.6 V *vs.* RHE, by surface reaction with the $-(OH)_{ads}$, leading to AA:

$$Pt-(CO-CH_3)_{ads} + Pt-(OH)_{ads} \rightarrow 2Pt + CH_3-COOH \qquad (1.56)$$

However, IR reflectance spectroscopy has clearly shown that Pt is able to break the C–C bond, leading to adsorbed CO species at relatively low anode potentials (from 0.3 V *vs.* RHE).[93] This was confirmed by Iwasita *et al.*,[95] who found some traces of CH_4 at potentials lower than 0.4 V *vs.* RHE, resulting from a surface reaction with some $-(H)_{ads}$ coming from water activation according to reaction (1.35). Thus:

$$Pt-(CO-CH_3)_{ads}+Pt \rightarrow Pt-(CO)_{ads}+Pt-(CH_3)_{ads} \quad \text{at } E>0.3 \text{ V } vs. \text{ RHE} \tag{1.57}$$

and

$$Pt-(CH_3)_{ads}+Pt-(H)_{ads} \rightarrow 2Pt+CH_4 \quad \text{at } E<0.4 \text{ V } vs. \text{ RHE} \tag{1.58}$$

At potentials higher than 0.6 V *vs.* RHE, the dissociative adsorption of water occurs on platinum to provide $-(OH)_{ads}$ adsorbed species, able to oxidize further the adsorption residues of ethanol. Then, oxidation of adsorbed CO species occurs, as was shown by IR reflectance spectroscopy and CO stripping experiments:[108]

$$Pt-(CO)_{ads}+Pt-(OH)_{ads} \rightarrow 2Pt+CO_2+H^++e^- \tag{1.40}$$

On the other hand, AAL can also be directly oxidized, leading to AA, as follows:

$$Pt-(CHO-CH_3)_{ads}+Pt-(OH)_{ads} \rightarrow 2Pt+CH_3-COOH+H^++e^- \tag{1.59}$$

Taking into account all the results from IR reflectance spectroscopy, together with those from the quantitative analysis of reaction products and by-products by DEMS and chromatographic techniques, a general mechanism can be proposed for the electrocatalytic oxidation of ethanol on a platinum electrode (Scheme 1.3).

The spectroelectrochemical study of the adsorption and oxidation of ethanol and the HPLC analyses of the reaction products and by-products underline the necessity to activate water molecules at lower potentials in order to increase the activity of the catalyst and the selectivity towards either acetic acid or CO_2 formation, which means the improvement of the voltage, ε_E, and faradic, ε_F, efficiencies. To perform this, the modification of platinum by other metals, such as tin, ruthenium, molybdenum, *etc.*, is necessary.

On PtSn, assuming that ethanol adsorbs only on platinum sites, the first step can be the same as for platinum alone. However, as was shown by SNIFTIRS experiments,[21,108] the dissociative adsorption of ethanol on a PtSn catalyst to form adsorbed CO species takes place at lower potentials than on a Pt catalyst, between 0.1 V and 0.3 V *vs.* RHE, whereas on a Pt catalyst the dissociative adsorption of ethanol takes place at potentials between 0.3 V and 0.4 V *vs.* RHE. Then, it can be stated that the same reactions occur at lower potentials and with a relatively fast rate. Once intermediate species like $Pt-(CO-CH_3)_{ads}$ and $Pt-(CO)_{ads}$ are formed, they can be oxidized at potentials close to 0.3 V *vs.* RHE, as confirmed by CO stripping experiments, because OH species are formed on tin at lower potentials:[110,111]

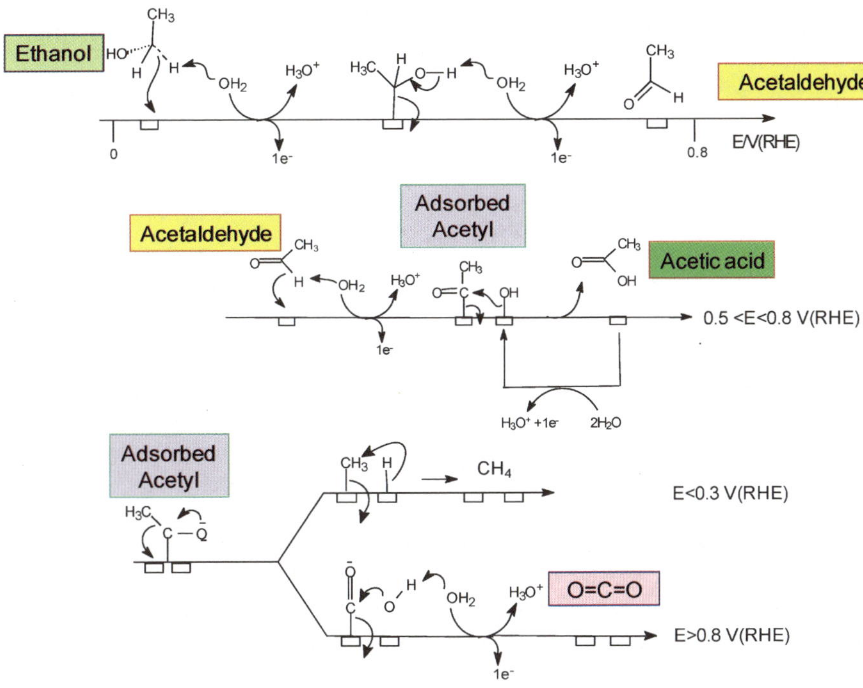

Scheme 1.3 Proposed mechanism for the electrocatalytic oxidation of ethanol on a smooth platinum electrode in acidic medium (all the species with coloured filling were detected either by IR reflectance spectroscopy or by chromatographic analysis).

$$Sn + H_2O \rightarrow Sn - (OH)_{ads} + H^+ + e^- \tag{1.60}$$

Then adsorbed acetyl species can react with adsorbed OH species to produce acetic acid according to:

$$Pt - (CO - CH_3)_{ads} + Sn - (OH)_{ads} \rightarrow Pt + Sn + CH_3 - COOH \tag{1.61}$$

Similarly, Pt–$(CO)_{ads}$ species are oxidized as follows:

$$Pt - (CO)_{ads} + Sn - (OH)_{ads} \rightarrow Pt + Sn + CO_2 + H^+ + e^- \tag{1.62}$$

These mechanisms explain also the higher efficiency of PtSn in forming AA compared to Pt at low potentials ($E < 0.35$ V *vs.* RHE), as was shown by electrolysis experiments. Indeed, adsorbed –$(OH)_{ads}$ species on Sn atoms allow oxidation of adsorbed –$(CO)_{ads}$ species to CO_2 or oxidation of adsorbed –$(CO-CH_3)_{ads}$ species to CH_3–COOH, according to the bifunctional mechanism.[80]

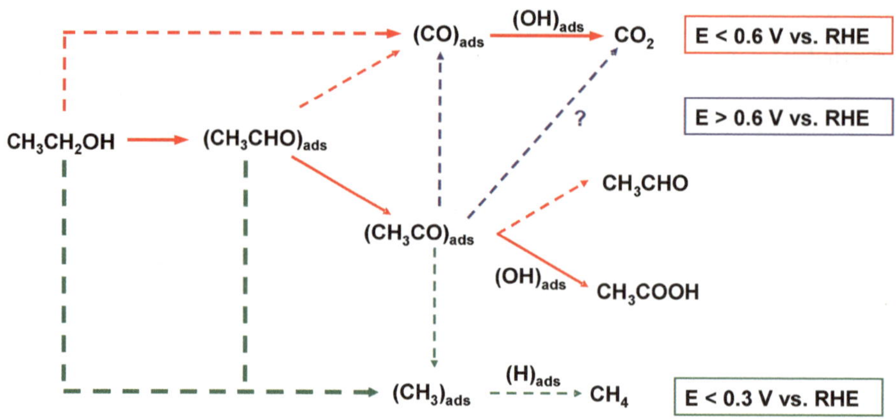

Scheme 1.4 Schematic representation of the mechanism of oxidation of –(CO–CH$_3$)$_{ads}$.

This reaction mechanism can be summarized by Scheme 1.4, where the key role of –(CO–CH$_3$)$_{ads}$ species (adsorbed acetyl) is indicated. In this mechanism the adsorbed acetyl plays a key role and its further oxidation is favoured by the addition to platinum of metal atoms more easily oxidized at low potentials, such as Ru, Sn, Mo, *etc.*

Other electrocatalysts were considered for the electro-oxidation of ethanol, such as rhodium, iridium[112] or gold,[102] leading to similar results in acid medium. The oxidation of ethanol on rhodium proceeds mainly through the formation of acetic acid and carbon monoxide, which is further oxidized to carbon dioxide when the rhodium surface begins to oxidize, at 0.5–0.7 V *vs.* RHE.[112] On gold in acid medium the oxidation reaction leads mainly to the formation of acetaldehyde.[102]

In conclusion, the generation of both poisoning species (mainly CO) and intermediate reaction products (AAL, AA) decreases correspondingly the useful energy density of the fuel, and also the power density of the fuel cell, since the oxidation current densities are lower than those obtained with the oxidation of methanol and above all of hydrogen. To improve the kinetics of ethanol oxidation would require the development of new electrocatalysts able to break the C–C bond at low temperatures and to oxidize adsorbed CO at lower potentials, *i.e.* to decrease the oxidation overpotential.

1.6.3 The Electro-oxidation of Polyols

The use of polyols as liquid fuels is an interesting alternative to hydrogen. Polyols such as ethylene glycol and glycerol are less toxic than methanol and display relatively high theoretical energy densities (5.2 and 5.0 kWh kg^{-1} for ethylene glycol and glycerol, respectively, *vs.* 6.1 and 8.0 kWh kg^{-1} for MeOH and EtOH, respectively[6,20]). Moreover, in these compounds, each carbon carries an alcohol group and as a consequence their partial oxidation to oxalic and mesoxalic acids in acid medium (according to reactions 1.63 and 1.64) and to oxalate and mesoxalate in alkaline medium (according to reactions 1.63′ and 1.64′), without C–C bond breaking, *i.e.* without CO_2 or CO_3^{2-} production, involves 8 and 10 exchanged electrons for ethylene glycol and glycerol, respectively, against 10 and 14, respectively, for their complete oxidation to CO_2 or CO_3^{2-} (according to reactions 1.65 and 1.65′ on one hand, and 1.66 and 1.66′ on the other hand):

$$CH_2OH-CH_2OH+2H_2O \rightarrow COOH-COOH+8H^+ +8e^- \quad (1.63)$$

$$CH_2OH-CH_2OH+10OH^- \rightarrow COO^- -COO^- +8H_2O+8e^- \quad (1.63')$$

$$CH_2OH-CHOH-CH_2OH+2H_2O \rightarrow COOH-CO-COOH+10H^+ +10e^- \quad (1.64)$$

$$CH_2OH-CHOH-CH_2OH+12OH^- \rightarrow COO^- -CO-COO^- +10H_2O+10e^- \quad (1.64')$$

$$CH_2OH-CH_2OH+2H_2O \rightarrow 2CO_2+10H^+ +10e^- \quad (1.65)$$

$$CH_2OH-CH_2OH+14OH^- \rightarrow 2CO_3^{2-} +10H_2O+10e^- \quad (1.65')$$

$$CH_2OH-CHOH-CH_2OH+3H_2O \rightarrow 3CO_2+14H^+ +14e^- \quad (1.66)$$

$$CH_2OH-CHOH-CH_2OH+20OH^- \rightarrow 3CO_3^{2-} +14H_2O+14e^- \quad (1.66')$$

Therefore, the possibility to oxidize these alcohols without breaking the C–C bonds could allow using 80% and 71.5% of the whole energy available, respectively.

1.6.3.1 Oxidation of Ethylene Glycol

The simplest alcohol, *i.e.* methanol, involves six electrons during its complete oxidation to CO_2, while other alcohols and polyols involve many more electrons. Therefore such oxidation reactions take place in a large number of steps, leading to several reaction products and by-products.

For example, the oxidation in acid medium of ethylene glycol to carbon dioxide, which needs 10 electrons, is a complex reaction including a parallel path, and involving several intermediate products. The reaction path is usually given as shown in Scheme 1.5.[6]

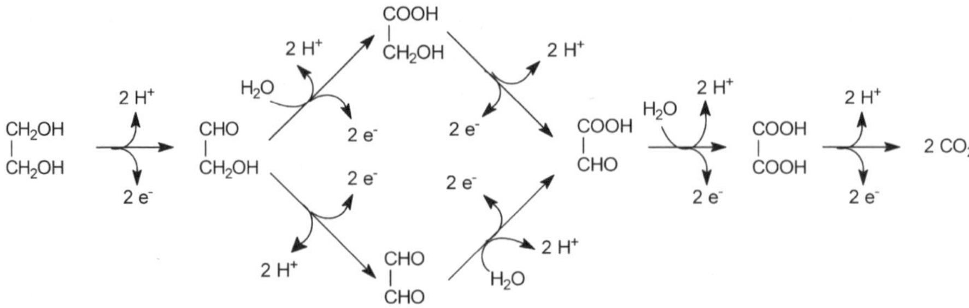

Scheme 1.5 Usual reaction mechanism proposed for the electro-oxidation of ethylene glycol.

Fourier transform IR transmission spectroscopy has been used as an analytical tool to identify the reaction products. The electrolysis of an alkaline solution (0.1 M NaOH) containing 0.1 M ethylene glycol has been carried out at room temperature on a smooth gold electrode (5 cm^2 surface area) at a constant controlled potential (1.2 V *vs.* RHE) in an electrochemical cell with two separated compartments.[113,114] The chosen potential corresponds to the maximum oxidation rate and the initial current density (10–12 mA cm^{-2}) stabilizes at around 3 mA cm^{-2} after 24 h. Samples of small volumes (0.1 cm^3) were taken out from the electrolysis compartment after different electrolysis times and analyzed by FTIRS with a Nicolet MX-10 spectrometer. The transmission spectra obtained, after subtraction of the background spectrum corresponding to the aqueous electrolyte alone, display different characteristic bands which may be assigned to different compounds (Figure 1.19). Except for the peaks at 1040 and 1085 cm^{-1}, which are always displayed and which correspond to the bands of ethylene glycol, there are three distinct peaks at 1325, 1350 and 1415 cm^{-1}, and a shoulder around 1380 cm^{-1}, the intensity of all of which continuously increases with the electrolysis time. It follows that small amounts of glycolaldehyde are detected (peak at 1383 cm^{-1}), that glycolate (or glyoxal due to the Cannizzaro reaction in alkaline medium) is clearly seen (peaks at 1077, 1327 and 1412 cm^{-1}), and that formate (peaks at 1350 and 1385 cm^{-1}) is also detected.

These preliminary investigations show that the reaction mechanism of ethylene glycol oxidation is much more complex than the one previously given, since formic acid is also detected as one of the main reaction products. This means that, at the relatively anodic potential where the reaction takes place on a gold electrode, the cleavage of the C–C bond may occur, even at room temperature. This result is at variance with what is usually found in the literature, *i.e.* the completion of the reaction in alkaline medium to oxalate.

In order to investigate further this electro-oxidation reaction in alkaline medium on catalyst nanoparticles, the "water-in-oil" microemulsion method was used to synthesize several nanostructured catalysts (Pt/C, PtPd/C, PtBi/C, PtPdBi/C). Characterization of the catalytic particles was carried out by different physical techniques. A Philips CM 120 microscope/EDX analyzer

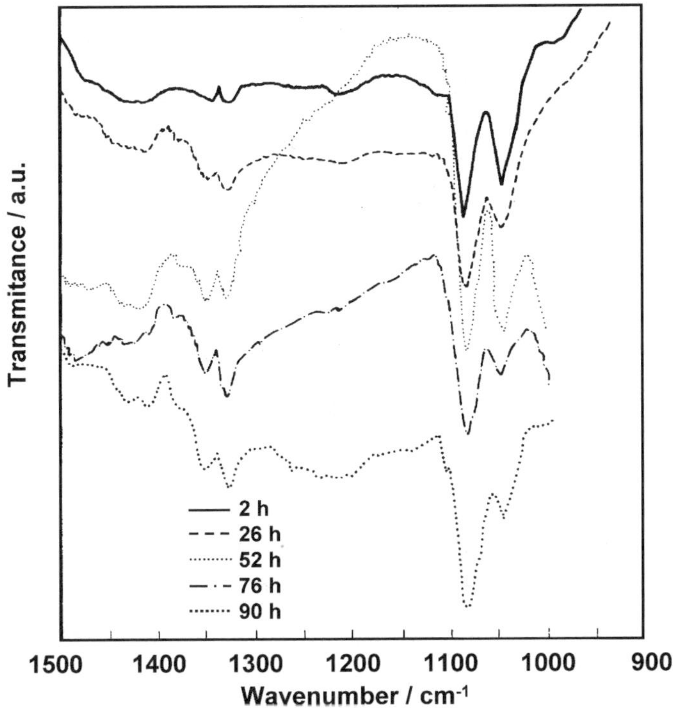

Figure 1.19 FTIR transmission spectra of the electrolyte solution (0.1 M ethylene glycol in 0.1 M NaOH) after different electrolysis times on a smooth gold electrode ($S \approx 5$ cm^2) at a controlled potential ($E = 1.2$ V *vs.* RHE).

equipped with a LaB_6 filament was used for transmission electron microscopy (TEM), allowing the morphological observation of the catalyst structure and the determination of the particle sizes, as well as an estimation of the particle composition by EDX. The confirmation of the catalyst composition was made by ICP-OES (Perkin–Elmer Optima 2000 DV). The results are given in Table 1.5.[115]

The addition of bismuth adatoms to a platinum surface has been studied by Kardigan *et al.*[54] These authors showed that, in alkaline medium, bismuth adatoms greatly enhance the catalytic activity towards ethylene glycol electro-oxidation. Polarization curves were then recorded with bismuth-containing nanostructured catalysts (Figure 1.20a). It can be seen that the lower Bi atomic ratio ($Pt_{0.9}Bi_{0.1}$/C catalyst) leads to the best activity in terms of higher achieved current densities at potentials higher than 0.45 V *vs.* RHE, *i.e.* the activity for ethylene glycol electro-oxidation at 0.6 V was twice as high as that obtained on a Pt/C 40 wt% catalyst, *i.e.* 8 mA cm^{-2} against 3.8 mA cm^{-2}. For higher bismuth content, a little increase of activity was revealed from 0.4 to 0.65 V, as shown by the higher current densities obtained with PtBi catalysts in this potential range. Addition of bismuth to platinum has a small effect on the

Table 1.5 Nominal metal loading, atomic composition determined by ICP-OES and particle size determined by TEM of Pt-based electrocatalysts supported on Vulcan XC-72 carbon used for the electro-oxidation of ethylene glycol in alkaline medium.

		Pt	Pt	$Pt_{0.5}Pd_{0.5}$	$Pt_{0.9}Bi_{0.1}$	$Pt_{0.8}Bi_{0.2}$	$Pt_{0.7}Bi_{0.3}$	$Pt_{0.45}Pd_{0.45}Bi_{0.1}$
Metal loading/wt%		20	50	20	50	50	50	50
ICP-OES composition/at%	Pt	–	–	59	88	78	68	43
	Pd	–	–	41	–	–	–	47
	Bi	–	–	–	12	22	32	10
d_{TEM}/nm		2.5	3.8	3.2	5.0	–	–	6.0

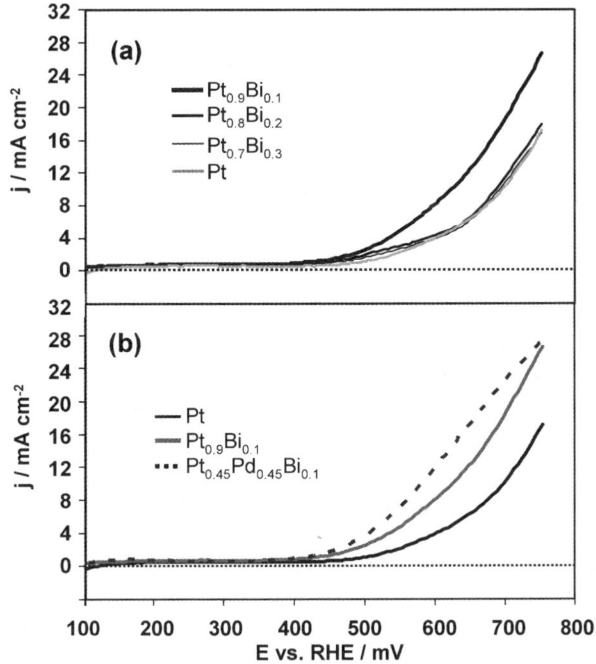

Figure 1.20 Polarization curves for the oxidation of ethylene glycol (EG) on several Pt-based electrocatalysts recorded in 0.2 M NaOH + 0.1M EG (v = 50 mV s^{-1}; 20 °C): (a) PtBi/C catalysts with different atomic compositions; (b) Pt/C, $Pt_{0.9}Bi_{0.1}$/C and $Pt_{0.45}Pd_{0.45}Bi_{0.1}$/C catalysts.

onset potentials of ethylene glycol electro-oxidation, which shifts between 0.47 and 0.4 V *vs.* RHE for Pt/C and PtBi/C catalysts, respectively.

Figure 1.20(b) compares the activity of Pt/C, PtBi/C and PtPdBi/C catalysts. Both bismuth-containing catalysts exhibit a better electroactivity towards ethylene glycol electro-oxidation than platinum alone for potentials higher than 0.39 V, as higher current densities are achieved. From 0.4 to 0.75 V the ternary catalyst leads to higher current densities than the bimetallic one, indicating its higher activity. The onset of the oxidation wave does not change between both catalysts; only the current densities of ethylene glycol electro-oxidation are increased.

To summarize, amongst the investigated catalysts, the electrocatalytic activity for ethylene glycol oxidation in alkaline medium was found to increase according to the following order: Pt/C 20 wt% < Pt/C 40 wt% < $Pt_{0.5}Pd_{0.5}$/C 20 wt% < $Pt_{0.9}Bi_{0.1}$/C 50 wt% < $Pt_{0.45}Pd_{0.45}Bi_{0.1}$/C 50 wt%. The difference observed for different metal loadings may be due to size effects and/or a change in the active surface area.

The oxidation products of ethylene glycol on Pt/C, PtBi/C and PtPdBi/C catalysts at a constant current density were qualitatively analyzed by liquid chromatography at the anode outlet of a solid alkaline membrane fuel cell

(SAMFC). For a current density of 20 mA cm^{-2}, the cell voltage is close to 0.33 V for the Pt/C anode, 0.51 V for the $Pt_{0.9}Bi_{0.1}$/C anode and 0.58 V for the $Pt_{0.45}Pd_{0.45}Bi_{0.1}$/C anode. Two different experiments were carried out: experiment 1 with a closed loop of fuel circulation and experiment 2 with an open loop circulation of fuel. In experiment 1, after 6 hours of product accumulation, glycolic acid, oxalic acid and some traces of formic acid were detected at the Pt/C anode, whereas glycolic acid, oxalic acid and some traces of glyoxylic acid were detected with the Bi-containing catalysts (Figure 1.21). The first important result is that platinum is able to break the C–C bond of ethylene glycol at ambient temperature. When Bi or PdBi is added to platinum, no formic acid is detected, which indicates that the dilution of platinum sites leads to a decrease in its ability to break the C–C bond. For experiment 2, glycolic acid and formic acid are again detected at the Pt/C anode outlet but no oxalic acid, whereas this latter is detected with bismuth-containing catalysts, but no glyoxylic acid.

In situ FTIR reflectance spectra (SPAIRS) of the adsorption and oxidation reaction of ethylene glycol in alkaline medium are given in Figure 1.22 for different electrodes: Pt/C, $Pt_{0.5}Pd_{0.5}$/C, $Pt_{0.9}Bi_{0.1}$/C and $Pt_{0.45}Pd_{0.45}Bi_{0.1}$/C.

For potentials as low as 0.23 V, negative absorption bands appear at wavenumbers close to 1577 and 1240 cm^{-1}, the latter being related to the C–O stretching mode of carboxylates.[116–119] These absorption bands can be related to those appearing at higher potentials, located close to 1357 cm^{-1} and assigned to adsorbed carboxylates.[118] Then, glycolate and/or oxalate and/or

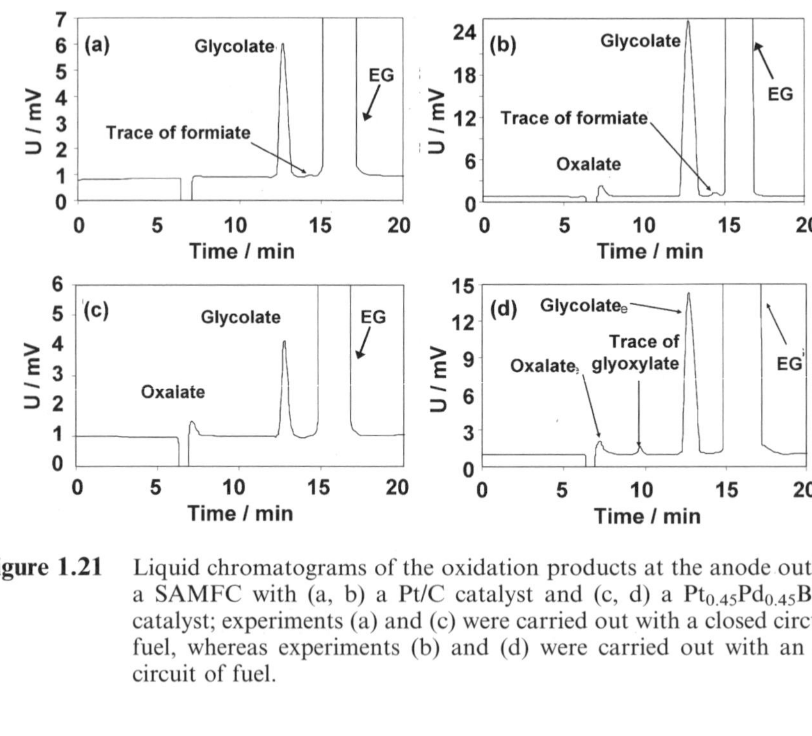

Figure 1.21 Liquid chromatograms of the oxidation products at the anode outlet of a SAMFC with (a, b) a Pt/C catalyst and (c, d) a $Pt_{0.45}Pd_{0.45}Bi_{0.1}$/C catalyst; experiments (a) and (c) were carried out with a closed circuit of fuel, whereas experiments (b) and (d) were carried out with an open circuit of fuel.

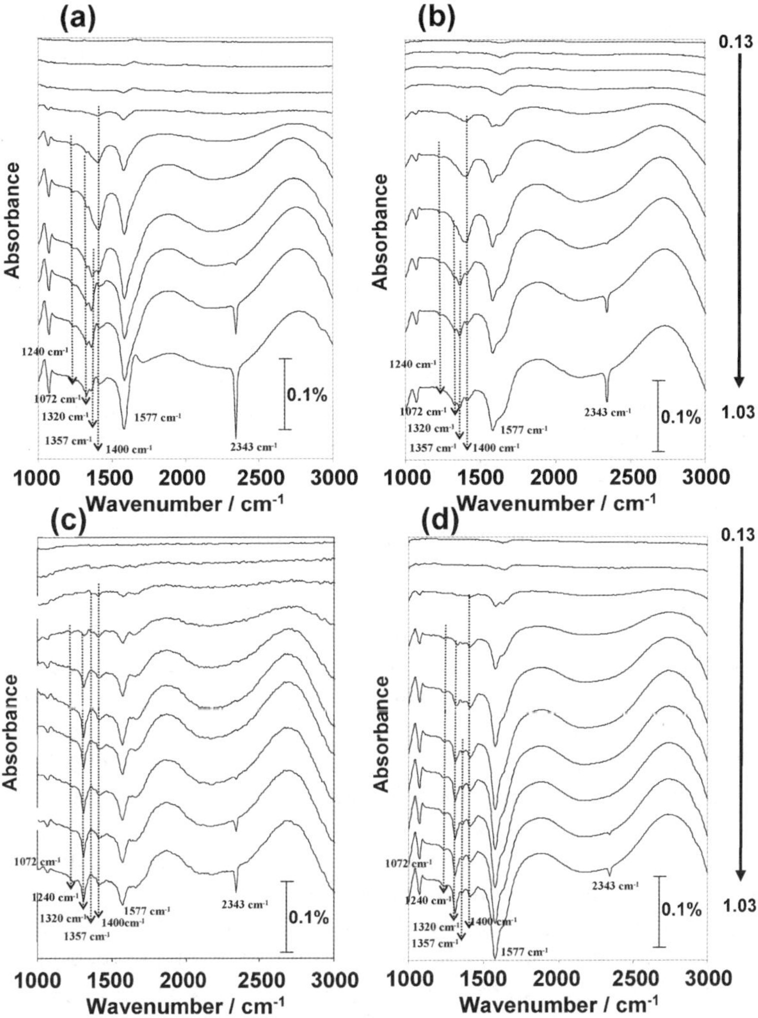

Figure 1.22 SPAIR spectra recorded in 0.2 M NaOH + 0.1 M ethylene glycol at (a) Pt/C, (b) $Pt_{0.5}Pd_{0.5}/C$, (c) $Pt_{0.9}Bi_{0.1}/C$ and (d) $Pt_{0.45}Pd_{0.45}Bi_{0.1}/C$ electrodes at various potentials (each 100 mV) from 0.130 to 1.030 V *vs*. RHE.

glyoxylate are formed.[119] However, according to Christensen and Hammett,[118] the presence of additional absorption bands at 1072 and 1320 cm^{-1} supports the formation of glycolate at low potentials at platinum. A broad band located close to 1400 cm^{-1} appears at 0.43 V, increases up to a potential of 0.63 V and seems then to decrease. This band can be due to the formation of carbonate (CO_3^{2-}) or glycolate. Christensen and Hammett[118] assigned the band located close to 1410 cm^{-1} to CO_3^{2-}. Finally, from 0.73 V, an absorption band located at 2343 cm^{-1} indicates unambiguously the production of CO_2.[21,24,118,119] The appearance of this band due to CO_2 coincides with the decrease of the band located at 1400 cm^{-1}. Considering at the same time the increase of the broad band between 2500 and 3000 cm^{-1}, which can be assigned to the consumption of OH^- (positive band),[118] it can be proposed that CO_2 appears as soon as CO_3^{2-} could not be further formed due to the lack of OH^-. Moreover, at potentials greater than 0.73 V, a shoulder appears in the band located at 1577 cm^{-1}. This shoulder, which can be due to the formation of some oxalate, can be correlated with the anode outlet analysis. It is proposed that oxalate is not a primary product of ethylene glycol oxidation at platinum, but that it comes from the further adsorption and oxidation of glycolate. Glycolate formed at the catalyst may readsorb and react to form oxalate. For $Pt_{0.5}Pd_{0.5}$/C, $Pt_{0.9}Bi_{0.1}$/C and $Pt_{0.45}Pd_{0.45}Bi_{0.1}$/C catalysts, some changes appear in the IR spectra, depending on the anode potential. The IR spectra of ethylene glycol adsorption and oxidation are very similar for these catalysts. In the spectra obtained with the $Pt_{0.45}Pd_{0.45}Bi_{0.1}$/C catalyst, a band located close to 1630 cm^{-1} appears at low potentials (0.13–0.23 V). For higher potentials, this band becomes a shoulder of an absorption band located close to 1572 cm^{-1}. This latter band is assigned, as for platinum, to the formation of glycolate and can be related to absorption bands appearing for higher potentials at 1350 and 1072 cm^{-1}. The former absorption band originates more likely from the formation of oxalate and can be related to the absorption band located close to 1307 cm^{-1}.[119] As for platinum, an absorption band located close to 1400 cm^{-1} appears but at a lower potential (from 0.33 V) and CO_2 formation occurs from 0.83 V. The relative intensity of the absorption bands related to glycolate is of the same order, indicating that the formation of this product occurs in an equivalent manner for both catalysts. On the other hand, the relative intensities of the CO_3^{2-} and CO_2 absorption bands are much higher at Pt/C than at other catalysts, indicating the higher ability of platinum to break the C–C bond and confirming the distribution products obtained under SAMFC working conditions. This can be explained by the fact that six to eight adjacent atoms of platinum are needed to realize the cleavage of the C–C bond occurring in the adsorption reaction of ethylene glycol. The dilution of surface platinum sites by addition of foreign metal atoms can decrease this ability. With the PtPdBi catalyst a small amount of CO_3^{2-} is produced, whereas oxalate is formed at lower potentials in a higher quantity. This can be explained by a bifunctional mechanism assuming that the presence of Bi provides extra OH species more easily to complete the oxidation reaction, as discussed in the next section.

1.6.3.2 Oxidation of Glycerol

The increasing world production of methyl esters as fuel additives to diesel oil has led to an increase in the production of glycerol (Gly), making it a cheap raw material, so that it can be considered now as a secondary primary energy source.[120–122]

In order to investigate the electrochemical oxidation of Gly on Pt- and Pd-based nanoparticles supported on Vulcan XC-72 carbon, several catalysts were prepared by the "water-in-oil" microemulsion method: Pd_xAu_{1-x}, Pd_xBi_{1-x}, Pt_xBi_{1-x} and $Pt_{0.45}Pd_{0.45}Bi_{0.1}$. Table 1.6 reports the physicochemical characterization of most of them. Metal loadings as determined by TGA and bulk composition as determined by ICP-OES are in agreement with the nominal ones. In the case of PdAu catalysts, XRD measurements indicate the formation of an alloy between both metals,[123] and TEM measurements show that the mean particle size increases with the Au content from 4.0 nm for pure Pd/C to 7.4 nm for pure Au/C. However, a palladium enrichment of the surface was evidenced by the method of Rand and Wood,[124] used for the determination of the surface composition. In the case of Bi-containing catalysts, no alloy was formed between Bi and Pt and/or Pd.[125] The addition of bismuth to Pd leads to an increase of the particle size, whereas with Pt it seems to have no significant effect on the value of the particle size.

The polarization curves for the oxidation of 0.1 M glycerol in 1.0 M NaOH were recorded at 10 mV s^{-1} on the different catalysts (Figure 1.23a). The platinum catalyst presents an onset potential for glycerol oxidation close to 0.4 V *vs.* RHE, which is approximately 0.15 V lower than that obtained with Pd/C and Au/C catalysts. Bimetallic PdAu catalysts lead to a shift of the onset potential by almost 0.1 V towards lower potentials compared to that obtained for Pd/C and Au/C catalysts (Figure 1.23b). The increase of the gold content leads to a slight increase in the activity between 0.45 and 0.7 V *vs.* RHE, as higher current densities are achieved. For higher potentials from 0.7 to 0.9 V *vs.* RHE, the palladium catalyst is the more active one. Obviously, different effects of the co-catalyst are involved in the mechanism of glycerol electro-oxidation at palladium-based materials. SPAIR spectra recorded during the oxidation of 0.1 M glycerol in 1.0 M NaOH aqueous solution on the monometallic Pt/C, Pd/C and Au/C catalysts are given in Figure 1.24.

Table 1.6 DTA-TGA, ICP-OES, TEM and XRD characterization of the different electrocatalysts used for the electro-oxidation of Gly in alkaline medium.

	Metal loading/*wt%*	Cell parameter/*nm*	Scherrer length, L_v/*nm*	Particle size, d_{TEM}/*nm*	Metal ratio/*at%*			
					Pd	Au	Pt	Bi
Pd/C	36	3.938	4.4	4.0	–	–	45	–
Au/C	39	4.067	–	7.4	–	–	–	–
Pt/C	37	3.942	5.5	5.3	–	–	–	–
$Pd_{0.7}Au_{0.3}$/C	37	3.986	–	4.9	69	31	–	–
$Pd_{0.5}Au_{0.5}$/C	37	4.001	–	5.0	50	50	–	–
$Pd_{0.3}Au_{0.7}$/C	40	4.018	–	5.1	31	69	–	–
$Pd_{0.9}Bi_{0.1}$/C	38	3.938	4.5	5.2	92	–	–	8
$Pt_{0.9}Bi_{0.1}$/C	36	3.943	5.7	4.7	90	–	–	10
$Pd_{0.45}Pt_{0.45}Bi_{0.1}$/C	38	3.943	4.7	4.5	43	–	–	12

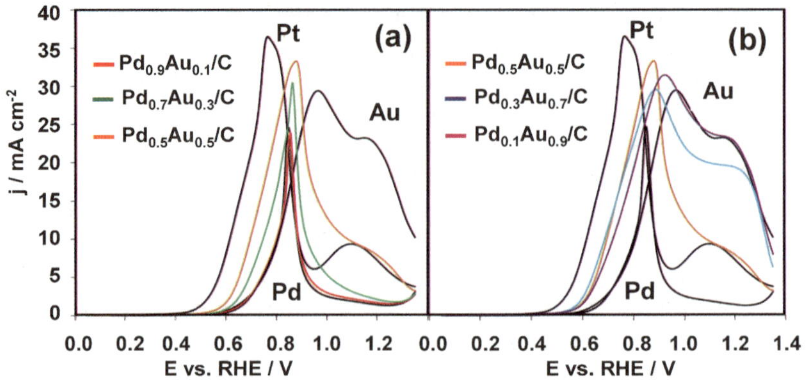

Figure 1.23 Oxidation curves recorded in a 1.0 M NaOH + 0.1 M glycerol solution on monometallic (Pt, Pd, Au) and bimetallic Pd_xAu_{1-x}/C catalysts (v = 10 mV s^{-1}; T = 20 °C).

The oxidation of glycerol leads to the formation of six main absorption bands in the 1100–1600 cm^{-1} range, identified as band A to band F in each figure. These six absorption bands appear with the three catalysts, indicating that the main reaction products may not differ irrespective of the catalytic material considered in this study. The main difference observed between these three catalysts is the presence of an absorption band at 1950 cm^{-1} with Pt/C and 1900 cm^{-1} with Pd/C. This absorption band is totally absent from IR spectra recorded with a gold catalyst. The position of this absorption band was

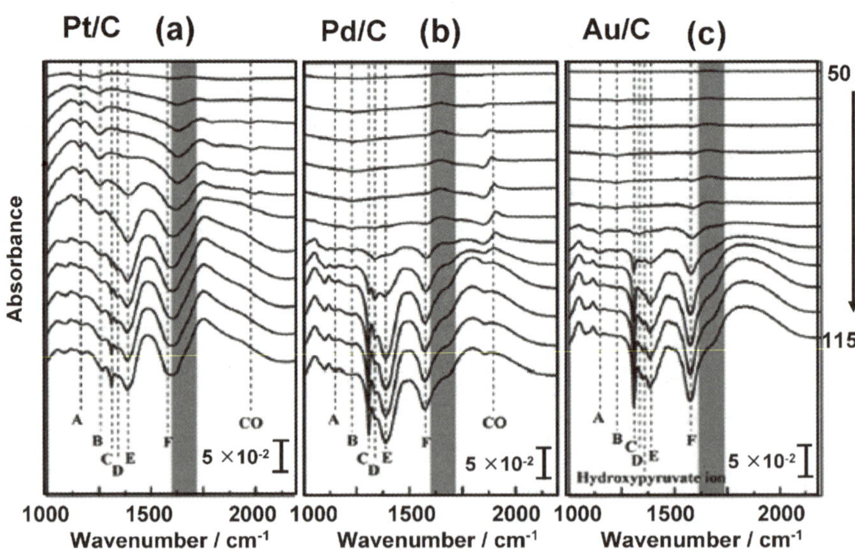

Figure 1.24 SPAIR spectra recorded during glycerol oxidation on (a) Pt/C, (b) Pd/C and (c) Au/C catalysts in a 1.0 M NaOH + 0.1 M glycerol solution at 20 °C. Scan rate: 1 mV s^{-1}; resolution 4 cm^{-1}.

attributed to carbon monoxide adsorbed on platinum (CO_L)[126] and palladium (CO_B).[127] The presence of adsorbed CO indicates that the dissociative adsorption of glycerol with the breaking of a C–C bond can occur at low potentials on both platinum and palladium nanoparticles. With a gold catalyst, this CO absorption band is not observed. This may support the hypothesis that glycerol adsorption does not occur on gold at low potentials or its adsorption does not lead to the breaking of a C–C bond. Indeed, Jeffery *et al.* demonstrated by using *in situ* FTIR the unambiguous presence of CO_2 as an electro-oxidation product of glycerol in alkaline media on a polycrystalline gold surface.[128] An absorption band which could be attributed to carbonates is slightly visible in Figures 1.24(a) and 1.24(b) for potentials close to 0.55 V *vs.* RHE, but it is then totally unobservable at higher potentials due to the appearance of the absorption band at 1385 cm^{-1} (named E in the spectra). A similar observation is made with adsorbed CO during the adsorption of propane-1,3-diol on a platinum catalyst in NaOH electrolyte, whereas no CO or CO_3^{2-} absorption bands were observed when the oxidation occured on a gold catalyst.[129] In the three spectra, two upward-going absorption bands are clearly visible in the 1000–1100 cm^{-1} range. They are attributed to glycerol consumption (these two peaks were identified independently by transmission IR spectroscopy with a 0.1 M glycerol + 1.0 M NaOH solution as being the two main absorption bands of glycerol). Considering the very low intensity of the absorption band characteristic of carbonates at 1400 cm^{-1} (which is not observed for potentials over 0.6 V *vs.* RHE), only molecules with three carbon atoms were considered (*i.e.* C–C bonds are considered as unbroken during the oxidation of glycerol). The absorption band located at 1575 cm^{-1} is typical of carboxylate ions, while other absorption band assignments are not straightforward. Notably, the weak peak located at 1335 cm^{-1} is absent from the IR spectra of most of the possible reaction products and could only be attributed to 1,3-dihydroxypropan-2-one.[130] IR spectra recorded during the oxidation of glycerol on gold nanoparticles (Figure 1.24c) present a particular absorption band at 1350 cm^{-1} in the potential domain over 0.7 V *vs* RHE. This absorption band is absent from spectra recorded with platinum and palladium catalysts and is characteristic of hydroxypyruvate ions.[131]

SPAIR spectra (Figure 1.25) show that the adsorption of glycerol was dissociative on Pd_xAu_{1-x} ($x > 0.3$) catalysts,[132] leading to the presence of adsorbed CO. The reaction mechanism seems very similar for the Pd- and/or Pt-based catalysts, while the gold monometallic catalyst presents a particular reaction product, *i.e.* the hydroxypyruvate ion. The fact that glycerol electro-oxidation at pure palladium and at pure gold catalysts starts at the same onset potential was unexpected. According to the bifunctional theory of electrocatalysis,[54,80] the oxidation of a primary alcohol to R–COOH and CO_2 (or R–COO^- and CO_3^{2-} in alkaline medium) needs the activation of water molecules to provide the OH adsorbed species at the surface to bring the extra oxygen atoms necessary to complete the oxidation reaction. Water activation is expected to occur at a palladium surface in the potential range from 0.5 to 0.6

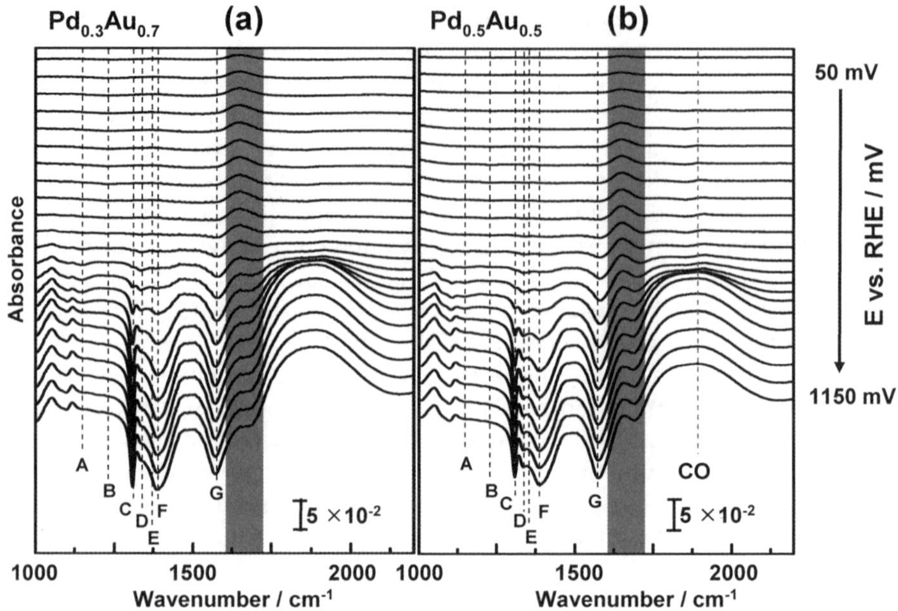

Figure 1.25 SPAIR spectra recorded during glycerol oxidation on (a) $Pd_{0.3}Au_{0.7}/C$ and (b) $Pd_{0.5}Au_{0.5}/C$ catalysts in a 1.0 M NaOH + 0.1 M glycerol solution at 20 °C. Scan rate: 1 mV s^{-1}; resolution 4 cm^{-1}.

V *vs.* RHE (just before PdO surface oxide formation). At a gold electrode, the electrochemically activated oxide layer formation starts at 1.2 V *vs.* RHE. Beden *et al.* proposed that the change in the catalytic behaviour for alcohol oxidation between acidic and alkaline media at a gold electrode is related to the existence of a sub-monolayer of adsorbed hydroxyl groups, the formation of which is favoured in alkaline medium at rather low potentials.[133]

Later, Chen and Lipkowski[134] evidenced, using *in situ* IR spectroscopy, that OH^- adsorption at a Au(111) single-crystal electrode in alkaline medium occurs at potentials as low as 0.6 V *vs.* RHE. Petrii *et al.*[135,136] studied the electro-oxidation of alcohols containing more than one carbon atom and the effect of adatoms on platinum. On the one hand, they concluded that their presence on the surface leads to significant effects on the amount and on the course of the electro-oxidation of strongly bonded species and that it is also possible that they affect the composition of chemisorbed species. Adžič[65] also proposed that some adatoms could prevent the adsorption and the formation of the strongly bound intermediates which occupy multiple surface sites. On the other hand, Petrii *et al.*[136] also proposed, in agreement with the bifunctional mechanism, that the considerable enhancement of the electrocatalytic activity due to the presence of adatoms in the electro-oxidation of alcohols may be explained assuming their "oxygen adsorbing" character, notably the formation of OH_{ads}. The first explanation is consistent with the behaviour of PdAu catalysts. The catalyst surface is palladium rich and gold may be considered mainly as an adatom. Of course, a bulk alloy effect cannot be discarded to explain the change in catalytic activity.[137] The increase of the gold coverage leads to a slight increase in the activity between 0.45 and 0.7 V *vs.* RHE. The adsorption of strongly bonded species at low potentials can be affected by the surface "dilution" of palladium by gold [according to Avramov-Ivič *et al.*, in the potential range before the Au(OH) formation there is no glycerol adsorption,[138] conversely to what happens on palladium], so that the coverage by the organic residues and by hydroxyl species are optimized, leading not only to the decrease of the onset of oxidation potential (by diminishing the poisoning effect) but also to the increase of the oxidation current densities.

Simoes *et al.*[125] found that bismuth-containing catalysts (Pd_xBi_{1-x}, Pt_xBi_{1-x} and $Pd_{0.45}Pt_{0.45}Au_{0.10}$) display a very high activity towards the glycerol electro-oxidation reaction, similar to or even higher than that of platinum alone.

The absorption band observed in the SPAIR spectra for all considered catalysts showed the occurrence of carboxylate ions among the reaction products. This molecular group involves the presence of the corresponding aldehyde as a reaction intermediate (or as a reaction product). While the presence of an aldehyde as an intermediate species is shown by the characteristic absorption band of carboxylate ions, their presence in the reaction products cannot be asserted since most of their IR absorption bands are overlapped with those of carboxylate ions. On palladium-based catalysts and platinum catalysts, the oxidation of glycerol may lead to three different reaction products, identified by their typical absorption band at 1575 cm^{-1}: glycerate ion, tartronate ion and mesoxalate ion. These three carboxylate ions involve different levels of oxidation of glycerol, with four exchanged electrons for glycerate ion formation, eight exchanged electrons for the formation of tartronate ions and 10 exchanged electrons to give mesoxalate ions. The latter molecule represents the last oxidation step before breaking the C–C bond. However, these three species cannot be distinguished by their IR spectra. The case of a gold catalyst is interesting, with an amount of hydroxypyruvate ion important enough to be observed by SPAIR spectroscopy. Furthermore, the absorption band named D, characteristic of 1,3-dihydroxypropan-2-one, is very weak in the SPAIR spectra recorded on a gold catalyst compared to that observed on platinum or palladium catalysts. Hence, it seems that in this particular case, 1,3-dihydroxypropan-2-one is only an intermediate species leading to the formation of the hydroxypyruvate ion, while in the case of the other catalytic materials the reaction stops at the 1,3-dihydroxypropan-2-one step.

Because *in situ* FTIR study does not allow determining all intermediate and/or reaction products, the elucidation of the oxidation mechanism has been completed by the analysis of reaction products by HPLC after long-term electrolysis at a given potential on Bi-containing catalysts. According to the work carried out by Bianchini's group on the determination of the reaction products of glycerol oxidation on Pd- and Pt-based anode catalysts in a

SAMFC, the main reaction products are glycerate, tartronate, glycolate, oxalate, formate and carbonate ions.[139,140]

The glycerol oxidation was performed at a constant electrode potential, for a long enough period of time to ensure a glycerol conversion yield of *ca.* 50%.[125] Table 1.7 gives the products detected by HPLC depending on the electrode potential for the different catalysts. The $Pd_{0.45}Pt_{0.45}Bi_{0.1}/C$ is the catalyst displaying the lower overpotential for the glycerol electro-oxidation reaction. The product distribution at 0.25 V *vs.* RHE (glycerate and dihydroxyacetone) indicates that glycerol oxidation occurs by an attack of the primary alcohol groups. No glyceraldehyde was detected, which seems to confirm that the isomerization reaction into dihydroxyacetone may occur at low potentials. However, the HPLC experimental conditions used made the separation of dihydroxyacetone and glycerate ions difficult owing to their very close retention times. Glycerate ion is also the main reaction product at low potentials for this long-term electrolysis experiment and traces of tartronate have been also detected. The increase of the electrode potential leads to a change in the reaction selectivity towards the production of tartronate and mesoxalate, although glycerate is detected in the whole studied potential range. It is then proposed that glycerol electro-oxidation occurs in several steps: the primary alcohol groups are first oxidized into carboxylates, and then the secondary alcohol group of tartronate is oxidized into mesoxalate. A similar mechanism seems to occur with $Pd_{0.9}Bi_{0.1}/C$ and $Pt_{0.9}Bi_{0.1}/C$ catalysts, but at higher potentials. However, experiments carried out at 0.85 V *vs.* RHE with these catalysts also indicate that the C–C bond of mesoxalate could be broken, forming oxalate and formate.

From *in situ* IR reflectance spectroscopy experiments and the determination of reaction products by HPLC after long-term electrolysis at different potentials, the mechanisms presented in Scheme 1.6 may be proposed for glycerol electro-oxidation, depending on the catalytic material used. The reaction pathway A first involves the oxidation of the primary alcohol group, leading to an aldehyde intermediate, whereas the reaction pathway B first implies the oxidation of the secondary alcohol group into a ketone function.

As a partial conclusion, the physical and electrochemical characterizations have shown that the atomic composition of the PdM/C catalysts can be controlled and that their structure depends on the nature of the catalytic metal added: an alloy is observed with gold and a non-alloyed bimetallic material is obtained with Bi. Their catalytic behaviour suggests an interaction between a Pd or a Pt phase and probably a $Bi(OH)_3$ phase. Electrocatalytic measurements indicate that the oxidation mechanism depends on the composition and structure of the PdM/C catalyst. *In situ* IR spectroscopy measurements have shown that the product distribution can be controlled by the catalyst composition and structure, and by the applied electrode potential. All these insights foresee the possibility to control the activity and selectivity of the catalysts towards a given reaction product. This may lead to the development of fuel cells allowing the cogeneration of energy and value-added chemicals.

Table 1.7 Products detected by HPLC and *in situ* spectroscopy for the different catalysts investigated used for the electro-oxidation of glycerol in alkaline medium.

Catalyst	Increasing potential →				
Pd/C	–	–	Dissociative adsorption → CO	Glyceraldehyde, dihydroxyacetone	Carboxylates
$Pd_{0.9}Bi_{0.1}/C$	–	Glyceraldehyde, dihydroxyacetone	Carboxylates		
$Pd_{0.45}Pt_{0.45}Bi_{0.1}/C$	Glyceraldehyde	Dihydroxyacetone, carboxylates	Carboxylates		
$Pt_{0.9}Bi_{0.1}/C$	Dihydroxyacetone, carboxylates	Carboxylates			
Pt/C	–	Dissociative adsorption → CO	Glyceraldehyde	Dihydroxyacetone, carboxylates	Carboxylates

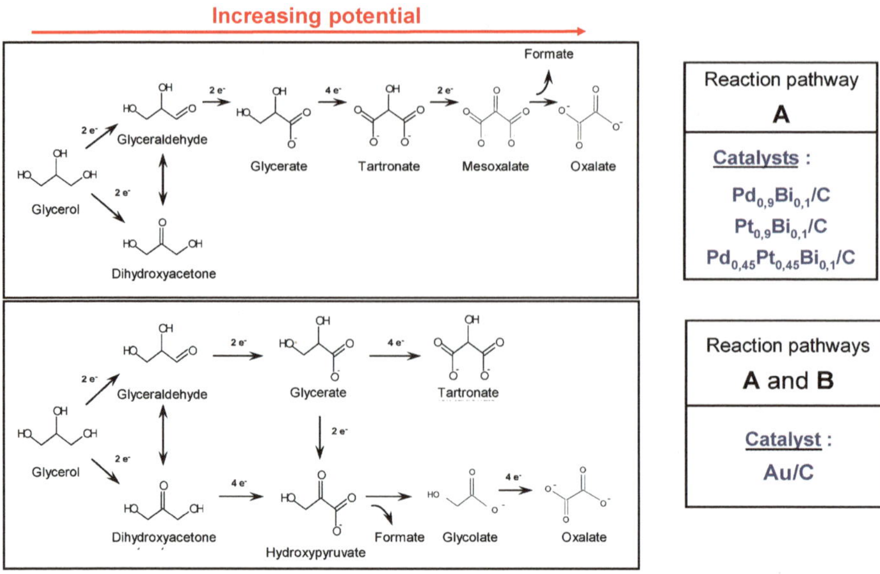

Scheme 1.6 Reaction mechanisms involved in the electro-oxidation of 0.1 M glycerol in 0.1 M NaOH.

1.7 Summary

The electrocatalytic oxidation of many alcohols has been investigated on several catalytic materials based on platinum group metals (PGMs). It has been demonstrated that the molecular reactivity depends both on the nature and structure of the metallic electrode and on the molecular structure of the alcohol.

The nature and structure of the catalytic electrodes have been varied using different synthesis methods to allow control of the particle size, the atomic composition and the metal loading of the fuel cell electrodes supported on large surface area carbon (*e.g.* Vulcan XC-72).

Several alcohols have been considered, such as aliphatic mono-alcohols (methanol, ethanol, *n*-butanol) and polyols (ethylene glycol, glycerol). Their oxidation mechanisms on catalytic electrodes of different nature and structure (smooth polycrystalline and single crystal electrodes, alloys and adatoms modified electrodes, dispersed nanoparticles, *etc.*) have been thoroughly investigated using different combined electrochemical and spectroscopic methods (IR reflectance spectroscopy, DEMS) and the analysis of reaction products and by-products by chromatographic techniques (GC and HPLC).

The detailed knowledge of the reaction mechanism is of prime importance to improve the catalytic properties of the electrode and to decrease its content in PGMs in order to reduce their cost for practical applications in fuel cell systems. This can be achieved by increasing the dispersion of the catalyst nanoparticles, *i.e.* by decreasing the particle size to a few nanometres and/or modifying the PGM catalysts by adding transition metals such as Sn, Mo, W, Ni, Co, Fe, *etc.*

From the mechanistic point of view, most of the alcohol oxidation reactions involve the formation, by partial oxidation of the primary alcoholic group, of an adsorbed aldehyde species $-(R-CHO)_{ads}$ which will lead to an adsorbed intermediate $-(R-CO)_{ads}$, the fate of which controls the rate of the overall reaction, since this aldehyde-like species may dissociate to $-(CO)_{ads}$ species and block the electrode active sites. Thus, the strategy to develop efficient electrocatalysts is to modify the Pt-based electrodes by adding some other metals able either to prevent the formation of $-(CO)_{ads}$ species by site dilution or to favour its oxidation at potentials much lower than those encountered on platinum.

References

1. B. D. McNicol, in *Catalysis*, ed. C. Kemball and D. A. Dowden, Royal Society of Chemistry, London, 1978, vol. 2, p. 243.
2. B. V. Tilak, R. S. Yeo, S. Srinivasan, in Comprehensive Treatise of Electrochemistry, J. O'M. Bockris, B. E. Conway, E. Yeager and R. E. White (Eds.), Plenum, New York, 1981, vol. 3, pp. 39–122.
3. C. Lamy, J.-M. Léger and S. Srinivasan, in *Modern Aspects of Electrochemistry*, ed. J. O'M. Bockris, B. E. Conway and R. E. White, Kluwer/Plenum, New York, 2001, vol. 34, p. 53.
4. A. Hamnett, in *Handbook of Fuel Cells: Fundamentals and Survey of Systems*, ed. W. Vielstich, A. Lamm and H. Gasteiger, Wiley, Chichester, 2003, vol. 1, p. 305.
5. A. Arico, V. Baglio and V. Antonucci, in *Electrocatalysis of Direct Methanol Fuel Cells*, ed. H. Zhang and H. Liu, Wiley-VCH, Weinheim, 2009, p. 1.
6. C. Lamy, E. M. Belgsir, in *Handbook of Fuel Cells: Fundamentals and Survey of Systems*, ed. W. Vielstich, A. Lamm and H. Gasteiger, Wiley, Chichester, 2003, vol. 1, pp. 323–334.
7. S. Rousseau, C. Coutanceau, C. Lamy and J.-M. Léger, *J. Power Sources*, 2006, **158**, 18.
8. C. Lamy, C. Coutanceau and J.-M. Léger, in *Catalysis for Sustainable Energy Production*, ed. P. Barbaro and C. Bianchini, Wiley-VCH, Weinheim, 2009, p. 3.
9. F. Kadirgan, B. Beden and C. Lamy, *J. Electroanal. Chem.*, 1982, **136**, 119.
10. L. Roquet, E. M. Belgsir, J.-M. Léger and C. Lamy, *Electrochim. Acta*, 1994, **39**, 2387.
11. R. S. Goncalves, J. M. Léger and C. Lamy, *Electrochim. Acta*, 1988, **33**, 1581.
12. D. Takky. B. Beden, J. M. Léger and C. Lamy, *J. Electroanal. Chem.*, 1983, **145**, 461.
13. D. Takky, B. Beden, J.-M. Léger and C. Lamy, *J. Electroanal. Chem.*, 1985, **193**, 159.

14. A. Capon and R. Parsons, *J. Electroanal. Chem.*, 1973, **45**, 205.
15. M. E. Vela, R. O. Lezna, N. R. de Tacconi, B. Beden, F. Hahn and C. Lamy, *J. Electroanal. Chem.*, 1992, **323**, 289.
16. C. Rice, S. Ha, R. I. Masel, P. Waszczuk, A. Wieckowski and T. Barnard, *J. Power Sources*, 2002, **111**, 83.
17. P. Olivi, L. O. S. Bulhoes, J. M. Léger, F. Hahn, B. Beden and C. Lamy, *J. Electroanal. Chem.*, 1994, **370**, 241.
18. B. Beden, C. Lamy, N. R. de Tacconi and A. J. Arvia, *Electrochim. Acta*, 1990, **35**, 691.
19. C. Roman-Martinez, D. Cazorla-Amoros, H. Yamashita, S. de Miguel and O. A. Scelza, *Langmuir*, 2000, **16**, 1123.
20. F. Vigier, C. Coutanceau, A. Perrard, E. M. Belgsir and C. Lamy, *J. Appl. Electrochem.*, 2004, **34**, 439.
21. F. Vigier, C. Coutanceau, F. Hahn, E. M. Belgsir and C. Lamy, *J. Electroanal. Chem.*, 2004, **563**, 81.
22. H. Bönnemann, W. Brijoux, R. Brinkmann, E. Dinjus, T. Joussen and B. Korall, *Angew. Chem. Int. Ed.*, 1991, **30**, 1312.
23. H. Bönnemann, W. Brijoux, R. Brinkmann, R. Fretzen, T. Joussen, R. Köppler, B. Korall, P. Neiteler and J. Richter, *J. Mol. Catal.*, 1994, **86**, 129.
24. L. Dubau, F. Hahn, C. Coutanceau, J.-M. Léger and C. Lamy, *J. Electroanal. Chem.*, 2003, **554/555**, 407.
25. L. Dubau, C. Coutanceau, J.-M. Léger and C. Lamy, *J. Appl. Electrochem.*, 2003, **33**, 419.
26. S. Brimaud, C. Coutanceau, E. Garnier, J.-M. Léger, F. Gérard, S. Pronier and M. Leoni, *J. Electroanal. Chem.*, 2007, **602**, 226.
27. M. Boutonnet, J. Kizling, P. Stenius and G. Maire, *Colloids Surf.*, 1982, **5**, 209.
28. F. Fievet, J. P. Lagier, B. Blin, B. Beaudoin and M. Fliglarz, *Solid State Ionics*, 1989, **32/33**, 198.
29. L. Ren and Y. Xing, *Electrochim. Acta*, 200, **53**, 5563.
30. C. Grolleau, C. Coutanceau, F. Pierre and J. M. Léger, *J. Power Sources*, 2010, **195**, 1569.
31. H. Yang, W. Vogel, C. Lamy and N. Alonso-Vante, *J. Phys. Chem. B*, 2004, **108**, 11024.
32. H. Yang, N. Alonso-Vante, J.-M. Léger and C. Lamy, *J. Phys. Chem. B*, 2004, **108**, 1938.
33. H. Yang, C. Coutanceau, J.-M. Léger, N. Alonso-Vante and C. Lamy, *J. Electroanal. Chem.*, 2005, **576**, 305.
34. V. S. Murthi, R. C. Urian and S. Mukerjee, *J. Phys. Chem. B*, 2004, **108**, 11011.
35. C. Coutanceau, A. Rakotondrainibe, A. Lima, E. Garnier, S. Pronier, J. M. Léger and C. Lamy, *J. Appl. Electrochem.*, 2004, **34**, 61.
36. P. Brault, A. Caillard, A. L. Thomann, J. Mathias, C. Charles, R. W. Boswell, S. Escribano, J. Durand and T. Sauvage, *J. Phys. D: Appl. Phys.*, 2004, **37**, 3419.

37. A. Caillard, P. Brault, J. Mathias, C. Charles, B. W. Boswell and T. Sauvage, *Surf. Coat. Technol.*, 2005, **200**, 391.
38. A. Caillard, C. Coutanceau, P. Brault, J. Mathias and J.-M. Léger, *J. Power Sources*, 2006, **162**, 66.
39. W. S. Rasband, ImageJ, US National Institutes of Health, Bethesda, Maryland, USA, http://rsbweb.nih.gov/ij/.
40. B. Beden, C. Lamy, A. Bewick and K. Kunimatsu, *J. Electroanal. Chem.*, 1981, **121**, 343.
41. B. Beden and C. Lamy, in *Spectroelectrochemistry: Theory and Practice*, ed. R. J. Gale, Plenum, New York, 1988, p. 189.
42. A. Bewick and B. S. Pons, in *Advances in Infrared and Raman Spectroscopy*, ed. R. J. H. Clark and R. E. Hester, Wiley-Heyden, London, 1985, vol. 12, p. 161.
43. D. S. Corrigan and M. J. Weaver, *J. Electroanal. Chem.*, 1988, **241**, 143.
44. K. Kunimatsu, *J. Electroanal. Chem.*, 1982, **140**, 205.
45. Z. Jusys, H. Massong and H. Baltruschat, *J. Electrochem. Soc.*, 1999, **146**, 1093.
46. E. M. Belgsir, E. Bouhier, H. Essis-Yei, K. B. Kokoh, B. Beden, H. Huser, J.-M. Léger and C. Lamy, *Electrochim. Acta*, 1991, **36**, 1157.
47. C. Lamy, J. M. Léger, J. Clavilier and R. Parsons, *J. Electroanal. Chem.*, 1983, **150**, 71.
48. B. Beden, C. Lamy and J.-M. Léger, in *Modern Aspects of Electrochemistry*, ed. J. O'M. Bockris, B. E. Conway and R. E. White, Plenum, New York, 1992, vol. 22, p. 97.
49. M. P. Hogarth and G. A. Hards, *Platinum Metal Rev.*, 1996, **40**, 150.
50. B. Beden, F. Hahn, S. Juanto, C. Lamy and J.-M. Léger, *J. Electroanal. Chem.*, 1987, **225**, 215.
51. E. Herrero, K. Franaszczuk and A. Wiekowski, *J. Phys. Chem.*, 1994, **98**, 5074.
52. A. Papoutsis, J.-M. Léger and C. Lamy, *J. Electroanal. Chem.*, 1993, **359**, 141.
53. R. R. Adžič, *Bull. Soc. Chim. Belgrad.*, 1983, **48**, 1.
54. F. Kadirgan, B. Beden and C. Lamy, *J. Electroanal. Chem.*, 1983, **143**, 135.
55. B. Beden, C. Lamy and J. M. Léger, *Electrochim. Acta*, 1979, **24**, 1157.
56. F. Kadirgan, B. Beden, J. M. Léger and C. Lamy, *J. Electroanal. Chem.*, 1981, **125**, 89.
57. M. M. P. Janssen and J. Moolhuysen, *Electrochim. Acta*, 1976, **21**, 869.
58. A. B. Anderson, E. Grantscharova and S. Seong, *J. Electrochem. Soc.*, 1996, **143**, 2075.
59. M. Watanabe and S. Motoo, *J. Electroanal. Chem.*, 1975, **60**, 267.
60. T. Iwasita, H. Hoster, A. John-Anaker, W. F. Lin and W. Vielstich, *Langmuir*, 2000, **16**, 522.
61. P. Waszczuk, G. Lu, A. Wieckowski, C. Lu, C. Rice and M. I. Masel, *Electrochim. Acta*, 2002, **47**, 3637.
62. A. Hamnett, *Catal. Today*, 1997, **38**, 445.

63. C. Ma and A.D. Kowalak. *Mater. Res. Soc. Proc.*, 1995, **368**, 145.
64. A. Lima, C. Coutanceau, J.-M. Léger and C. Lamy, *J. Appl. Electrochem.*, 2001, **31**, 379.
65. R. R. Adžič, in *Advances in Electrochemistry and Electrochemical Engineering*, ed. H. Gerisher and C. W. Tobias, Wiley-Interscience, New York, 1984, vol. 13, p. 159.
66. S. Rousseau, PhD Thesis, University of Poitiers, France, 2004.
67. K. Kinoshita, *J. Electrochem. Soc.*, 1990, **137**, 845.
68. A. Kabbabi, F. Gloaguen, F. Andolfatto and R. Durand, *J. Electroanal. Chem.*, 1994, **373**, 251.
69. B. J. Kennedy and A. Hamnett, *J. Electroanal. Chem.*, 1990, **283**, 271.
70. M. Watanabe, S. Saegusa and P. Stonehart, *J. Electroanal. Chem.*, 1989, **271**, 213.
71. T. Frelink, W. Visscher and J. R. van Veen, *J. Electroanal. Chem.*, 1995, **382**, 65.
72. F. Gloaguen, J. M. Léger and C. Lamy, *J. Appl. Electrochem.*, 1997, **27**, 1052.
73. D. M. Kolb, in *Spectroelectrochemistry: Theory and Practice*, ed. R. J. Gale, Plenum, New York, 1988, p. 87.
74. D. A. Buttry, in *Electroanalytical Chemistry*, ed. A. J. Bard, Dekker, New York, 1991, vol. 17, pp. 1–86.
75. G. Horányi, in *Interfacial Electrochemistry: Theory, Experiment, and Applications*, ed. A. Wieckowski, Dekker, New York, 1999, p. 477.
76. S. Juanto, B. Beden, F. Hahn, J.-M. Léger and C. Lamy, *J. Electroanal. Chem.*, 1987, **237**, 119.
77. B. Beden, S. Juanto, J.-M. Léger and C. Lamy, *J. Electroanal. Chem.*, 1987, **238**, 323.
78. T. Iwasita-Vielstich, in *Advances in Electrochemical Science and Engineering*, VCH, Weinheim, 1990, vol. 1, p. 127.
79. E. M. Belgsir, H. Huser, J.-M. Léger and C. Lamy, *J. Electroanal. Chem.*, 1987, **225**, 281.
80. M. Watanabe and S. Motoo, *J. Electroanal. Chem.*, 1975, **60**, 275.
81. H. A. Gasteiger, N. Markovic, P. N. Ross and E. J. Cairns, *J. Phys. Chem.*, 1993, **97**, 12020.
82. H. A. Gasteiger, N. Markovic, P. N. Ross and E. J. Cairns, *J. Electrochem. Soc.*, 1994, **141**, 1795.
83. T. Frelink, W. Visscher, A. P. Cox and J. A. R. van Veen, *Electrochim. Acta*, 1995, **40**, 1537.
84. J. G. Love and A. J. McQuillan, *J. Electroanal. Chem.*, 1989, **274**, 263.
85. B. Beden, F. Hahn, J.-M. Léger, C. Lamy and M.-I. Lopez, *J. Electroanal. Chem.*, 1989, **258**, 463.
86. T. Iwasita, F. C. Nart, B. Lopez and W. Vielstich, *Electrochim. Acta*, 1992, **37**, 2361.
87. A. Kabbabi, R. Faure, R. Durand, B. Beden, F. Hahn, J.-M. Léger and C. Lamy, *J. Electroanal. Chem.*, 1992, **444**, 41.

88. Z. Jusys, T. J. Schmidt, L. Dubau, K. Lasch, L. Jörissen, J. Garche and R. J. Behm, *J. Power Sources*, 2007, **105**, 297.
89. C. Lamy, A. Lima, V. Le Rhun, F. Delime, C. Coutanceau and J.-M. Léger, *J. Power Sources*, 2002, **105**, 283.
90. B. V. Tilak, B. E. Conway and H. Angerstein-Kozlowska, *J. Electronal. Chem.*, 1973, **48**, 1.
91. N. H. Li, S. G. Sun and S. P. Chen, *J. Electroanal. Chem.*, 1997, **430**, 57.
92. W. Chrzanowski and A. Wieckowski, *Langmuir*, 1998, **14**, 1967.
93. J.-M. Perez, B. Beden, F. Hahn, A. Aldaz and C. Lamy, *J. Electroanal. Chem.*, 1989, **262**, 251.
94. H. Hitmi, E. M. Belgsir, J.-M. Léger, C. Lamy and R. O. Lezna, *Electrochim. Acta*, 1994, **39**, 407.
95. T. Iwasita and E. Pastor, *Electrochim. Acta*, 1994, 39, 531547.
96. S. C. Chang, L. W. Leung and M. J. Weaver, *J. Phys. Chem.*, 1990, **94**, 6013.
97. X. H. Xia, H. D. Liess and T. Iwasita, *J. Electroanal. Chem.*, 1997, **437**, 233.
98. T. Iwasita, B. Rasch, E. Cattaneo and W. Vielstich, *Electrochim. Acta*, 1989, **34**, 1073.
99. C. Rice, Y. Y. Tong, E. Oldfield, A. Wieckowski, F. Hahn, F. Gloaguen, J.-M. Léger and C. Lamy, *J. Phys. Chem. B*, 2000, **104**, 5803.
100. F. Delime, J.-M. Léger and C. Lamy, *J. Appl. Electrochem.*, 1999, **29**, 1249.
101. H. Wang, Z. Jusys and R. J. Behm, *J. Power Sources*, 2006, **154**, 351.
102. G. Tremiliosi-Filho, E. R. Gonzalez, A. J. Motheo, E. M. Belgsir, J.-M. Léger and C. Lamy, *J. Electroanal. Chem.*, 1998, **444**, 31.
103. V. M. Schmidt, R. Ianniello, E. Pastor and S. Gonzalez, *J. Phys. Chem.*, 1996, **100**, 17901.
104. E. Stenhagen, S. Abrahamsson and F. W. McLafferty, *Atlas of Mass Spectral Data*, Interscience, New York, 1969.
105. J. Wang, S. Wasmus and R. F. Savinell, *J. Electrochem. Soc.*, 1995, **142**, 4218.
106. J. Souza, F. J. B. Rabelo, I. R. de Moraes and F. C. Nart, *J. Electroanal. Chem.*, 1997, **420**, 17.
107. C. Lamy, E. M. Belgsir and J.-M. Léger, *J. Appl. Electrochem.*, 2001, **31**, 799.
108. C. Lamy, S. Rousseau, E.-M. Belgsir, C. Coutanceau and J.-M. Léger, *Electrochim. Acta*, 2004, **49**, 3901.
109. P. Liu, A. Logadottir and J. K. Nørskov, *Electrochim. Acta*, 2003, **48**, 3731.
110. Y. Morimoto and E. B. Yeager, *J. Electroanal. Chem.*, 1998, 441, 77.
111. T. E. Shubina and M. T. M. Koper, *Electrochim. Acta*, 2002, 47, 3621.
112. N. R. de Tacconi, R. O. Lezna, B. Beden, F. Hahn and C. Lamy, *J. Electroanal. Chem.*, 1994, **379**, 329.
113. B. Beden, F. Kadirgan, A. Kahyaoglu and C. Lamy, *J. Electroanal. Chem.*, 1982, **135**, 329.
114. F. Kadirgan, E. Bouhier-Charbonnier, C. Lamy, J.-M. Léger and B. Beden, *J. Electroanal. Chem.*, 1990, **286**, 41.
115. L. Demarconnay, S. Brimaud, C. Coutanceau and J.-M. Léger, *J. Electroanal. Chem.*, 2007, **601**, 169.

116. R. B. de Lima, V. Paganin, T. Iwasita and W. Vielstich, *Electrochim. Acta*, 2007, **49**, 85.
117. L.-W. H. Leung and M. J. Weaver, *J. Phys. Chem.*, 1988, **92**, 4019.
118. P. A. Christensen and A. Hammett, *J. Electroanal. Chem.*, 1989, **260**, 347.
119. A. Dailey, J. Shin and C. Korzeniewski, *Electrochim. Acta*, 1998, **44**, 1147.
120. S. S. Yazadani and R. Gonzalez, *Curr. Opin. Biotechnol.*, 2007, **18**, 213.
121. M. A. Dasari, P. P. Kiatsimkul, W. R. Sutterlin and G. J. Suppes, *Appl. Catal. A*, 2005, **281**, 225.
122. J.-M. Clacens, Y. Pouilloux and J. Barrault, *Appl. Catal. A*, 2002, **227**, 181.
123. M. Simões, S. Baranton and C. Coutanceau, *J. Phys. Chem. C*, 2009, **113**, 13369.
124. A. J. Rand and R. Woods, *J. Electroanal. Chem.*, 1972, **36**, 57.
125. M. Simoes, S. Baranton and C. Coutanceau, *Appl. Catal. B*, 2011, **110**, 40.
126. A. Couto, A. Rincón, M. C. Pérez and C. Gutiérrez, *Electrochim. Acta*. 2001, **46**, 1285.
127. Y.-X. Jiang, S.-G. Sun and N. Ding, *Chem. Phys. Lett.*, 2001, **344**, 463.
128. D. Z. Jeffery and G. A. Camara, *Electrochem. Commun.*, 2010, **12**, 1129.
129. M. El Chbihi, D. Takky, F. Hahn, H. Huser, J.-M. Léger and C. Lamy, *J. Electroanal. Chem.*, 1999, **463**, 63.
130. C. J. Pouchert, in *The Aldrich Library of Infrared Spectra*, Aldrich, Milwaukee, 3rd edn., 1981.
131. F. Dickens and D. H. Williamson, *Biochem. J.*, 1958, **68**, 74.
132. M. Simoes, S. Baranton and C. Coutanceau, *Appl. Catal. B*, 2010, **93**, 354.
133. B. Beden, A. Cetin, A. Kahyaoglu, D. Takky and C. Lamy, *J. Catal.*, 1987, **104**, 37.
134. A. Chen and C. Lipkowski, *J. Phys. Chem. B*, 1999, **103**, 682.
135. B. I. Podlovchenko, O. A. Petrii, A. N. Frumkin and H. Lal, *J. Electronal. Chem.*, 1966, **11**, 12.
136. N. W. Smirnova, O. A. Petrii and A. Grzejdziak, *J. Electroanal. Chem.*, 1988, **251**, 73.
137. H. Igarashi, T. Fujino, Y. Zhu, H. Uchida and M. Watanabe, *Phys. Chem. Chem. Phys.*, 2001, **3**, 306.
138. M. L. Avramov-Ivić, J.-M. Léger, C. Lamy, V. D. Jović and S. D. Petrović, *J. Electroanal. Chem.*, 1991, **308**, 309.
139. V. Bambagioni, C. Bianchini, A. Marchionni, J. Filippi, F. Vizza, J. Teddy, P. Serp and M. Zhiani, *J. Power Sources*, 2009, **190**, 241.
140. C. Bianchini and P. K. Shen, *Chem. Rev.*, 2009, **109**, 4183.

CHAPTER 2

Nanoalloy Electrocatalysts for Alcohol Oxidation Reactions

JUN YIN, BRIDGID WANJALA, BIN FANG, JIN LUO, RAMESHOWRI LOUKRAKPAM, LEFU YANG, SHIYAO SHAN, MING NIE AND CHUAN-JIAN ZHONG*

Department of Chemistry, State University of New York at Binghamton, Binghamton, New York 13902, USA
*E-mail: cjzhong@binghamton.edu

2.1 Introduction

Alcohol fuel cells, such as direct methanol fuel cell (DMFCs) and direct ethanol fuel cells (DEFCs), become attractive because of high conversion efficiency, low pollution, light weight, high power density, and applications from small power supplies for electronic devices such as PCs, notebooks, and cellular phones. In a typical alcohol fuel cell (Scheme 2.1), the oxidation of methanol (for DMFC) or ethanol (for DEFC) occurs, where the oxygen reduction occurs at the air cathode. In DMFCs, the oxidation of methanol on a catalyst layer in the presence of water forms carbon dioxide. Protons (H^+) are transported across the proton exchange membrane (*e.g.*, Nafion) to the cathode, where they react with oxygen to produce water. Electrons (six in total) are transported through an external circuit, providing power to the connected load. The thermodynamic potential is 1.21 V. Both the methanol oxidation reaction at the anode and the oxygen reduction reaction at the cathode are highly irreversible and thus there is a loss of about 0.2 V at the anode for DMFC under open-circuit conditions, and an enhanced loss of

RSC Energy and Environment Series No. 6
Catalysts for Alcohol-Fuelled Direct Oxidation Fuel Cells
Edited by Zhen-Xing Liang and Tim S. Zhao

Published by the Royal Society of Chemistry, www.rsc.org

about 0.1 V at the oxygen electrode because of the crossover of methanol from the anode to the cathode.[1] In DEFCs, ethanol is used instead of the more toxic methanol. Ethanol is an attractive alternative to methanol. In comparison with methanol, ethanol is a hydrogen-rich liquid and it has a higher energy density (8.0 kWh kg^{-1}) compared to methanol (6.1 kWh kg^{-1}). Ethanol also has a 33% higher energy density than methanol. Ethanol can be obtained from renewable resources like sugar cane, wheat, or corn. Bio-ethanol is thus attractive and has the potential to significantly increase fuel utilization and fuel cell performance, lower costs, and improve safety.

The catalyst is therefore one of the key components in alcohol fuel cells.[2–4] According to the cost breakdown of fuel cell components,[5] the cost of catalysts in manufacturing fuel cells is the highest (~30%) for small production volumes, and remains very high with increasing production volumes. Currently, the low activity, poor durability, and high cost of platinum-based anode and cathode catalysts in DMFCs constitute some of the major barriers to commercialization of fuel cells. In addition, the durability of Pt-based catalysts can be compromised by sintering and dissolution of the catalysts in fuel cells. There is a major gap in the development of catalyst technology between the laboratory test and the practical application, which largely lies at the lack of abilities to engineer the size, composition, and stability. A key challenge to the commercialization of fuel cells is the development of active, robust, and low-cost catalysts.[6,7] In contrast to most existing catalyst preparations based on traditional co-precipitation or impregnation methods, the foundation of the nano-engineering of multimetallic alloy catalysts[8–15] has many important attributes to bridge the existing gap and lead to new opportunities in creating a synergistic balance of activity and stability of the

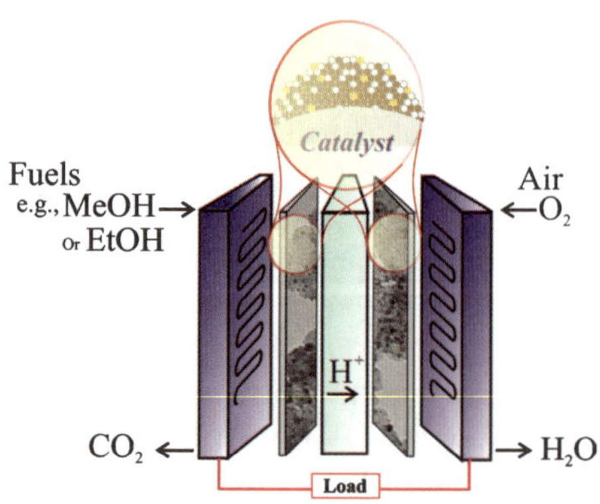

In DMFC:		In DEFC:	
Anode	$CH_3OH + H_2O \rightarrow CO_2 + 6H^+ + 6e^-$	Anode	$CH_3CH_2OH + 3H_2O \rightarrow 2CO_2 + 6H^+ + 12e^-$
Cathode	$\frac{3}{2}O_2 + 6H^+ + 6e^- \rightarrow 3H_2O$	Cathode	$3O_2 + 12H^+ + 12e^- \rightarrow 6H_2O$
Overall	$CH_3OH + \frac{3}{2}O_2 \rightarrow CO_2 + 2H_2O$	Overall	$CH_3CH_2OH + 3O_2 \rightarrow 2CO_2 + 3H_2O$

Scheme 2.1 A schematic illustration of a DMFC and a DMFC, and the fuel cell reactions.

catalysts.[3] For example, TiO_2-supported bimetallic PtFe nanoparticles of different compositions have been demonstrated to exhibit an enhanced electrocatalytic activity for methanol oxidation in acidic electrolytes,[16,17] that is comparable with those for PtRu catalysts.

The development of DEFCs has drawn a surge of interest recently due to its biofuel characteristic, its ability to eliminate the toxicity issue of methanol as in DMFCs, and its high energy density. Carbon-supported PtRu, a well-known catalyst for DMFCs, was naturally studied for DEFCs.[18] Recently, a palladium-based catalyst has been identified as a promising area of research. The mechanism of the ethanol oxidation reaction on palladium in alkaline media has been studied, which showed the rate-determining step to be the removal of the adsorbed ethoxy group by the adsorbed hydroxyl.[19,20] The combination of Pd with another metal has been shown to enhance both the catalytic activity and stability of Pd for the ethanol oxidation reaction (EOR) in alkaline media, including bimetallic and trimetallic catalysts such as PdIr/C, PdPt/C, PdNi/CNF, PtSnPd/C, *etc.*[21–25] Enhanced electrocatalytic activity has been reported for ethanol oxidation using a PdAu alloy supported on carbon[26] and tungsten carbide.[27]

While various nanoparticle catalysts have been reported for both DMFCs and DEFCs, the understanding of the nanoscale properties of the catalysts is rather limited, which constitutes a major hurdle for the rationale design of active, selective, robust, and cost-effective catalysts. Nanoscale materials differ from their bulk counterparts in many profound ways, which has become a topic of increasing interest in recent years, especially in fuel cell related catalytic reactions. One major area of interest involves the study of nanoparticles with multimetallic compositions such as bimetallic and trimetallic nanoparticles, where the alloying or phase segregation is dramatically dependent on the particle size. The ability to control such nanoscale phase properties is important for the exploration of the bimetallic or trimetallic nanoparticles for the design of advanced catalysts. One significant example that is highlighted in this article is the recent progress in the study of gold–platinum (AuPt) nanoparticles. In contrast to the miscibility gap in a wide bimetallic composition range well known for bulk gold–platinum systems,[28,29] our work in 2005 demonstrated the single-phase alloy properties for bimetallic AuPt nanoparticles (2–5 nm) prepared by two-phase synthesis and thermal activation on carbon support materials.[30] Because AuPt nanoparticles exhibit many interesting properties, especially electrocatalytic activities for fuel cell reactions such as the methanol oxidation reaction (MOR) and oxygen reduction reactions (ORRs),[2,3,8,9,14,32–34] the understanding of the nanoscale phase properties and their correlation with their electrocatalytic properties constituted an important driving force for a rapid growth of research interest in chemistry, materials, and theory in the last five years.[3,8,9,14,30–53] This article will highlight the recent insights into the nanoscale AuPt system, focusing mainly on findings from our laboratory,[3,8,9,14,30–34,40–47] and some relevant experimental and theoretical studies from other research laboratories.[35–39,48,51–53]

The early finding that AuPt nanoparticles supported on carbon exhibit single-phase alloy properties[2] represented the first example demonstrating the difference of nanoscale materials from their bulk counterparts. The finding was supported by a theoretical study based on a thermodynamic model and an analytically embedded atom method,[35] which showed a negative heat of formation for AuPt particles smaller than ~6 nm, and several experimental studies of different AuPt systems.[36–39,51–53] Theoretically or experimentally, both alloying and phase-segregation are possible, depending on the preparation conditions and the theoretical model. For example, thin films of AuPt prepared by co-sputtering were shown to exhibit alloy properties dependent on surface composition,[38] whereas thin films of AuPt prepared by crossed-beam pulsed laser deposition showed the formation of metastable AuPt alloys over the entire composition range.[51] AuPt/C prepared by a microemulsion method showed segregation of Pt on the surface for low-Au particles, but single-phase alloy properties on high-Au particles;[49] however, AuPt/C prepared *via* a polyol method appeared to be unalloyed.[50] It is now recognized that the alloying or phase-segregation depends on a number of parameters, including the preparation method and support properties; the latter was supported by insight from a recent *in situ* X-ray diffraction study of AuPt nanoparticles on different supports[40] and molecular dynamic simulations of AuPt on supports of different surface properties.[53]

There are also relevant examples demonstrating the ability to control the bimetallic phase properties, *e.g.* preparation of Au-core/Pt-shell, Pt-core/Au-shell, and other core/shell nanoparticles.[46,47,54,55] Some of these systems were exploited for the electrocatalytic methanol oxidation reaction[56,57] and the oxygen reduction reaction, which include Au-core/Pt-shell catalysts,[46,47,58,59] partially alloyed AuPt catalysts,[48] and Pt nanoparticles modified with gold clusters.[52] The importance of the relative surface alloying or layering arrangements has also been recognized for metals on single-crystal substrates, including the "near surface alloy" model of metal adlayer on metal substrate for the ORR.[58,59] However, the correlation between the synergistic surface properties and the spatial arrangement for different components in metal or oxide nanoparticles remains elusive.

In experimental and theoretical studies aimed at establishing the correlation between the nanoscale phase properties and the catalytic activities, a fundamental question is how the nanoscale alloying and phase segregation can be controlled. For Au_nPt_{100-n} nanoparticles supported on a certain type of support material, the nanoscale alloying or phase segregation could lead to the formation of several possible states. Some of these states have been revealed in our work to be obtainable, depending on the composition, substrate, and thermal treatment temperature.[30,42] There are currently two aspects of the fundamental insights that are considered in supporting some of these alloying or phase-segregation states. One involves the theoretical finding of a negative heat of formation for AuPt nanoparticles smaller than ~6 nm, which constituted as a thermodynamic basis for favoring the formation of alloyed phase.[35] The other involves the theoretical finding of a possible phase segregation of AuPt particles as a result of the combination of a larger surface free energy of Pt (2.48 J m^{-2}) than Au (1.50 J m^{-2})[35] and the lower melting point of smaller-sized nanoparticles compared with larger-sized particles ($\Delta T \propto 1/r_{(\text{particle radius})}$).[60–62] These two aspects are discussed based on the latest experimental findings on the phase properties of supported Au_nPt_{100-n} nanoparticles. The insights into the composition, morphology, phase, and surface properties have provided not only the experimental confirmation of the theoretical findings, but also the first experimental demonstration of the control of the nanoscale alloying and phase segregation for this bimetallic system.

While significant progress has been documented in technological areas of alcohol fuel cells, including hardware design, membrane, catalysts, and theoretical modeling, it is the catalytic aspect that constitutes the focus of this article. In this chapter, some recent results in our investigations of nano-engineered alloy catalysts for the oxidation reactions involved in DMFCs and DEFCs are discussed. Selected examples from bimetallic nanoparticle catalysts, especially AuPt catalysts, are highlighted in terms of their structural and catalytic properties.

2.2 Preparation of Nanoalloy Catalysts

The preparation method of catalysts has been an important focus for various types of catalytic reactions. Traditional approaches to preparing supported nanoparticle catalysts involve co-precipitation, deposition precipitation, ion exchange, impregnation, successive reduction and calcination, *etc.*, which have been widely used for preparing noble metal catalysts.[63] While a variety of supported Pt-group binary or ternary catalysts have been prepared by traditional methods,[63–70] the ability to control the size and composition is limited owing to the propensity of aggregation of metals at the nanoscale. Aggregation of nanoparticles leads to an eventual loss of the nanoscale catalytic activity in practical applications. The preparation of nanoparticles capped in monolayers, polymers, or dendrimers is rapidly emerging, demonstrating remarkable parallels to catalytic activities for supported nanoparticles.[71–75] Among many emerging approaches to the preparation of nanoparticles or nanostructures, one particular class of nanoparticles with core–shell-type structures is beginning to attract interest for addressing some of the challenges in nanoscale catalyst preparation.[76,77] The core–shell-type nanomaterials can be broadly defined as a core and shell of different matter in close interaction, including inorganic/organic and inorganic/inorganic combinations.[71,76–79] There has been an increasing volume of studies in recent years aimed at synthesizing metal nanoparticles in the presence of organic capping agents.[71,77b,79–83]

In contrast to traditional approaches to the preparation of supported catalysts, the molecular encapsulation-based synthesis and processing strategy

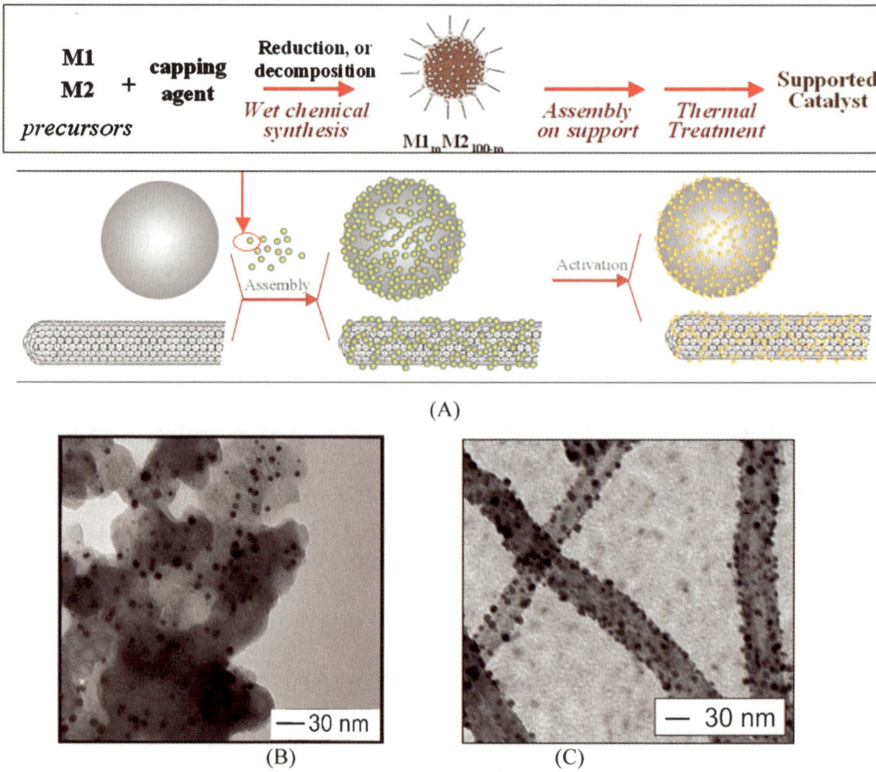

Figure 2.1 (A) A general scheme showing a molecularly engineered synthesis of bimetallic nanoparticles capped with a monolayer shell of oleic acid/oleylamine, and the preparation of bimetallic nanoparticles supported on carbon powders or carbon nanotubes by assembly and activation. (B, C) Two sets of TEM images showing $Au_{22}Pt_{78}$ nanoparticles supported on carbon black (B) and Au nanoparticles supported on carbon nanotubes (C).

involves a sequence of three steps for the preparation of multimetallic catalysts: (1) chemical synthesis of the metal nanocrystal cores with molecular encapsulation, (2) assembly of the encapsulated nanoparticles on support materials (*e.g.*, carbon powders or carbon nanotubes), and (3) thermal treatment of the supported nanoparticles.[8,10,84] Figure 2.1 depicts the general concept and some examples of gold and gold–platinum supported on carbon blacks and carbon nanotubes.

Using a modified two-phase method,[8,9,31,33,85] Au or AuPt alloy nanoparticles of 2 nm core size and different compositions encapsulated with organic shells were synthesized by controlling the feed ratios of the two metal precursors (*e.g.*, $AuCl_4^-$ and $PtCl_6^{2-}$).[8,9] Bimetallic nanoparticles in which the nanocrystal core consists of one metal core and another metal shell (core@shell), such as Au@Pt and Pt@Au, were also synthesized by a seeded growth method.[46] Such nanoparticles were assembled onto carbon black materials, and the carbon-supported nanoparticles were then subjected to thermal treatment under a controlled atmosphere which involved the removal of organic shells and calcination of the alloy nanoparticles. The as-synthesized Au_mPt_{100-m} nanoparticles with different compositions were capped with thiol/amine monolayer shells. Varying the feeding ratio of the metal precursors used in the synthesis controlled the bimetallic composition and sizes of the nanoparticles.

2.3 Electrocatalytic Activity of Bimetallic Catalysts

Different bimetallic alloy catalysts have investigated in assessing the electrocatalytic activities for the MOR and EOR.[9,31,32,44,86] In the following subsections, examples of AuPt and PdCo catalysts for the MOR or EOR are discussed.

2.3.1 AuPt Alloy and Core–Shell Nanoparticle Catalysts

Nanoalloy and core–shell types of AuPt nanoparticles supported on carbon represent one group of bimetallic alloy catalysts that have been demonstrated to exhibit synergistic properties for the MOR. Figure 2.2 shows a representative set of cyclic voltammetric curves comparing the electrocatalytic MOR characteristics in alkaline solution between different catalysts, *i.e.* Pt/C, PtRu/C, $Au_{82}Pt_{18}$/C, and $Au_{72}Pt_{28}$/C catalysts, in comparison with those from well-documented Pt and PtRu catalysts (*e.g.*, E-tek's Pt/C and PtRu/C). The data for Au/C catalysts[31] are also included in Figure 2.2 for comparison. To ensure the quantitative comparison of the data under a relatively comparable loading condition, we also measured the peak current *vs.* the amount of catalysts loaded onto the electrode surface. The observed linear relationship for our catalyst loading range[31] indicates comparability of data because the catalyst loading on the electrode surface falls in a similar range.

The general electrocatalytic characteristics for AuPt/C catalysts is quite similar to those observed for the Pt/C and PtRu/C catalysts. A close examination of the peak potentials and currents reveal subtle differences. The peak potential for $Au_{82}Pt_{18}$/C catalysts (~ -170 mV) is more positive by +10 mV than that for the Pt/C (~ -180 mV) and by +80 mV than that for PtRu/C (~ -250 mV). The peak current density for the $Au_{82}Pt_{18}$/C catalyst [$\sim$8500 mA cm^{-2} (mg AuPt)$^{-1}$], after being normalized to the total metal loading, is larger than that for PtRu/C catalyst [$\sim$7600 mA cm^{-2} (mg PtRu)$^{-1}$] and slightly smaller than Pt/C catalyst [$\sim$8900 mA cm^{-2} (mg Pt)$^{-1}$]. This observation indicates that there is a major improvement in comparison with that of Au/C catalysts in terms of the peak potential (by ~ -600 mV) and the peak current (by $\sim 25\times$). The presence of a small fraction of Pt in the Au-based bimetallic nanoparticles significantly modified the catalytic properties.

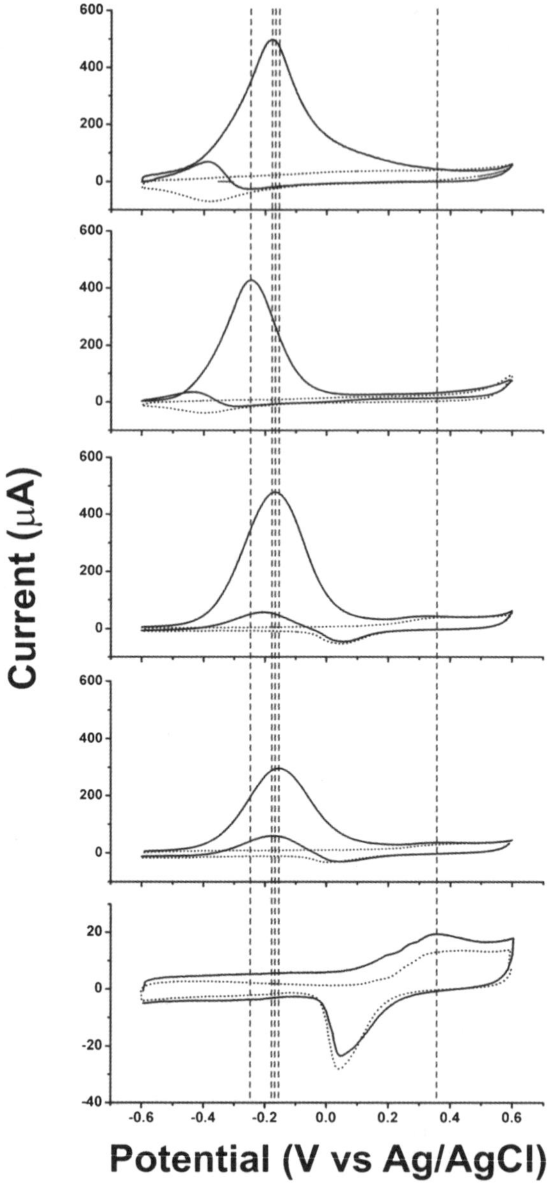

Figure 2.2 Cyclic voltammetric curves for several catalysts (on a GC electrode, 0.07 cm^2) in 0.5 M KOH with (*thick curves*) and without (*thin curves*) 0.5 M methanol: (A) Pt/C, (B) PtRu/C, (C) $Au_{82}Pt_{18}$/C, (D) $Au_{72}Pt_{28}$/C, and (E) Au/C. All data were obtained with 20% metal loading. Scan rate: 50 mV s^{-1}. Reprinted with permission from Luo *et al.*[9] Copyright (2006) American Chemical Society.

The catalytic modification of the bimetallic composition is further reflected by the remarkable difference of the voltammetric characteristics observed in the reverse scan, especially in the alkaline electrolyte. For Pt/C and PtRu/C, the reverse wave for the alkaline electrolyte occurs at a potential less positive than the forward wave by ~200 mV. In contrast, the reverse wave for $Au_{82}Pt_{18}$/C occurs at a potential which differs from the potential for the wave in the forward sweep by only ~20 mV. The relative peak current of the reverse/forward wave is also found to be dependent on Au% in the bimetallic nanoparticle. The oxides formed on the catalyst surface at the potential beyond the anodic peak potential in the positive sweep are reduced in the reverse sweep.[87] Poisonous CO species formed on the Pt surface can also be removed in the reverse sweep. The observation of the more positive potential for the reverse wave likely reflects the bimetallic effect on the re-activation of the catalyst surface after the anodic sweep, a scenario that is under further investigation using Fourier transfer infrared (FTIR) spectroscopic techniques. The re-activation of the surface catalytic sites after the anodic sweep is likely modified by the presence of Au in the catalyst, which leads to the shift of the peak potential of the reverse wave to a more positive potential (by ~200 mV) for $Au_{82}Pt_{18}$/C than for Pt/C.

In comparison with the redox waves for the monometallic Au and Pt catalysts (in 0.5 M KOH), which exhibit redox waves at –0.4 to 0.0 V for Pt and 0.0 to 0.4 V for Au,[31] the detection of the redox waves corresponding to Au and Pt in the bimetallic $Au_{72}Pt_{28}$/C and $Au_{35}Pt_{65}$/C catalysts demonstrate the presence of a bimetallic surface composition on the catalysts. This is consistent with the bimetallic nanoparticle core composition. In the absence of methanol, the catalysts exhibit redox waves corresponding to gold and gold oxide on the surface. The gold oxidization wave was found at 0.3 V, whereas the reduction wave was around 0.06–0.07 V for both catalysts. The methanol oxidation activity was also found to be dependent on both composition and treatment temperature.

The correlation between the composition and the electrocatalytic activity for the above catalysts is assessed by analyzing the mass activity (*i.e.*, the peak current density per unit total mass of metals) and the kinetic parameters (Tafel slope). Figure 2.3 summarizes a representative set of results showing the dependence of MOR mass activity on the bimetallic composition. One of the significant findings is that the mass activity for the alkaline condition appears to exhibit a maximum around the composition of 65–85% Au, which is remarkably close to the composition range (~65% Au) observed for the transition of the band features. While the presence of Au in Pt increases the lattice distance of Pt, the higher electronegativity of Au than Pt could cause an increase in the amount of change being transferred from Pt to Au, which was in fact supported by high-resolution XPS data showing an Au $4f_{7/2}$ binding energy of 83.32 eV for Au/Pt and of 83.87 eV for bulk-like Au atoms,[88] and consequently an increase of the d-orbital vacancy in AuPt. The composition was found to significantly modify the electrocatalytic properties of both Au

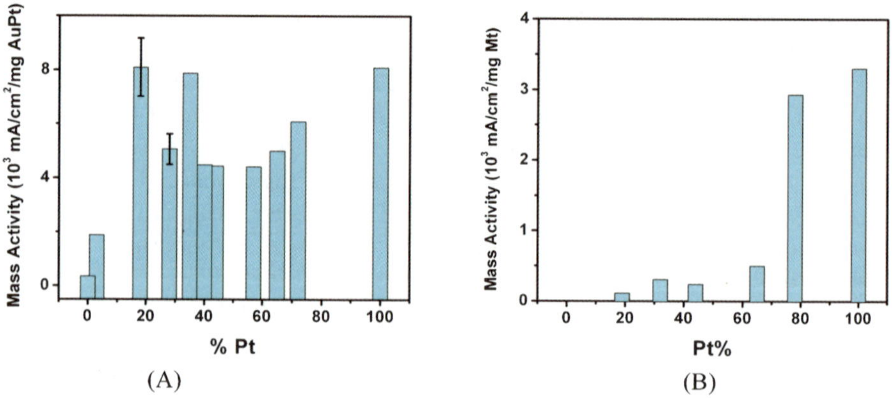

Figure 2.3 MOR mass activities *vs.* bimetallic composition for AuPt/C in (A) 0.5 M KOH and in (B) 0.5 M H_2SO_4 with 0.5 M MeOH (based on maximum current density in voltammetric curve at 50 mV s^{-1}).

and Pt. The finding that the mass activity in the alkaline electrolyte exhibits a maximum around 65–85% Au is in contrast to the small and gradual increase from no activity of Au to high activity of Pt in the acidic electrolyte. The variation of the Tafel slope appears to be relatively small ($\sim$120 mV dec^{-1}). The display of the maximum mass activity for AuPt/C catalysts, comparable or higher than that for Pt/C catalyst in the alkaline electrolyte, is remarkable in view of the fact that only 15–35% Pt was present in the catalyst.

To understand the origin of unique activity characteristic, recent modeling results based on a DFT calculation for CO adsorption on small clusters of AuPt[89,90] are considered. The modeling results showed that the CO adsorption energy increases with the % Pt, which maximizes at $\sim$30% Pt for CO adsorption on Au atoms, and increases with the % Pt for CO adsorption on Pt atoms for all compositions. This is important because it implies that gold atoms surrounding Pt atoms in the AuPt alloy with a relatively high percentage of Au may have played an important role in either removing the intermediate CO-like species or providing oxygenated species in the methanol oxidation process. This assessment is consistent with the known facts that the nanoscale Au is catalytically highly active for CO oxidation,[91–93] and Au is known to be capable of producing surface oxygenated species in basic electrolytes.[94,95]

We believe that a bifunctional electrocatalytic property may be operative for the bimetallic catalysts in the alkaline electrolyte. The similarity of the mass activity for the bimetallic catalyst with 65–85% Au to those for pure Pt catalyst is suggestive of the participation of Au in the catalytic reaction of Pt. The alloy character for both the surface and the core of the AuPt/C catalysts is supported by X-ray diffraction (XRD) data for the phase properties and FTIR data of CO adsorption on the catalysts. Thus, the Pt atom for an AuPt alloy with $\sim$75% Au would be practically surrounded by Au atoms. Scheme 2.2 illustrates the proposed reaction pathway for the conversion of methanol to carbonate ion mediated by a gold–platinum bimetallic catalyst.

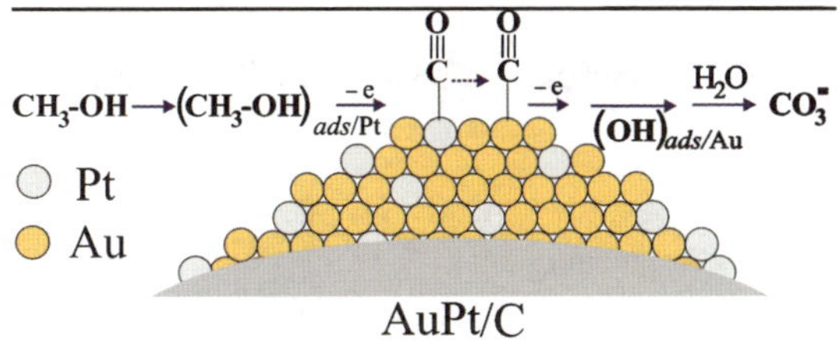

Scheme 2.2 A schematic illustration of the electrocatalytic oxidation of methanol on a AuPt/C catalyst in an alkaline electrolyte. Reprinted from Mott *et al.*[44] Copyright (2007), with permission from Elsevier.

While a detailed delineation of the proposed electrocatalytic mechanism is part of our on-going work, several important insights are supportive of the mechanistic view. The formation of intermediate CO_{ad} species on Pt is a well-known fact. The transfer of the intermediate CO_{ad} species from Pt atop sites to neighboring Au atop sites is possible in view of the favorable adsorption of CO on Au nanoparticles known from both experimental measurements[91] and theoretical calculations.[89,90] The formation of Au–OH_{ad} or surface oxides on gold in an alkaline electrolyte was in fact proposed to explain some of the electrocatalytic properties observed for a gold electrode (*e.g.*, incipient hydrous oxide/adatom mediator model[94]). Our previous *in situ* measurements of the interfacial mass change accompanying the electrocatalytic oxidation of methanol also indicated the formation of Au oxides [Au_2O_3, AuOH, or $Au(OH)_3$] on gold nanoparticle surfaces.[95] The bifunctional activity in the alkaline electrolyte involves the participation of CO and OH adsorption on Au sites in the catalytic reaction of Pt in the alloy *via* a combination of reaction steps (see Scheme 2.2), including the adsorption of MeOH on Pt followed by dehydrogenation, the formation of intermediate CO_{ad}/Pt, the transfer of CO_{ad}/Pt to neighboring Au atop sites forming CO_{ad}/Au, and the reactions of Pt–CO_{ad} + Au–OH_{ad} and Au–CO_{ad} + Au–OH_{ad} towards the final product (CO_3^-).

The understanding of the synergistic electrocatalytic activity of AuPt/C in the MOR and ORR (Figure 2.4) has been an important focus of interest in this bimetallic system.[3,4,8,31,35] For the MOR, one of the most significant findings is that the mass activity in the alkaline electrolyte exhibits a maximum around 65–85% Au (Figure 2.4). Considering the modeling results for CO adsorption on small clusters of AuPt and the FTIR results for CO adsorption, the Au atoms surrounding Pt atoms in the AuPt alloy are believed to play an important role in either removing the intermediate CO-like species and/or providing oxygenated species in the methanol oxidation process (see scheme in Figure 2.4). This assessment is consistent with the known facts that the nanoscale Au is highly active for CO oxidation, and produces surface oxygenated species in basic electrolytes.

The above conclusions on the surface properties due to alloying and phase segregation as a function of the bimetallic composition (Figure 2.4) and the

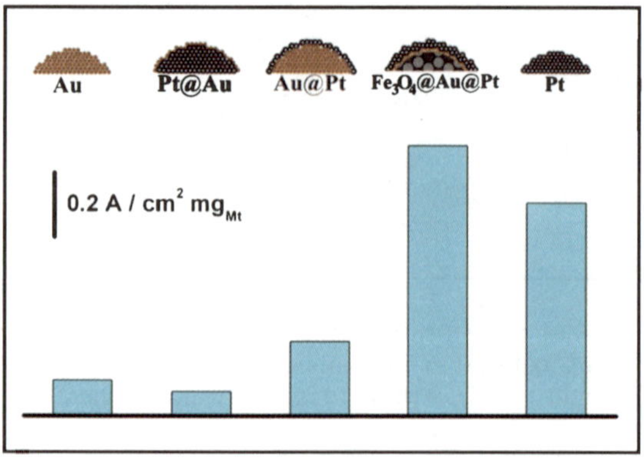

Figure 2.4 Plots of activities for the MOR (*bottom*) for different catalysts in 0.5 M H_2SO_4 saturated with O_2. The inserted upper scheme illustrates the corresponding nanoparticle surface structures by design.[46]

thermal temperature were further assessed in terms of MOR activities in acidic electrolytes for carbon-supported core@shell nanoparticles such as Au@Pt and Pt@Au.[46] In contrast to the absence of activities for Au and Au-coated nanoparticles in acidic solution, the increased activities for Pt-coated catalysts (Figure 2.4) are characteristic of the electrocatalytic properties of the Pt component in Au@Pt and Fe3O4@Au@Pt catalysts. The fact that the mass activity for Fe3O4@Au@Pt is close to or higher than Pt demonstrates the feasibility of producing the synergistic catalytic effect by a metal oxide core–shell nanostructure.

2.3.2 PdCo Alloy Nanoparticle Catalysts

Palladium has attracted recent interest in developing Pt-free catalysts for alcohol fuel cells. In Figure 2.5, typical cyclic voltammetric (CV) curves for a Pd/C catalyst, which was synthesized by the method described earlier, for methanol (A) and ethanol (B) oxidation (0.5 M) in an alkaline electrolyte (0.5 M KOH) are shown. In the case of the MOR, the forward peak was at ~ -0.17 V and the backward peak was located at ~ -0.35 V. The current intensity of the forward peak was about four times larger than the backward peak. In the case of the EOR, the anodic current in the reversed sweep (at -0.32 V), which is two times higher than the forward one, is attributed to the oxidation of ethanol upon the removal of adsorption and interactions of various intermediate species, such as CO and OH^-, on the catalyst surface formed in the anodic sweep.

Different Pd-based bimetallic catalysts have been studied in our recent work. One example involves PdCo catalyst which was synthesized by the method described earlier. In Figure 2.5(A), a typical CV curve of a PdCo/C catalyst for methanol oxidation (0.5 M) in an alkaline electrolyte (0.5 M KOH) is shown. The forward peak was at ~ -0.18 V and the backward peak was located at

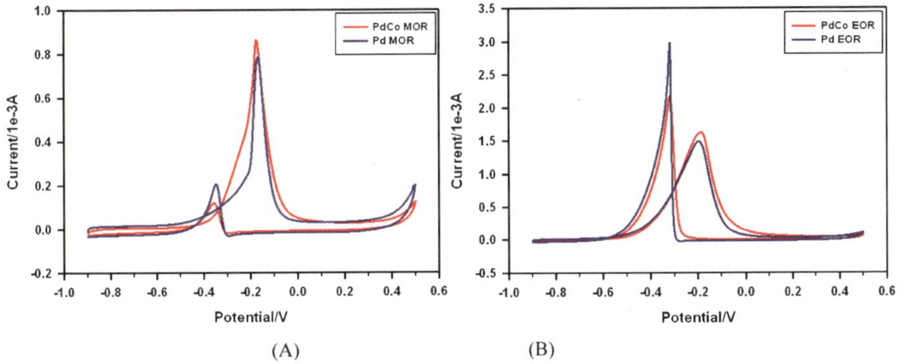

Figure 2.5 Cyclic voltammetric curves for a PdCo/C catalyst (*red curve*) and a Pd/C catalyst (*blue curve*) (on a GC electrode, 0.196 cm^2) 0.5 M KOH electrolyte with 0.5 M methanol (A) and 0.5 M ethanol (B). Scan rate: 20 mV s^{-1}.

~ −0.36 V. The current intensity of the forward peak was about seven times larger than the backward peak. Figure 2.5(B) shows a typical CV curve of a PdCo/C catalyst for ethanol oxidation (0.5 M) in an alkaline electrolyte (0.5 M KOH). The forward peak was at ~ −0.19 V and the backward peak was located at ~ −0.32 V. The current intensity of the backward peak was about the same as the forward one. In comparison with the data for Pd/C, the relative current intensity of the backward peak is apparently weaker. These subtle differences are believed to reflect the operation of some synergistic activity of the bimetallic catalysts in the MOR and EOR, which calls for further in-depth investigations.

2.4 Phase and Surface Properties of Bimetallic Nanoparticle Catalysts

To understand the synergistic properties of the bimetallic nanoalloys on the electrocatalytic activity, the phase, structure, and surface properties of the nanocatalysts have been investigated using different techniques.[9,31,44,86] In the following subsection, examples of AuPt catalysts in correlation with MOR catalytic activity are discussed.

2.4.1 Bimetallic Phase Properties

The nanoscale phase of AuPt nanoparticles is a very important parameter for the exploration of nanoscale bimetallic functional properties. In our earlier work,[30,44,45] the lattice constants of the bimetallic nanoparticles were found not only to scale linearly with the % Pt (Figure 2.6), in contrast to the bulk AuPt counterparts which display a miscibility gap at 20–90% Au,[28,29] but also to be smaller than those of the bulk counterparts. This finding was the first example demonstrating that the nanoscale AuPt nanoparticles could exhibit single-phase character and reduced inter-atomic distances in the entire bimetallic composition range, in contrast to the bulk counterparts.[30,44,45] This finding is supported by the result from density functional theory (DFT) computation for small AuPt nanoparticles (~1.2 nm), which revealed a difference in the lattice constant–composition relationship between the nanoscale and the bulk systems, in agreement with the experimental data.[42]

The XRD examination of the bimetallic catalysts as a function of thermal treatment temperature has revealed a gradual splitting of a single-peak pattern at low temperature into a two-peak overlapping pattern upon increasing the temperature (Figure 2.7A).[42] The peak splitting translates to an evolution of the lattice parameter, as shown by the temperature dependence of the lattice parameters derived from XRD peak deconvolution (Figure 2.7B). It is interesting that the lattice parameter evolves from a single-phase alloy (0.4029 nm) at low temperature to the directions corresponding to single-phase Au and Pt, *i.e.* 0.4075 nm for Au/C and 0.3915 nm for Pt/C.[42] In other words, while the data for the 400 °C treated samples were largely characteristic of alloys with an intermittent lattice parameter (*i.e.*, between 0.3915 and 0.4075 nm), the data for the higher-temperature treated samples showed two lattice parameters, with one closely matching the lattice parameter of Au and the other shifting toward 0.3915 nm for Pt/C. This finding demonstrates that the phase properties are dependent on the thermal treatment temperature.

There are several fundamental questions about the alloying and phase segregation properties. The first question is why alloying occurs at the nanoscale, in contrast to the miscibility gap well established for bulk AuPt

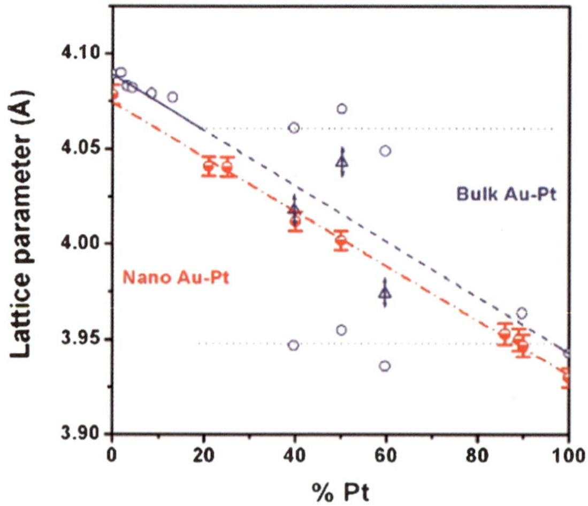

Figure 2.6 The lattice parameters *vs.* Pt% for AuPt nanoparticles () and bulk AuPt () (triangular points represent a frozen state for the bulk metals). Reprinted from Mott *et al.*[44] with permission of Elsevier.

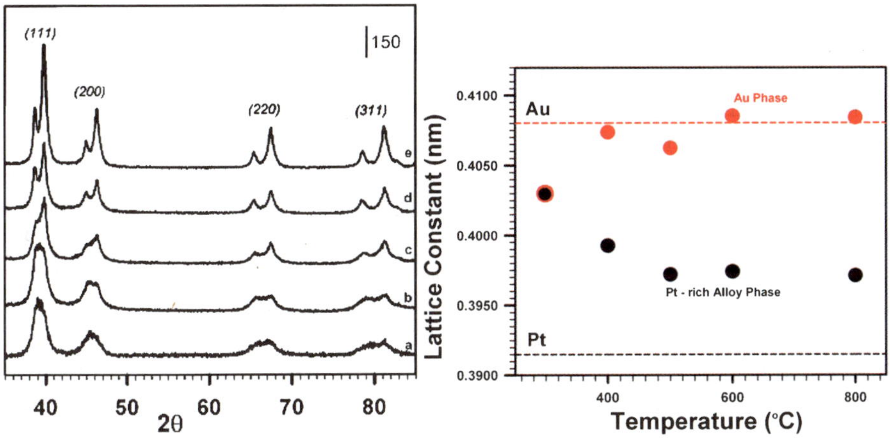

(A) (B)

Figure 2.7 (A) XRD patterns for $Au_{22}Pt_{78}/C$ catalysts after organic shell removal at 280 °C for 60 min and different annealing temperatures: (a) 300, (b) 400, (c) 500, (d) 600, and (e) 800 °C.[42] (B) Lattice parameters of $Au_{22}Pt_{78}/C$ catalysts treated at different temperatures; the *upper* and *bottom* dotted horizontal lines represent the values of the lattice parameters for Au/C and Pt/C obtained from XRD deconvolution for $Au_{22}Pt_{78}/C$. Reprinted from Wanjala *et al.*[42] with permission of the American Chemical Society.

materials.[28,29] This question is theoretically addressed by the demonstration of a negative heat of formation for AuPt particles smaller than ~6 nm (Figure 2.8A).[35] This energy difference thermodynamically favors the formation of an alloyed phase.[5] The second question is why alloying and phase segregation are dependent on the thermal treatment temperature. This question can be addressed by considering both thermodynamic analysis and DFT modeling results. The thermodynamic analysis result indicates that there is a clear difference in surface free energy between these two metals, *i.e.* a larger surface free energy for Pt (2.48 J m^{-2}) and a smaller one for Au (1.50 J m^{-2}). This difference favors the formation of an Au shell, as illustrated by the evolution of the metal distribution as a function of temperature (Figure 2.8B).[35] This conclusion is consistent with the fact that Au has a lower surface free energy than Pt, as shown in the earlier molecular dynamics simulation for AuPt nanoparticles of various sizes. There are two important findings. First, the heat of formation is shown to be not only dependent on the composition, but also on the particle size (Figure 2.8A). For Au_mPt_{100-m} nanoparticles <6 nm, there is a negative heat of formation, which is thermodynamically favorable, in contrast to larger sized or bulk AuPt materials. Second, the results from the thermodynamic computation revealed a gradual evolution of the distribution of the two metals in the nanoparticles as the temperature is increased. For alloy AuPt nanoparticles of ~3.6 nm, upon increasing temperature from 1000 to 1400 K (Figure 2.8B) there is a redistribution of Au and Pt in the nanoparticle, which evolves in such a way that the Au concentration enriches in the region near the outer shell whereas the Pt concentration enriches in the core region, *i.e.* near the center of mass. This finding is illustrated by the model in Figure 2.8(B) in terms of three regions (a, b and c) as a function of distance from the nanoparticle center of mass. At temperatures below 600 K, the distributions of Au and Pt are identical in region a, or show small differences in regions b and c, exhibiting largely single-phase alloy character. However, at temperatures above 800 K the distributions of Au and Pt are very different, depending on the regions in the nanoparticle. In regions a and b, the % Au decreases whereas the % Pt increases with temperature. In region c the evolution is the opposite of regions a and b, showing a % Pt decrease and a % Au increase with temperature. The nanoparticle separates out into two phases with a Pt-rich core and an Au-rich shell, forming core–shell-like structures. These theoretical findings are consistent with our experimental findings.[11a]

Similar theoretical findings were also obtained by molecular dynamic simulations of bimetallic AuPt nanoparticles supported by graphite and bundles of carbon nanotubes.[53] For AuPt clusters of different compositions, the atomic distributions of Au and Pt are also shown to be dependent on the distance from the center of mass in the nanoparticle (Figure 2.9). In contrast to an ideal alloy distribution represented by the dashed line regardless of the regions, the computation results revealed that the % Au in region a is consistently lower than the dashed line, whereas the % Au in regions b and c are higher than the dashed

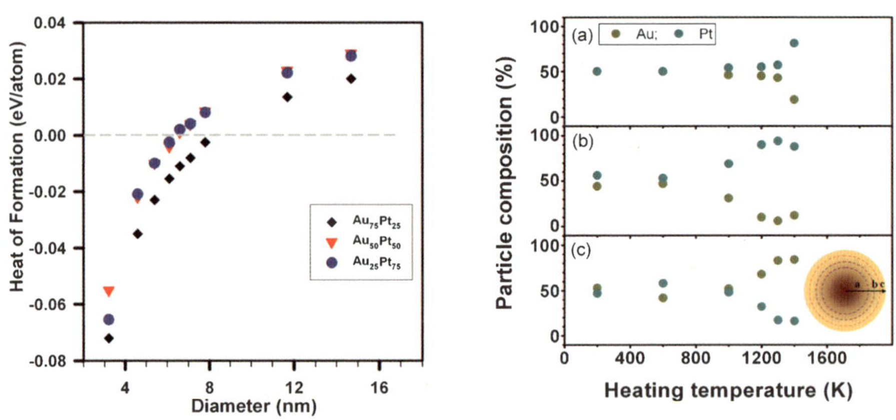

(A) (B)

Figure 2.8 (A) The calculated heat of formation for a solid solution of AuPt particles with different composition as a function of particle diameter. For bulk metals: $\Delta E = 0.048$ ($Au_{75}Pt_{25}$), 0.052 ($Au_{50}Pt_{50}$), and 0.065 ($Au_{25}Pt_{75}$) eV atom^{-1}. (B) The calculated atomic distributions of Au and Pt in a nanoparticle (size ~3.5 nm) at different distances from the center of mass as a function of temperature [*inset*: model for a nanoparticle of radius *r*: (a) <1.0 nm (57% *r*), (b) 1.3 nm (71% *r*), and (c) 1.5 nm (86% *r*)]. For both (A) and (B) the data were extracted from Xiao *et al.*[35]

line. There seems to be a trend of the enrichment of Au in the shell region and enrichment of Pt in the core region. $Au_{75}Pt_{25}$ and $Au_{25}Pt_{75}$ form a mixed AuPt core surrounded by an Au shell and a Pt core covered with an AuPt shell, respectively. The support was also shown to affect the atomic distribution, as indicated by the decrease of the number of Pt atoms on the surface for $Au_{25}Pt_{75}$ when bundles of carbon nanotubes are used as the support, and the change in atomic distribution when the support geometry is changed.

The phase properties are indeed quite complex, as recently evidenced by an *in situ* time-resolved XRD study of phase change kinetics at high temperature.[40] For example, AuPt/SiO_2 nanoparticles were shown to exhibit an alloyed structure that is preserved even after extensive annealing at temperatures as high as 700 °C, whereas the AuPt/Al_2O_3 nanoparticles exhibit a rich phase evolution that is sensitive to alloy composition and the details of the synthesis process.[40] By comparing the phase and morphology transformations in Au, Pt, and AuPt metal nanoparticles on silica substrates, different mechanisms are believed to be operative for thermal-driven melting and sintering in the nanoparticle shaping and growth.[41] The AuPt nanoparticle structure and size depend on the details of the heating process as well as the particle density. In contrast to the behavior of bulk AuPt systems[28] and other experimental observations of AuPt nanoparticles on different substrates,[50,54,96] AuPt nanoparticles on silicon undergo only partial phase segregation upon extensive annealing above 700 °C.[41] We also note that the phase behavior could be different for nanoparticles in the solution phase, where an evolution

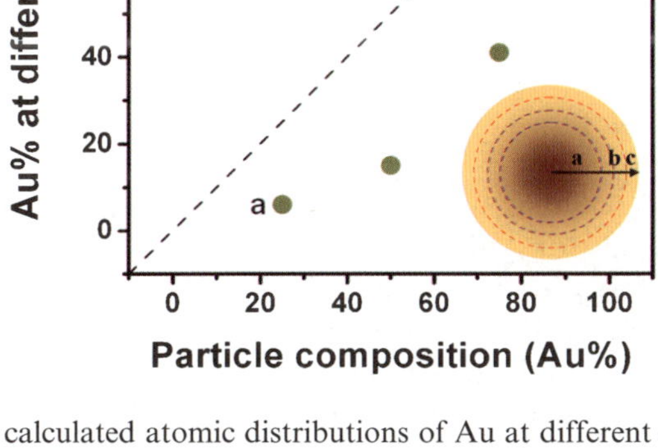

Figure 2.9 The calculated atomic distributions of Au at different distances from the center of mass in a nanoparticle (size ~2.7 nm) as a function of the homogeneous composition: (a) <0.8 nm (57% *r*); (b) 1.0 nm (71% *r*); and (c) 1.2 nm (86% *r*). Data were extracted from Morrow and Striolo.[53]

of the surface plasmon band of gold is possible, as shown for concentrated AuPt nanoparticle solutions under thermal evolution conditions.[62]

2.4.2 Bimetallic Surface Properties

Having determined the nanoscale alloying and phase segregation in terms of the evolution of the atomic distribution as a function of temperature, the next question is how the nanoscale phase evolution translates to surface alloying or segregation. This question is addressed by a number of different approaches, including FTIR, X-ray photoelectron spectroscopy (XPS), and electrochemical characterization.[42,44,46] The FTIR probing of CO adsorption on the surface of the AuPt nanoparticles represents one important approach to understanding the surface binding sites.[44] There are two important findings from the FTIR spectral evolution for CO adsorption on AuPt/SiO_2 as a function of bimetallic composition (Figure 2.10). First, the 2115 cm^{-1} band observed for Au/SiO_2 displays a clear trend of diminishing absorbance as the Pt concentration increases in the bimetallic catalysts. This band becomes insignificant or even absent at $>\sim45\%$ Pt. Secondly, the ~2050 cm^{-1} band shows a clear trend of shift towards that for the Pt-atop-CO band observed for Pt/SiO_2 as the Pt concentration increases. This finding, together with the XRD data for the nanocrystal core properties,[30] demonstrated that both the core and the surface of the bimetallic nanoparticles exhibit bimetallic alloy properties. The detection of both Au-atop- and Pt-atop-CO bands on the surface of alloy nanoparticles of a wide range of bimetallic composition can be correlated with the electronic effect as a result of the d-band shift of Pt in the bimetallic nanocrystals. There exists a stronger electron donation to the CO band by a Pt-atop site surrounded by Au atoms in the bimetallic alloy surface than that from the monometallic Pt surface as a consequence of the upshift in d-band center of Pt atoms surrounded by Au atoms, which explains the preference of Pt-atop-CO over the Au-atop-CO adsorption. The observed decrease of the Pt-atop-CO band frequency with increasing Au concentration is in agreement with the d-band theory for the bimetallic system.[89]

The assessment of the nanoscale surface properties can be further substantiated by comparing CO adsorption on silica-supported Au, Pt, Au@Pt, Pt@Au, and Fe3O4@Au@Pt nanoparticles (Figure 2.11).[46] The adsorption of CO on Pt is characterized by a major band at 2090 cm^{-1} that is attributed to CO on Pt-atop sites, whereas CO on Au is characterized by a band at 2115 cm^{-1} that is attributed to CO linearly adsorbed on Au-atop sites.[97] The CO bands for AuPt alloy catalysts were detected at 2115 and 2066 cm^{-1}. The key observation is that the main CO bands for Au@Pt and Pt@Au nanoparticles resemble Pt and Au nanoparticles, respectively. The subtle difference for Pt@Au and Pt@Au, *i.e.* the shift of the Au-atop CO band to 2130 cm^{-1} and the presence of the alloy-like band at 2065 cm^{-1}, could be due to the sites in the Au surface adlayer and its effect on the Pt sites.

For the AuPt/C catalysts thermally treated at different temperatures, the relative change of Au and Pt distribution on the particle surface was also

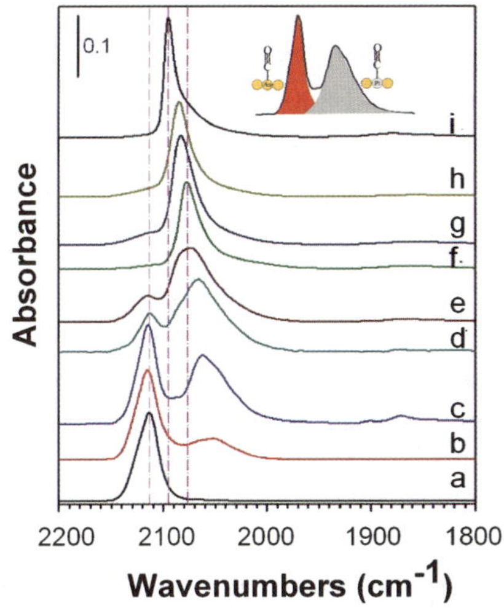

Figure 2.10 FTIR spectra of CO adsorption on SiO_2-supported AuPt catalysts of different composition (% Au): (a) Au; (b) $Au_{96}Pt_4$; (c) $Au_{82}Pt_{18}$; (d) $Au_{72}Pt_{28}$; (e) $Au_{65}Pt_{35}$; (f) $Au_{56}Pt_{44}$; (g) $Au_{43}Pt_{57}$; (h) $Au_{35}Pt_{65}$; and (i) Pt. Reprinted from Mott *et al.*[44] with permission of Elsevier.

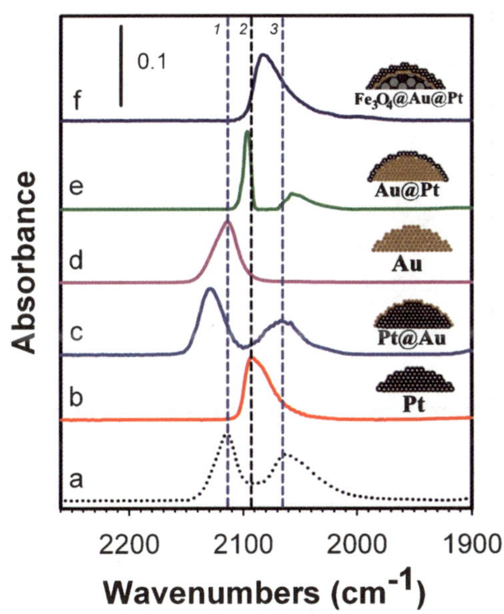

Figure 2.11 FTIR spectra comparing CO adsorption on SiO_2-supported (a) AuPt, (b) Pt, (c) Pt@Au, (d) Au, (e) Au@Pt, and (f) Fe3O4@Au@Pt.[13a] The vertical dashed lines, "1", "2", and "3", represent CO stretching bands for Au-atop, Pt-atop, and Pt sites in the alloy, respectively.[44]

examined as a function of the treatment temperature using the XPS technique.[42] The comparison of XPS spectra in the Pt 4f and Au 4f regions of $Au_{22}Pt_{78}/C$ treated at 400 and 800 °C (Figure 2.12) serves as an example to illustrate how surface properties change. For $Au_{22}Pt_{78}/C$ treated under O_2 for 60 min followed by annealing under H_2 at different temperatures, the Pt $4f_{7/2}$ and Pt $4f_{5/2}$ bands were observed at 71.90 and 75.25 eV for 400 °C, and 71.75 and 75.10 eV at 800 °C, whereas the Au $4f_{7/2}$ and Au $4f_{5/2}$ bands were observed at 84.45 and 88.10 eV for 400 °C, and 84.53 and 88.30 eV at 800 °C, respectively. In addition to a subtle increase in the Au 4f peak by 0.2 eV from 400 to 800 °C, a decrease of the Pt/Au ratio by a factor of 4 was also observed.

By considering electron attenuation theory to model the relative changes of the two metal components, the relative change of the Au/Pt atomic ratio was modeled based on an ideal Pt(core)/Au(shell) nanoparticle considering the electron attenuation effect.[98,99] It revealed that the surface treated at 800 °C is enriched with Au in comparison with that treated at 400 °C, which is consistent with partial phase segregation revealed by the XRD data. In contrast to the

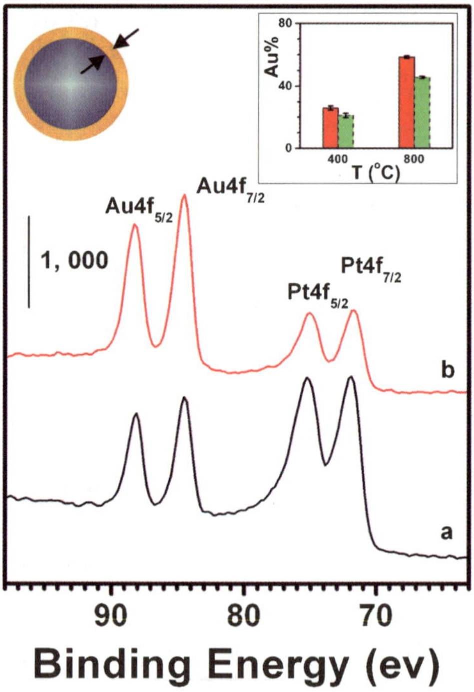

(A) (B) (C)

Figure 2.12 XPS spectra comparing $Au_{22}Pt_{78}/C$ catalysts (60 min thermal removal under O_2) by thermal treatment under H_2 at (a) 400 (b) and 800 °C. *Insets*: comparison of the XPS-determined % Au (*red*) and the % Au calculated (*green*) based on an idealized Pt(core)/Au(shell) nanoparticle. Reprinted in part from Wanjala *et al.*[42] with permission of the American Chemical Society.

largely alloyed character for those treated at 300–400 °C, the 800 °C treated catalysts were shown to consist of a Pt-rich alloy core and an Au shell or phase-segregated Au domains enriched on the surface. The interaction remains between Au and Pt domains through alloying bonding or phase boundaries.[42]

2.5 Summary

In summary, this chapter has shown that both synthetic and processing parameters in the preparation of nanoalloy catalysts can have a significant impact on the electrocatalytic activities. This impact stems from the synergistic correlation between the structure and the activity of the alloys at the nanoscale. While the fundamental understanding along this direction holds promise for the development of alcohol fuel cell catalysts, the ability to produce the catalyst by design remains a key challenging area. Opportunities to address this challenge are emerging in the field of nanostructured materials and nanotechnology. The understanding of how the atomic-scale core structures and surface binding sites evolve in the nanoalloys in the catalytic reactions will undoubtedly help us in achieving catalyst-by-design capabilities, which is part of our on-going efforts in this exciting endeavor of fuel cells and other energy-related catalysis explorations. DEFCs have recently been receiving increased attention. Most studies in acidic media have showed only 2 or 4 electrons per ethanol molecule rather than the 12 electrons as expected when ethanol is fully oxidized. The recent development of alkaline membranes that conduct hydroxyl ions makes alkaline membrane fuel cells very attractive, because the rapid kinetics allows the use of catalysts with lower noble metal content feasible and potentially enables the 12-electron oxidation, which is also an important direction of current research.

Acknowledgements

The authors express their gratitude to former and current members of the Zhong Research Group and collaborators who have made contributions to the work described in this article. The research work is supported currently or in the past by the National Science Foundation (CBET 0709113, CHE 0848701, CHE 0316322), NYSERDA, DOE Basic Science, ACS Petroleum Research Funds, NYSTAR, and the World Gold Council.

References

1. A. K. Shukla and R. K. Raman, *Annu. Rev. Mater. Res.*, 2003, **33**, 155.
2. B. Wanjala, J. Luo, B. Fang, D. Mott and C. J. Zhong, *J. Mater. Chem.*, 2011, **21**, 4012.
3. C. J. Zhong, J. Luo, B. Fang, B. Wanjala, P. Njoki, R. Loukrakpam and J. Yin, *Nanotechnology*, 2010, **21**, 062001.

4. C. J. Zhong, J. Luo, P. Njoki, D. Mott, B. Wanjala, R. Loukrakpam, S. Lim, L. Wang, B. Fang and Z. Xu, *Energy Environ. Sci.*, 2008, **1**, 454.
5. H. Stone, *Economic Analysis of Stationary PEMFC Systems*, DOE Hydrogen Program FY, US Department of Energy, Washington, 2005.
6. H. A. Gasteiger, S. S. Kocha, B. Sompalli and F. T. Wagner, *Appl. Catal. B*, 2005, **56**, 9.
7. T. E. Mallouk and E. S. Smotkin, in *Handbook of Fuel Cells: Fundamentals, Technology and Applications*, ed. W. Vielstich, A. Lamm, and H. A. Gasteiger, Wiley, Chichester, 2005.
8. J. Luo, P. Njoki, Y. Lin, L. Wang, D. Mott and C. J. Zhong, *Electrochem. Commun.*, 2006, **8**, 581.
9. J. Luo, P. Njoki, Y. Lin, L. Wang and C. J. Zhong, *Langmuir*, 2006, **22**, 2892.
10. J. Luo, N. Kariuki, L. Han, L. Y. Wang, C. J. Zhong and T. He, *Electrochim. Acta*, 2006, **51**, 4821.
11. J. Luo, L. Wang, D. Mott, P. N. Njoki, N. Kariuki, C. J. Zhong and T. He, *J. Mater. Chem.*, 2006, **16**, 1665.
12. H. Bonnemann and K. S. Nagabhushana, *J. New Mater. Electrochem. Syst.*, 2004, **7**, 93.
13. C. J. Zhong, J. Luo, M. M. Maye, L. Han, N. Kariuki and T. He, *U.S. Pat.* 7 053 021, 2006.
14. C. J. Zhong, J. Luo, M. M. Maye and N. Kariuki, *U.S. Pat.* 7 208 439, 2007.
15. T. He, C. J. Zhong, J. Luo, M. M. Maye, L. Han, N. Kariuki and T. He, *U.S. Pat.* 7 335 245, 2008.
16. M. Hepel, I. Kumarihamy and C. J. Zhong, *Electrochem. Comm.*, 2006, **8**, 1439.
17. M. Hepel, I. Dela, T. Hepel, J. Luo and C. J. Zhong, *Electrochim. Acta*, 2007, **52**, 5529.
18. W. Zhou, B. Zhou, W. Li, P. Tsiakaras, W. J. Zhou, B. Zhou, W. Z. Lia, Z. H. Zhou, S. Q. Song, G. Q. Sun, Q. Xin, S. Douvartzidesc, M. Goulac and P. Tsiakaras, *J. Power Sources*, 2004, **126**, 16.
19. Z. Liang, T. Zhao, J. Xu and L. D. Zhu, *Electrochim. Acta*, 2009, **54**, 2203.
20. M. M. Dimos and G. J. Blanchard, *J. Phys. Chem. C*, 2010, **114**, 6019.
21. S. Shen, T. Zhao, J. Xu and Y. Li, *Energy Environ. Sci.*, 2011, **4**, 1428.
22. S. Mahapatra, A. Dutta and J. Datta, *Electrochim. Acta*, 2010, **55**, 9097.
23. L. Jou, J. Chang, T. Twhang and I. W. Sun, *J. Electrochem. Soc.*, 2009, **156**, D193.
24. T. Maiyalagan and K. Scott, *J. Power Sources*, 2010, **195**, 5246.
25. E. Lee, I. Park and A. Manthiram, *J. Phys. Chem. C*, 2010, **114**, 10634.
26. Y. Lee, M. Kim, Y. Kim, S. W. Kang, J. H. Lee and S. W. Han, *J. Phys. Chem. C*, 2010, **114**, 7689.
27. M. Nie, H. Tang, Z. Wei, S. P. Jiang and P. K. Shen, *Electrochem. Commun.*, 2007, **9**, 2375.
28. *Catalysis by Gold*, ed. G. C. Bond, C. Louis and D. T. Thompson, Imperial College Press, London, 2006.

29. V. Ponec and G. C. Bond, *Catalysis by Metals and Alloys*, Elsevier, Amsterdam, 1995.
30. J. Luo, M. M. Maye, V. Petkov, N. N. Kariuki, L. Wang, P. Njoki, D. Mott, Y. Lin and C. J. Zhong, *Chem. Mater.*, 2005, **17**, 3086.
31. J. Luo, M. M. Maye, N. N. Kariuki, L. Wang, P. Njoki, Y. Lin, M. Schadt, H. R. Naslund and C. J. Zhong, *Catal. Today*, 2005, **99**, 291.
32. Y. Lou, M. M. Maye, L. Han, J. Luo and C. J. Zhong, *Chem. Commun.*, 2001, **5**, 473.
33. M. M. Maye, N. N. Kariuki, J. Luo, L. Han, P. Njoki, L. Wang, Y. Lin, H. R. Naslund and C. J. Zhong, *Gold Bull.*, 2004, **37**, 217.
34. C. J. Zhong, J. Luo, L. Wang, P. Njoki and D. Mott, *U.S. Pat.* 7 687 428, 2010.
35. S. Xiao, W. Hu, W. Luo, N. N. Kariuki, Y. Wu, X. Li and H. Deng, *Eur. Phys. J. B*, 2006, **54**, 479.
36. P. Hernandez-Fernandez, S. Rojas, P. Ocon, J. L. G. de la. Fuente, J. S. Fabian, J. Sanza, M. A. Pena, F. J. Garcia-Garcia, P. Terreros and J. L. G. Fierro, *J. Phys. Chem. C*, 2007, **111**, 2913.
37. G. Selvarani, S. V. Selvaganesh, S. Krishnamurthy, G. V. M. Kiruthika, P. Sridhar, S. Pitchumani and A. K. Shukla, *J. Phys. Chem. C*, 2009, **113**, 7461.
38. B. Brown, S. D. Wolter, B. R. Stoner and J. T. Glass, *J. Electrochem. Soc.*, 2008, **155**, B852.
39. J. H. Choi, K. J. Jeong, Y. Dong, J. Han, T. H. Lim, J. S. Lee and Y. E. Sung, *J. Power Sources*, 2006, **163**, 71.
40. O. Malis, M. Radu, D. Mott, B. Wanjala, J. Luo and C. J. Zhong, *Nanotechnology*, 2009, **20**, 245708.
41. O. Malis, C. Byard, D. Mott, B. Wanjala, R. Loukrakpam, J. Luo and C. J. Zhong, *Nanotechnology*, 2011, **22**, 025701.
42. B. Wanjala, J. Luo, R. Loukrakpam, B. Fang, D. Mott, P. Njoki, M. Engelhard, H. R. Naslund, J. K. Wu, L. Wang, O. Malis and C. J. Zhong, *Chem. Mater.*, 2010, **22**, 4282.
43. B. Fang, B. Wanjala, X. Hu, J. Last, R. Loukrakpam, J. Yin, J. Luo and C. J. Zhong, *J. Power Sources*, 2011, **196**, 659.
44. D. Mott, J. Luo, P. Njoki, Y. Lin, L. Wang and C. J. Zhong, *Catal. Today*, 2007, **122**, 378.
45. D. Mott, J. Luo, A. Smith, P. Njoki, L. Wang and C. J. Zhong, *Nanoscale Res. Lett.*, 2007, **2**, 12.
46. J. Luo, L. Wang, D. Mott, P. N. Njoki, Y. Lin, T. He, Z. Xu, B. Wanjala, I. I. S. Lim and C. J. Zhong, *Adv. Mater.*, 2008, **20**, 4342.
47. L. Wang, H. Y. Park, S. I. I. Lim, M. J. Schadt, D. Mott, J. Luo and C. J. Zhong, *J. Mater. Chem.*, 2008, **18**, 2629.
48. Hitachi Maxell Ltd., presented at the 101st Catalysis Conference, Tower Hall, Funabori, Tokyo, Japan, March 29, 2008.
49. A. Habrioux, W. Vogel, M. Guinel, L. Guetaz, K. Servat, B. Kokoh and N. Alonso-Vante, *Phys. Chem. Chem. Phys.*, 2009, **11**, 3573.
50. S. S. Kumar and K. L. N. Phani, *J. Power Sources*, 2009, **187**, 19.

51. E. Irissou, F. Laplante, S. Garbarino, M. Chaker and D. Guay, *J. Phys. Chem. C*, 2010, **114**, 2192.
52. J. Zhang, K. Sasaki, E. Sutter and R. R. Adzic, *Science*, 2007, **315**, 220.
53. B. H. Morrow and A. Striolo, *Phys. Rev. B*, 2010, **81**, 155437.
54. S. Wang, N. Kristian, S. Jiang and X. Wang, *Nanotechnology*, 2009, **20**, 025605.
55. N. Braidy, G. R. Purdy and G. A. Botton, *Acta Mater.*, 2008, **56**, 5972.
56. W. Tang, S. Jayaraman, T. F. Jaramillo, G. D. Stucky and E. W. McFarland, *J. Phys. Chem. C*, 2009, **113**, 5014.
57. N. Kristian and X. Wang, *Electrochem. Commun.*, 2008, **10**, 12.
58. J. Zhang, M. B. Vukmirovic, K. Sasaki, A. U. Nilekar, M. Mavrikakis and R. R. Adzic, *J. Am. Chem. Soc.*, 2005, **127**, 12480.
59. J. L. Zhang, M. B. Vukmirovic, Y. Xu, M. Mavrikakis and R. R. Adzic, *Angew. Chem. Int. Ed.*, 2005, **44**, 2132.
60. M. M. Maye, W. X. Zheng, F. L. Leibowitz, N. K. Ly and C. J. Zhong, *Langmuir*, 2000, **16**, 490.
61. M. J. Schadt, W. Cheung, J. Luo and C. J. Zhong, *Chem. Mater.*, 2006, **18**, 5147.
62. M. M. Maye and C. J. Zhong, *J. Mater. Chem.*, 2000, **10**, 1895.
63. K. J. Klabunde, *Nanoscale Materials in Chemistry*, Wiley, New York, 2001.
64. U. A. Paulus, A. Wokaun, G. G. Scherer, T. J. Schmidt, V. Stamenkovic, V. Radmilovic, N. M. Markovic and P. N. Ross, *J. Phys. Chem. B*, 2002, **106**, 4181.
65. (a) E. Antolini, *Mater. Chem. Phys.*, 2003, **78**, 563; (b) C. R. K. Rao and D. C. Trivedi, *Coord. Chem. Rev.*, 2005, **249**, 613.
66. H. Yang, W. Vogel, C. Lamy and N. Alonso-Vante, *J. Phys. Chem. B*, 2004, **108**, 11024.
67. D. L. Feldheim and C. A. Foss, Jr., *Metal Nanoparticles: Synthesis, Characterization, and Applications*, Dekker, New York, 2002.
68. (a) P. Waszczuk, G. Q. Lu, A. Wieckowski, C. Lu, C. Rice and R. I. Masel, *Electrochim. Acta*, 2002, **47**, 3637; (b) K. W. Park, J. H. Choi, B. K. Kwon, S. A. Lee, Y. E. Sung, H. Y. Ha, S. A. Hong, H. Kim and A. Wieckowski, *J. Phys. Chem. B*, 2002, **106**, 1869.
69. T. J. Schmidt, H. A. Gasteiger and R. J. Behm, *Electrochem. Commun.*, 1999, **1**, 1.
70. (a) R. Raja, T. Khimyak, J. M. Thomas, S. Hermans and B. F. G. Johnson, *Angew Chem. Int. Ed.*, 2001, **40**, 4638; (b) F. E. Jones, S. B. Milne, B. Gurau, E. S. Smotkin, S. R. Stock and C. M. Lukehart, *J. Nanosci. Nanotechnol.*, 2002, **2**, 81; (c) A. J. Dickinson, L. P. L. Carrette, J. A. Collins, K. A. Friedrich and U. Stimming, *Electrochim. Acta*, 2002, **47**, 3733.
71. C. J. Zhong and M. M. Maye, *Adv. Mater.*, 2001, **13**, 1507.
72. (a) A. Crown, H. Kim, G. Q. Lu, I. R. de Moraes, C. Rice and A. Wieckowski, *J. New Mater. Electrochem. Syst.*, 2000, **3**, 275; (b) J D. Aiken III and R. G. Finke, *J. Mol. Catal. A.*, 1999, **145**, 1.

73. (a) H. Li, Y.-Y. Luk and M. Mrksich, *Langmuir*, 1999, **15**, 4957; (b) R. S. Ingram and R. W. Murray, *Langmuir*; 1998, **14**, 4115.
74. (a) X. Peng, M C. Schlamp, A. V. Kadavanich and A. P. Alivisatos, *J. Am. Chem. Soc.*, 1997, **119**, 7019; (b) S. J. Oldenburg, R. D. Averitt, S. L. Westcott and N. Halas, *J. Chem. Phys. Lett.*, 1998, **288**, 243.
75. T. H. Galow, U. Drechsler, J. A. Hanson and V. M. Rotello, *Chem. Commun.*, 2002, 1076.
76. (a) M. Brust, M. Walker, D. Bethell, D. J. Schiffrin and R. Whyman, *J. Chem. Soc., Chem. Commun.*, 1994, 801; (b) A. C. Templeton, W. P. Wuelfing and R. W. Murray, *Acc. Chem. Res.*, 2000, **33**, 27.
77. (a) G. Schmid, V. Maihack, F. Lantermann and S. Peschel, *J. Chem. Soc., Dalton Trans.*, 1996, 589; (b) U. A. Paulus, U. Endruschat, G. J. Feldmeyer, T. J. Schmidt, H. Bonnemann and R. J. Behm, *J. Catal.*, 2000, **195**, 383.
78. R. L. Whetten, J. T. Khoury, M. M. Alvarez, S. Murthy, I. Vezmar, Z. L. Wang, P. W. Stephens, C. L. Cleveland, W. D. Luedtke and U. Landman, *Adv. Mater.*, 1996, **8**, 428.
79. (a) M. A. El-Sayed, *Acc. Chem. Res.*, 2001, **34**, 257; (b) C. J. Kiely, J. Fink, J. G. Zheng, M. Brust, D. Bethell and D. J. Schiffrin, *Adv. Mater.*, 2000, **12**, 640.
80. (a) F. Caruso, *Adv. Mater.*, 2001, **13**, 11; (b) J. J. Schneider, *Adv. Mater.*, 2001, **13**, 529; (c) W. Schärtl, *Adv. Mater.*, 2000, **12**, 1899.
81. (a) J. J. Storhoff and C. Mirkin, *Chem. Rev.*, 1999, **99**, 1849; (b) J. K. N. Mbindyo, B. D. Reiss, B. R. Martin, C. D. Keating, M. J. Natan and T. E. Mallouk, *Adv. Mater.*, 2001, **13**, 249. (c) R. M. Crooks, M. Q. Zhao, L. Sun, V. Chechik and L. K. Yeung, *Acc. Chem. Res.*, 2001, **34**, 181.
82. (a) S. H. Sun, C. B. Murray, D. Weller, L. Folks and A. Moser, *Science*, 2000, **287**, 1989; (b) S. H. Sun, E. E. Fullerton, D. Weller and C. B. Murray, *IEEE Trans. Mag.*, 2001, **37**, 1239.
83. M. Chen and D. E. Nikles, *Nano Lett.*, 2002, **2**, 211.
84. L. Han, W. Wu, F. L. Kirk, J. Luo, M. M. Maye, N. Kariuki, Y. Lin, C. Wang and C. J. Zhong, *Langmuir*, 2004, **20**, 6019.
85. M. J. Hostetler, C. J. Zhong, B. K. H. Yen, J. Anderegg, S. M. Gross, N. D. Evans, M. D. Porter and R. W. Murray, *J. Am. Chem. Soc.*, 1998, **120**, 9396.
86. J. Luo, M. M. Maye, Y. Lou, L. Han, M. Hepel and C. J. Zhong, *Catal. Today*, 2002, **77**, 127.
87. T. Page, R. Johnson, J. Hormes, S. Noding and B. Rambabu, *J. Electroanal. Chem.*, 2000, **485**, 34.
88. C. Berg, H. J. Venvik, F. Strisland, A. Ramstad and A. Borg, *Surf. Sci.*, 1998, **409**, 1.
89. M. Ø. Pedersen, S. Helveg, A. Ruban, I. Stensgaard, E. Lægsgaard, J. K. Nørskov and F. Besenbacher, *Surf. Sci.*, 1999, **426**, 395.
90. C. Song, Q. Ge and L. Wang, *J. Phys. Chem. B*, 2005, **109**, 22341.
91. M. Haruta, *Nature*, 2005, **437**, 1098.
92. M. Haruta, *Catal. Today*, 1997, **36**, 153.

93. M. Haruta and M. Date, *Appl. Catal. A*, 2001, **222**, 427.
94. L. D. Burke, *Gold Bull.*, 2004, **37**, 125.
95. J. Luo, M. M. Maye, L. Han, C. J. Zhong and M. Hepel, *J. New Mater. Electrochem. Syst.*, 2000, **5**, 237.
96. Z. Peng and H. Yang, *Nano Today*, 2009, **4**, 143.
97. S. Strbac and R. R. Adzic, *J. Electroanal. Chem.*, 1996, **403**,169.
98. T. Tunc, S. Suzer, M. A. Correa-Duarte and L. M. Liz-Marzan, *J. Phys. Chem. B*, 2005, **109**, 7597.
99. D. Q. Yang, J. N. Gillet, M. Meunier and E. Sacher, *J. Appl. Phys.*, 2005, **97**, 024303.

CHAPTER 3

Theoretical Studies of Formic Acid Oxidation

WANG GAO AND TIMO JACOB*

Institut für Elektrochemie, Universität Ulm, Albert-Einstein-Allee 47, D-89081 Ulm, Germany
*E-mail: timo.jacob@uni-ulm.de

3.1 Introduction

Formic acid (HCOOH) oxidation and analogous reactions have received great interest over the past decades, both for their status as prototype reactions for the electrochemical oxidation of small organic molecules as well as for their relevance in low-temperature fuel cells.[1,2] Several different metal catalysts such as Pt, Pd and Au, as well as related alloy catalysts, have been widely used in HCOOH oxidation.

To shed light on this fundamental process, the commonly used Pt catalyst is often applied as a model system. Experimentally, researchers choose gas-phase HCOOH oxidation on the Pt surface as the starting point for probing mechanisms that may unlock low-temperature HCOOH fuel cells. Ultra-high-vacuum (UHV) studies revealed that HCOOH oxidation depends on both the system temperature (T) as well as the HCOOH surface coverage.[3] Between 80–100 K, HCOOH resides on Pt(111) as molecular monomers and/or dimers. Between 150–170 K, HCOOH decomposes into formate (HCOO) and atomically adsorbed hydrogen. A further increase of the temperature up to T = 260 K causes HCOO to preferentially oxidize to CO_2, while higher coverages of HCOOH cause desorption rather than oxidization. Another work found that HCOOH exclusively dissociates to CO_2 and H_2 at low exposures

RSC Energy and Environment Series No. 6
Catalysts for Alcohol-Fuelled Direct Oxidation Fuel Cells
Edited by Zhen-Xing Liang and Tim S. Zhao

Published by the Royal Society of Chemistry, www.rsc.org

and the yields of CO_2 and H_2 increase with increasing exposure up to 4 Torr·s, reaching their saturation between 4 and 6 Torr·s and following with an additional molecular desorption.[4] Notably, it was also found that co-adsorbed water does not hinder formic acid adsorption and decomposition in any measurable way.

Electrochemical HCOOH oxidation on platinum electrodes is difficult to characterize experimentally; however, it is generally believed to proceed *via* a dual-path mechanism consisting of indirect and direct paths (see Figure 3.1).[1c] The indirect path has HCOOH being first converted to adsorbed carbon monoxide (CO*) and then to CO_2, while the direct path has HCOOH reacting to give CO_2 *via* a reactive intermediate.

Numerous experimental studies have presented different perspectives on what may be the relevant intermediates in HCOOH electro-oxidation. *In-situ* IR reflection absorption spectroscopy (IRAS) experiments identified CO*, a product of HCOOH dehydration, as a poisonous species, evidencing the indirect path.[1d,e] However, the reactive intermediate in the direct path has been controversial for a long time. The intermediates from both the indirect and direct paths compete with one another for adsorption sites and the opportunity to react with oxidizers (*e.g.* OH*) on the surface. This competition couples these reaction paths kinetically, hampering the elucidation of their individual reaction mechanisms.[5,6]

Wilhelm and co-workers initially suggested that either COH or CHO would be the reactive intermediate for the direct path,[1f] while others have long assumed it is COOH.[1c,h,2c] Recently, Osawa *et al.* found that formate (HCOO)

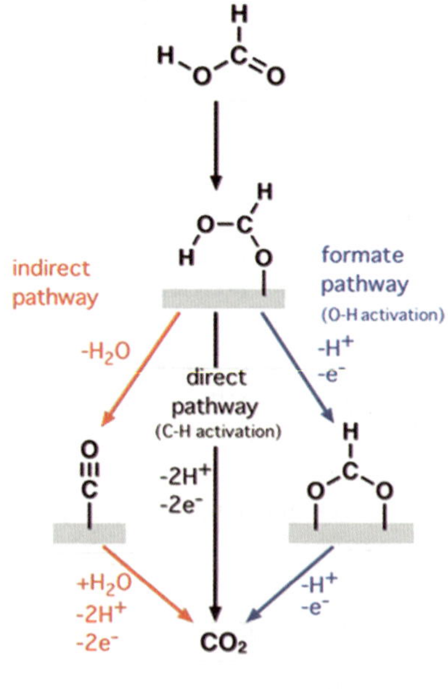

Figure 3.1 Mechanistic scheme for electrochemical formic acid oxidation.

is the reactive intermediate and that oxidation of HCOO to CO_2 is rate determining.[5–10] In contrast, Behm *et al.* argued that the weakly adsorbed HCOOH could be the intermediate in the dominant reaction path, leaving HCOO to be just a spectator.[11–13] The electrochemical and spectral data obtained under static and flow conditions, which provide the basis for these latter two proposed reactive intermediates, are essentially identical. However, different interpretations of the non-linear relationship between the measured current and the formate coverage lead to different conclusions.[6]

Cyclic voltammograms (CVs) of HCOOH oxidation show that the current (*I*) first peaks around 0.6 V as the potential (*U*) increases. The current then remains stable or decreases between 0.6 and 0.8 V in what is termed the negative differential resistance (NDR) region. A sharp increasing peak around 0.95 V follows the NDR region.[5,7,12,14–16] During this process, IR spectroscopy measurements reveal that the polycrystalline Pt surface is covered with a relatively constant coverage of CO below 0.8 V. Above this potential, the coverage of adsorbed CO rapidly decreases due to oxidation, while HCOO quickly increases with increasing potential, until the surface is nearly saturated with HCOO at above 0.9 V. Once CO is almost completely removed from the surface around 0.95 V, the coverage of HCOO rapidly decreases with increasing potential up to 1.2 V.[5]

Thus, these IR measurements suggest that the CV curve can be understood in terms of CO adsorption, desorption and oxidation by OH. Nevertheless, the detailed roles of adsorbed CO* and OH* have yet to be elucidated, substantially hindering our understanding of the mechanism of this fundamental reaction.

First-principles simulations should help to clarify the complicated mechanism of HCOOH oxidation. Unfortunately, only a few theoretical studies exist on the Pt/H_2O system, since modeling the electrocatalytic system, especially solid/liquid interfaces, and the applied potential are quite challenging.

Different results suggested that the most stable structure of HCOOH on Pt(111) in the gas phase has the carbonyl oxygen's lone pair of HCOOH binding to an atop site while the O–H group of HCOOH points towards the surface.[17,18] In contrast, others found that HCOOH prefers to adsorb in an upright fashion, binding to Pt(111) with the oxygen lone pairs of both its O–H and carbonyl groups.[19] Depending on the latter observation, simulations by Neurock *et al.* proposed that COOH is the reactive intermediate while HCOO is just a spectator for HCOOH oxidation on Pt(111) in the presence of a surrounding solution and an external electrode potential, supporting experiments by Behm *et al.* Also, they found that an applied potential has little effect on the barriers of formate oxidation, but influences the oxidation of CO significantly. Wang and Liu have used continuum solvation approaches to model the Pt/H_2O interface, finding that a weakly adsorbed HCOOH* with its C–H bond in a "down" configuration is the reactive intermediate.[20] Their results also suggested that an applied potential has minor effects on HCOOH oxidation barriers.

Recent theoretical investigations have reported that the initial adsorption structure of a reactant might influence or even change the entire catalytic reaction pathway. To best discern the gas-phase HCOOH oxidation mechanism, all reasonable adsorption configurations of HCOOH (as well as different decomposition intermediates) should be investigated. This might then lead to a complicated collection of possible reaction processes, some of which may have already been ruled out at certain conditions by previous calculations and/or experiments. However, using this collection as the road map for a full mechanistic study, insight on the actual reaction processes can later come to light by explicitly finding how water solvation, ion concentration, electrode potential, temperature and pressure all affect these mechanisms. Therefore, as a starting point to fundamentally understand the HCOOH oxidation, we modeled the gas-phase reaction on Pt(111) using density functional theory (DFT) since these results are readily comparable to UHV experiments.[21]

Then we tried to include the effects of solvation in the reaction process, which is crucial for electrocatalysis and has not been addressed yet. To elucidate the controversial mechanisms of HCOOH oxidation and to investigate how different explicit water models affect these mechanisms, we used two different explicit solvation models to simulate the aqueous interface, which is the site where HCOOH oxidation occurs. A large number of unique reaction pathways involving subtly different reaction intermediates and transition states were obtained explicitly. The effects of applied potentials were later included to investigate different preferred electrocatalytic pathways. By comparing these results to analogous simulations in the gas phase, the critical role of explicit water molecules and the applied potential on HCOOH oxidation can be clarified.[22]

Further, one has to consider the complicated role of co-adsorbed CO* and OH* in the electro-oxidation of HCOOH. We included a fractional CO* and OH* coverage to simulate different electrochemical interfaces, which can clarify the individual effect of CO* and/or OH* on the HCOOH oxidation. By considering the effects of applied potentials on the different pathways, our results substantially reveal the detailed process of HCOOH oxidation and clarify the mechanism for this fundamental electrocatalytic reaction.[23]

3.2 Methods

All calculations were performed using DFT implemented in CASTEP,[24] with Vanderbilt-type ultrasoft pseudopotentials[25] and the generalized gradient approximation (GGA) exchange–correlation functional proposed by Perdew, Burke and Ernzerhof (PBE).[26] Intermediate surface coverages of $^1/_4$ and $^1/_9$ ML (ML = monolayer) on Pt(111) were studied within (2 $\times$ 2) and (3 $\times$ 3) surface unit cells, respectively. A five-layered slab represented the Pt(111) surface, and a vacuum width >13 Å minimized interactions between neighboring slabs in the supercell geometry. For each slab the topmost three surface layers were allowed to relax, while the bottom two layers were fixed to the calculated crystal structure. Integrations in reciprocal space used a 2 × 2 Monkhorst–Pack *k*-point grid after test calculations with the 4 × 4 *k*-point grid amounted to energy differences <0.04 eV. A plane-wave basis set with an energy cutoff of 400 eV was used in all calculations. To investigate the reaction pathways, the transition state (TS) search procedure in the CASTEP code was used, which employs a combination of LST/QST (linear and quadratic synchronous transit; see Govind *et al.*[27] for details) algorithms with subsequent conjugate gradient methods.[24]

We define the binding energy (*i.e.* adsorption energy) E_{bind} for co-adsorbed H_2O and HCOOH on Pt(111) as:

$$E_{bind} = -\left(E_{total} - E_{surf} - E_{HCOOH} - nE_{H_2O}\right) \tag{3.1}$$

where E_{total} is the energy of the H_2O/HCOOH complex on Pt(111), E_{surf} is the energy of the clean Pt(111) surface, E_{HCOOH} is the energy of an HCOOH molecule in vacuum, and nE_{H_2O} is the energy of a single H_2O molecule in vacuum times the number of water molecules in the simulation. All other adsorption energies are defined in a similar way. Applied electrode potentials are incorporated by explicitly shifting Fermi energies as calculated by DFT. More details of this approach can be found in Keith *et al.*[28]

3.3 Results and Discussion

3.3.1 Gas-Phase Reaction

We start the procedure of stepwise increasing the complexity of the model and thus becoming more realistic by first studying the oxidation of HCOOH under gas-phase conditions. Figure 3.2(a) shows the most stable configuration of HCOOH, where the carbonyl oxygen binds to an atop surface site with the –OH group pointing towards the surface. Our results are consistent with other theoretical studies reported in the literature.[17,18] We find the binding energies for this structure are dependent on the surface coverage: 0.34 eV with $^1/_4$ ML and 0.40 eV with $^1/_9$ ML coverage. This agrees with values reported by Bakó and Pálinkás, who found 0.42 eV with $^1/_6$ ML,[17] as well as values by Hartnig *et al.*[18] of 0.40 eV with $^1/_{16}$ ML. These results at different coverage show that at coverages $\leq ^1/_9$ ML the adsorbate–adsorbate interactions are negligible, and that $^1/_9$ ML can already be interpreted as a low-coverage limit.

Regarding the HCOOH structure on the Pt(111) surface, the bond distance between the oxygen atom O1 and its underlying Pt atom (d_{Pt-O1}) is 2.27 Å on the *p*(2 × 2) adlayer configuration and 2.24 Å on the *p*(3 × 3) adlayer configuration. We find that HCOOH adsorption in an upright fashion (as shown in Figure 3.2b) is less stable and has almost no favorable surface binding, *i.e.* $E_{bind} < 0.01$ eV.

We then investigated different decomposition pathways of HCOOH based on those adsorption structures, leading to the energy diagrams in Figure 3.3.

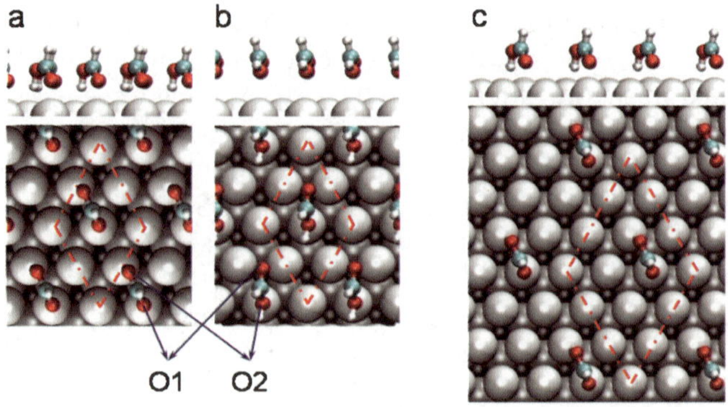

Figure 3.2 Side and top views of HCOOH adsorbed on Pt(111) at $^{1}/_{4}$ (a and b) and $^{1}/_{9}$ ML (c) coverage. Structure (a) indicates the most stable adsorbate configuration, while (b) shows the upright adsorption that is sometimes discussed in the literature. The O atom of the C=O bond is denoted as O1, the oxygen of the OH group as O2. The dashed box indicates the surface unit cells.

We found that two different pathways are possible, one eventually leading to CO_2 + 2H and the other leading to CO + H_2O. In the former case, three different pathways can be distinguished: (1) the COOH pathway, (2) the $HCOO^-$ pathway and (3) the direct pathway. In case (1), the COOH pathway on the (2 × 2) unit cell proceeds by HCOOH first breaking its C–H bond, a process that is exothermic and has a barrier of E_{act} = 1.58 eV. This barrier is substantially higher than that found for HCOOH when absorbed in a tilted configuration (E_{act} = 0.74 eV).[29] Not surprisingly, the difference between these two barriers is due to the different adsorption structures of HCOOH. After breaking the C–H bond, the resulting COOH prefers to be adsorbed at an

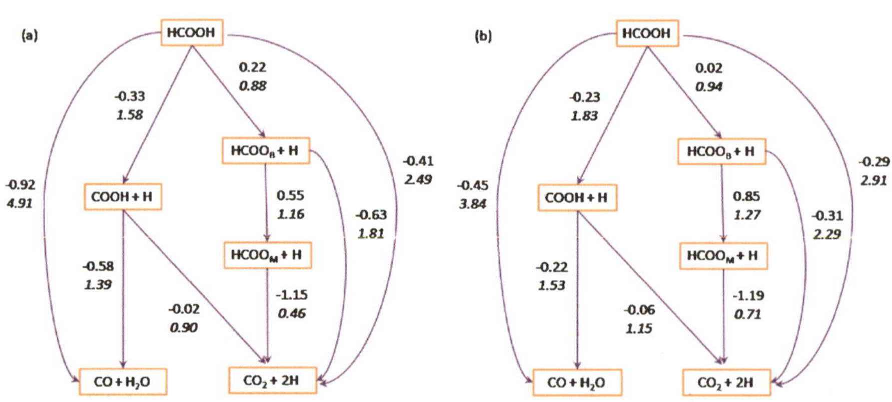

Figure 3.3 Reaction mechanisms for the HCOOH oxidation without water at (a) θ = $^{1}/_{4}$ ML [(2 × 2) unit cell] and (b) θ = $^{1}/_{9}$ ML [(3 × 3) unit cell]. All energy changes (upper numbers) and activation barriers (lower italic numbers) are in eV.

Table 3.1 Adsorption energies of HCOOH and different reaction intermediates along its oxidation on Pt(111).[a]

		E_{bind} (eV)					
		This work	Literature				
Adsorbate[b]	Periodicity	PBE	PBE	PW91	RPBE	GGA+U	Exp.
HCOOH	(2 × 2)	0.34	–	–	–	–	–
	(3 × 2$\sqrt{3}$)	–	0.42[17]	–	–	–	–
	(3 × 3)	0.40	–	–	–	–	–
	(4 × 4)	–	–	0.37[18]	–	–	–
COOH	(2 × 2)	2.31	–	2.29,[30] 2.28[31]	1.76[31]	–	–
	(3 × 3)	2.30	–	–	–	–	–
$HCOO_B$	(2 × 2)	2.50	3.73[29]	2.31[31]	1.65[31]	–	–
	(3 × 3)	2.32	–	–	–	–	–
$HCOO_M$	(2 × 2)	1.56	2.83[29]	–	–	–	–
	(3 × 3)	1.46	–	–	–	–	–
CO	(2 × 2)	1.66	1.56[33]	1.82[31]	1.68,[34] 1.34[35]	1.33,[31] 1.72,[36] 1.39,[37] 1.44[38]	1.66[39]
	(3 × 3)	1.61	–	–	–	–	–

[a]For each system, (2 × 2) and (3 × 3) adlayers, representing $\theta = {}^1/_4$ and ${}^1/_9$ ML, were considered. Furthermore, available theoretical (with the used exchange–correlation functionals being indicated) and experimental literature values are given. [b]$HCOO_B$ = bidentate configuration; $HCOO_M$ = monodentate structure.

atop site with E_{bind} = 2.31 eV. Note that this is consistent with PW91 calculations (~2.29 eV) but not with the RPBE calculated value (1.76 eV; see Table 3.1).[30,31] COOH can then further decompose to CO_2 with a barrier of E_{act} = 0.90 eV, a value in reasonable agreement with the PW91 calculated barrier of 1.02 eV reported by Gong *et al.*[30]

Breaking the O–H bond of HCOOH has a relatively small barrier of E_{act} = 0.88 eV. This energy is much lower than the 1.63 eV required to dissociate the O–H bond when HCOOH is adsorbed in the tilted configuration.[29] The resulting HCOO intermediate binds in a bidentate configuration ($HCOO_B$), where the molecule is adsorbed on a Pt bridge site *via* two Pt–O bonds with a binding energy of E_{bind} = 2.50 eV. This value is in agreement with PW91 calculations by Mavrikakis and co-workers (E_{bind} = 2.31 eV),[31] though not with the same group's RPBE calculations (E_{bind} = 1.65 eV) and nor with previous PBE calculations by Jiang and co-workers (E_{bind} = 3.73 eV).[29] Needless to say, the large range of calculated values shown in Table 3.1 means future studies on this species should deserve careful attention. If $HCOO_B$ then decomposes to CO_2 and 2H directly, the barrier is E_{act} = 1.81 eV, for which Gabow *et al.* calculated a value of only 1.04 eV.[31] Owing to the present orientation of $HCOO_B$, forming CO_2 directly in the gas phase is difficult since adjacent molecules are nearby to receive the H atom. However, forming CO_2

and 2H is much easier if one first breaks one Pt−O bond, yielding a HCOO intermediate species with a monodentate structure ($HCOO_M$). The barrier associated with this transition is E_{act} = 1.16 eV. The resulting $HCOO_M$ is adsorbed at an atop site with a binding energy of 1.56 eV. Since breaking the C−H bond of $HCOO_M$ is both exothermic by ΔE = –1.15 eV and involves only a relatively small activation barrier (E_{act} = 0.46 eV), one would expect this process to be both kinetically and thermodynamically favorable compared to other processes so far considered. Indeed, it agrees with other DFT results by Gao *et al.* of ΔE = –1.35 eV and E_{act} = 0.59 eV.[29] Moreover, $HCOO_M$ was also suggested to be the intermediate for the reaction of HCOO → CO_2 in UHV experiments on TiO_x/Pt(111) surfaces.[32]

Although the direct pathway, HCOOH → CO_2 + 2H, is exothermic by −0.41 eV, it is kinetically hindered by having a large activation barrier of E_{act} = 2.49 eV. Similarly, if HCOOH directly decomposes to CO and H_2O instead, the process is exothermic by –0.92 eV but requires overcoming an extremely high barrier of 4.91 eV. Experimental studies have not found evidence for this process, though it was speculated that while the barrier should be high, it might occur on steps or defect sites.[19] Moreover, $E_{bind}{}^{CO}$ is 1.66 eV, which is in good agreement with other simulations giving a binding energy range of 1.34–1.82 eV, as well as with experiments.[33–39] For COOH + H → CO + H_2O, the activation energy (E_{act} = 1.39 eV) is higher than that for COOH + H → CO_2 + 2H (E_{act} = 0.90 eV). Thus, we predict that HCOOH preferentially reacts to give CO_2 + 2H, where, based on our previous discussion of energy barriers, the minimum energy pathway (MEP) would be the pathway with the lowest barriers, and those barriers are on the pathway involving HCOO. Thus, we expect HCOO should be the reactive intermediate for HCOOH decomposition in the gas phase.

With $^1/_9$ ML HCOOH the pathways are nearly the same as those obtained with $^1/_4$ ML (see Figure 3.3b). The pathway based on HCOO is also the kinetically most favored one for forming CO_2 + 2H, while the CO pathway is still thermodynamically favored. The correlated barriers along the CO_2 pathways increase as the coverage decreases, which is consistent with the tendency of the adsorption energies that are influenced by adsorbate−adsorbate repulsions. Along the HCOO pathway, E_{act} = 0.71 eV for $HCOO_M$ → CO_2 + H is higher than E_{act} = 0.42 eV for $HCOO_M$ → $HCOO_B$. Therefore, $HCOO_M$ is easier to first react to give $HCOO_B$ than to give CO_2, which is different to the behavior on the higher coverage (2 × 2) unit cell. Therefore, $HCOO_B$ might accumulate on the surface until higher coverage is reached, supporting the follow-up reaction steps to CO_2. In the CO pathway, E_{act} = 3.84 eV for HCOOH → CO + H_2O is even 1.07 eV lower than the corresponding value obtained for $\theta = {}^1/_4$ ML. As a result, for the gas phase system it seems that increasing the coverage from $^1/_9$ to $^1/_4$ ML supports the oxidation of HCOOH to CO_2 + 2H. This agrees with experimental UHV measurements that proposed HCOO as active intermediate by analyzing the corresponding vibrational frequencies.[3,4]

In conclusion, our calculations show that HCOO is the active intermediate of the HCOOH oxidation in gas phase, which is consistent with corresponding experimental studies under UHV conditions.

3.3.2 Influence of Water Solvation

In the next step we included the presence of the surrounding solvent by considering co-adsorbed water. We successively increased the number of water molecules per unit cell, starting with one and two water molecules (model A) up to the full water bilayer (model B). While the details of the first systems can be found in Gao *et al.*,[22] here we focus on model B, which has a water bilayer consisting of a fully periodic ice-like network of water molecules. Figure 3.4(a) depicts waters in the ordered ice-like structure where the outer waters have their OH bonds pointing toward the surface.[40] When HCOOH is incorporated into this surface, its size and structure permits substitution of two water molecules in the bilayer. An HCOOH/H_2O cluster thereby forms with two neighboring H_2O molecules, all of which fits in the H-bonding network spanning the entire unit cell. With this observation, we found two stable ring-network structures distinguishable by the direction of HCOOH, both with similar binding energies: E_{bind} = 2.48 eV (B1; see Figure 3.4b) and 2.60 eV (B2; see Figure 3.4c). In B1, the O−H bond in HCOOH forms an H-bond with a water molecule that lies parallel to Pt(111). In B2, the O−H also forms an H-bond, but with a water oriented perpendicularly to Pt(111). Based on observed atomic distances tabulated in Table 3.2, HCOOH in B1 more strongly

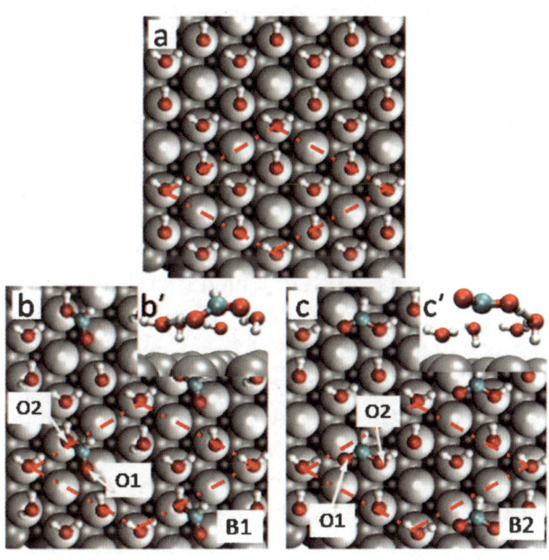

Figure 3.4 (a) Top view of the water bilayer structure on Pt(111). (b) B1 and (c) B2 are different HCOOH structures with four waters on Pt(111). Insets (b′, c′) denote side views of the interface.

Table 3.2 Geometric data for different HCOOH structures on the (3 × 3) Pt(111) surface unit cell.[a]

Adsorbate	d_{O1-Pt}	d_{O2-Pt}	d_{O1-H}	d_{O2-H}	d_{Pt-H_2O}	$\Theta^1_{H_2O}$	$\Theta^2_{H_2O}$
HCOOH + H_2O	2.24	3.13	1.79	–	2.09	106.7	–
A1	2.70	3.18	1.74	–	2.19	110.4	–
A2	2.52	3.15	1.84	–	2.22	104.7	–
B1	3.55	2.93	1.97	2.05	–	102.7	107.9
B2	3.87	3.68	1.86	1.99	–	108.8	101.6

[a]HCOOH + H_2O denotes structural information for HCOOH solvated by a single water molecule. Structures A1 and A2 are models for HCOOH solvated with two waters, while structures B1 and B2 show HCOOH solvated with a water bilayer. The CO oxygen in HCOOH is denoted O1 while the OH oxygen is denoted O2. The distances d_{O1-Pt} and d_{O2-Pt} are the H-bond lengths between HCOOH and the adjacent water molecule; d_{Pt-H_2O} denotes the distance between the water molecule adjacent to O1 to the nearest Pt atom. The angles $\Theta^1_{H_2O}$ and $\Theta^2_{H_2O}$ are the intramolecular bond angles of waters coordinating to O1 and O2, respectively. Distances are in Å and angles in degrees.

interacts with the Pt surface, while HCOOH in B2 more strongly interacts with the co-adsorbates. As a result, HCOOH in B2 practically lies flat to the surface to maximize the number of H-bonds to adjacent water molecules. Wang and Liu reported a different structure for HCOOH whereby HCOOH binds with four H_2O molecules and adsorbs on the surface with its C–H bond pointing towards the surface.[20] Direct comparison to that work is not possible, but in our own calculations the CH-down structure is >0.6 eV less stable (E_{bind} = 1.84 eV) than either structure we calculated based on models B1 or B2.

Using solvation model B, the direct pathway involving simultaneous C−H and O−H activation from HCOOH* to give CO_2 has lower barriers (B1: E_{act} = 1.68 eV; B2: E_{act} = 1.43 eV) compared with those with model A (E_{act} = 2.5−3.0 eV), indicating that the water bilayer indeed catalyzes this process. Despite this catalytic effect, the barriers remain relatively high and will be later ruled out as possible reaction processes on Pt(111) terraces.

In the B1 and B2 models, C−H activation from HCOOH* leads to two different products as well. In the B1 model, C−H activation leads to a spontaneous O−H bond-breaking step, which then leads to an adsorbed CO_2* intermediate. This intermediate will be discussed in more detail later. Interestingly, other calculations on O−H bond activation from COOH* into an aqueous phase indicate this to be a barrier-less process.[19,29,30,41] The concerted reaction step from the B1 structure is overall exothermic by –0.33 eV and has a barrier of E_{act} = 0.79 eV. In the B2 model, O−H bond breaking does not accompany C−H bond breaking and so COOH* is formed instead. Here, the reaction step is slightly endothermic by +0.07 eV and has a barrier of E_{act} = 1.63 eV.

The B1 pathway has the lower barrier and also permits simultaneous C−H and O−H bond activation. This particular reactivity is most likely due to the combination of two aspects. First, the water bilayer in the B1 model promotes

O−H bond dissociation. Second, the B1 model has HCOOH* binding more strongly to the surface, so CO_2* can remain bound. In previous calculations by Neurock *et al.* on the upright HCOOH*, C−H activation came with a barrier of 0.50 eV.[19] A similar barrier of E_{act} = 0.45 eV was proposed in the CH-down configuration by Wang and Liu.[20] In comparison, our B1 reaction barrier is slightly higher, probably because we initiated our studies with a more stable HCOOH* configuration. An H-bond network due to explicit waters stabilizes the CO_2* intermediate. The CO_2* intermediate is metastable as it is –0.33 eV more stable than the preceding HCOOH* intermediate in B1. This means that, under specific configurations of water molecules, highly stable molecules such as CO_2 can be confined close to a surface with van der Waals interactions, as has been seen with organic molecular porous materials.[42] Naturally, these CO_2* intermediates on Pt(111) are expected to be short lived.

Comparing calculations from models A to those from models B, one easily recognizes the importance of the water bilayer in describing the O−H activation steps. Without the extensive H-bonding network, barriers for O−H bond breaking can be more than two times greater than without the catalytic effects from nearby waters. Furthermore, the different orientations of the HCOOH* and COOH* can lead to different pathways. Overall, the most relevant pathway to be considered for all the direct mechanisms, however, is that with the lowest energy: the B1 pathway.

For the formate pathway, as B1 and B2 have different structures embedded in the water bilayer, the calculated values are slightly different, but not substantially. However, the final C−H activation steps of $HCOO_M$* with the B1 and B2 models are exothermic by –0.85 and –0.46 eV, respectively, with activation barriers of E_{act} = 0.58 and E_{act} = 1.10 eV. In general, the B2 calculations match fairly well with the model A calculations, although the final CO_2 molecule is even less stable when sharing the interface with additional water molecules. In B1 calculations, the orientation of $HCOO_M$* within the bilayer facilitates both easier C−H bond breaking and more favorable bonding with CO_2 at the interface.

In the indirect pathway, the reactions for models B1 and B2 are exothermic by –0.43 eV and –0.44 eV, respectively, with barriers of E_{act} = 3.32 eV and E_{act} = 3.47 eV. Based on these calculations alone, it would appear that CO* should not be formed in HCOOH oxidation at all. However, the presence of CO* is of course well established to be found on Pt(111) electrodes, particularly at relatively low applied potentials. Feliu and co-workers have shown that the presence of CO* seems to be proportional to the amount of step-edges and other defect sites on Pt(111).[43] Since our results show dehydrogenation processes leading to CO* are always high-barrier processes on idealized Pt(111) terraces, our results are still in agreement with experiment, assuming the presence of CO* is related to defects or low-coordinated surface sites. Given the high barriers to CO* formation, we did not calculate further oxidations to CO_2*, although this topic has been studied by DFT in previous work.[44,45]

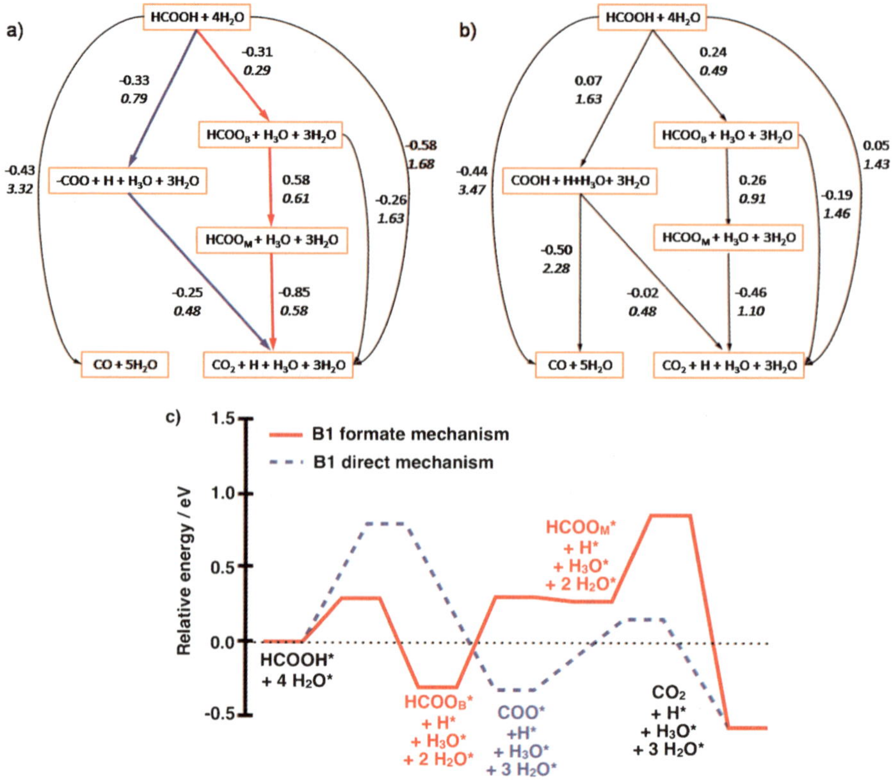

Figure 3.5 Langmuir–Hinshelwood HCOOH oxidation reaction mechanisms using the water bilayer solvation model starting from different orientations of HCOOH*: (a) B1 and (b) B2. The lowest energy pathway is shown in *red*, while the next-lowest energy pathway is shown in *blue*. Individual ΔE for intermediates (*in bold*) and activation barriers (*in italics*) are given in eV. (c) Potential energy diagram comparing the two lowest reaction pathways.

Using solvation model B, the MEP is again the formate pathway; however, the next lowest energy process is the direct pathway involving the COO* intermediate. Both of these pathways were found using model B1, and a summary of these data are presented in Figure 3.5. Using the bilayer model, the presence of a dual-pathway mechanism is at least more possible.

3.3.3 Eley–Rideal Mechanisms

Until now, our discussion of HCOOH oxidation has assumed that H atoms removed from HCOOH during oxidation remain at the electrochemical interface, either as a hydronium molecule or as an explicitly bound H* species. Instead, we can assume that each hydrogen atom removed from HCOOH* enters the electrolyte as a proton. Since the proton is removed into the bulk electrolyte, this would be an Eley–Rideal (ER) mechanism rather than the previously discussed Langmuir–Hinshelwood mechanism. Such ER mechanisms have been observed to be preferred in electrocatalytic oxygen reduction on Pt(111).[28] If one then assumes the proton immediately participates in the hydrogen electrode reaction shown in eqn (3.2), it is then a coupled-electron-proton-transfer process (CPET), and one can use this coupling to implicitly model applied electrode potentials (*vs.* the RHE) *via* explicit shifts of calculated Fermi energies for different intermediates.[46] In other words, for every transferred H atom, the energy of the intermediate following this transfer is shifted by the desired electrode potential. Potential-dependent reaction barriers for these processes can also be approximated with an extension to this model, and further details of this approach can be found in Keith *et al.*[28] and Marenich *et al.*[47]

$$H^+_{(aq)} + e^- \rightarrow \tfrac{1}{2}H_{2(g)} \tag{3.2}$$

Previous DFT studies on the electrocatalytic HCOOH oxidation using different models found that an applied potential has little effect on oxidation barriers.[19,20] However, we have already shown that the bilayer solvation model sometimes substantially influences reaction barriers and intermediates. Thus, in the following we will concentrate on the water solvated system (bilayer water model) and will discuss ER mechanisms (Figure 3.6).

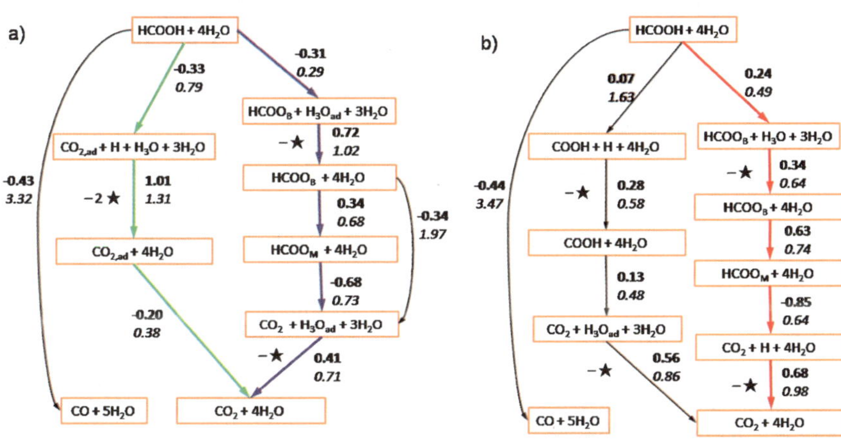

Figure 3.6 Eley–Rideal HCOOH oxidation reaction mechanisms with the water bilayer solvation model starting from different orientations of HCOOH*: (a) B1 and (b) B2. The lowest energy pathway is shown in *red*, the next-lowest energy in *blue*, while the third-lowest energy pathway is in *green*. Reaction steps marked with a star (★) denote a CPET process between the interface and the electrolyte. Although the pathway denoted in (b) has the lowest individual barrier of 0.04 eV, its final product state is thermodynamically 0.56 eV less stable than the product states in (a). Therefore, only pathways in (a) are considered further.

The ER pathways involving CO formation are the same as the LH pathways because dehydrogenation leads to a fifth water molecule per (3 $\times$ 3) surface unit cell and no charge is transferred during CO oxidation. Thus, the reaction barriers for these processes remain very high ($E_{act} > 3.32$ eV), and this result remains in good agreement with electrochemical experiments.[7,8,11–13,43]

ER pathways involving CO_2 formation do change because intermediates formed from a CPET no longer involve an explicit proton within the simulation cell. In general, all of the CO_2 pathways in both B1 and B2 models become thermodynamically more favorable as applied potentials increase, consistent with the general trends found by Neurock *et al.*[22] Figure 3.6 shows that the final product obtained using the B2 model is 0.56 eV higher than the product found in the B1 model. Since the relative energies of the final products are all influenced by the applied potential equally, we have to consider the potential dependence on the barriers for all of these processes.

We now report individual reaction processes, including transition states at different applied potentials (see Figure 3.7). At 0.0 V (*vs.* RHE), the MEP is the formate pathway in the B2 model: $E_{act} = 0.98$ eV and $\Delta E = 1.04$ eV, overall. The highest barrier process for this mechanism is the potential-dependent CPET process removing H*. The next lowest energy mechanism is the formate pathway in the B1 model: $E_{act} = 1.02$ eV and $\Delta E = 0.60$ eV, overall with respect to the more stable HCOOH* structure calculated in model B2. The highest barrier process for this mechanism is the CPET process after formation of the $HCOO_B$* intermediate. Lastly, the third lowest energy pathway is the direct pathway in the B1 model: $E_{act} = 1.31$ eV and its final product is also 0.6 eV endothermic overall. Here, the highest barrier process is removal of two hydrogen atoms after formation of the CO_2* intermediate, a double-CPET process.

Although the formate pathway from the B2 model is calculated to have the lowest barrier of all, it is only slightly lower than the barrier for the formate pathway from the B1 model. Both formate pathways should be considered competitive, but the B1 pathway is more relevant as it leads directly to a lower-energy final product structure. For now, we only show both the formate and the direct pathways *via* the CO_2* intermediate as calculated with the B1 model in Figure 3.7. Comparing the steps with the highest barrier in both mechanisms shows that at low electrode potentials (0.0 V *vs.* RHE) the formate pathway is strongly preferred on Pt(111) terraces over the direct pathway due to the high barrier associated with the high-energy double-CPET process in the latter mechanism.

At 0.5 V the relative energies between the two B1 pathways change since the formate pathway has two separate CPET processes (after $HCOO_B$* and CO_2 formation), while the direct pathway has a double-CPET process (after CO_2* formation). The highest barrier in the formate pathway is now $E_{act} = 0.73$ eV ($HCOO_M \rightarrow CO_2$), while the highest barrier in the direct pathway is 0.79 eV (simultaneous C−H and O−H bond breaking). Note that the highest barrier processes at 0.5 V are now potential-independent processes. Overall, the energy

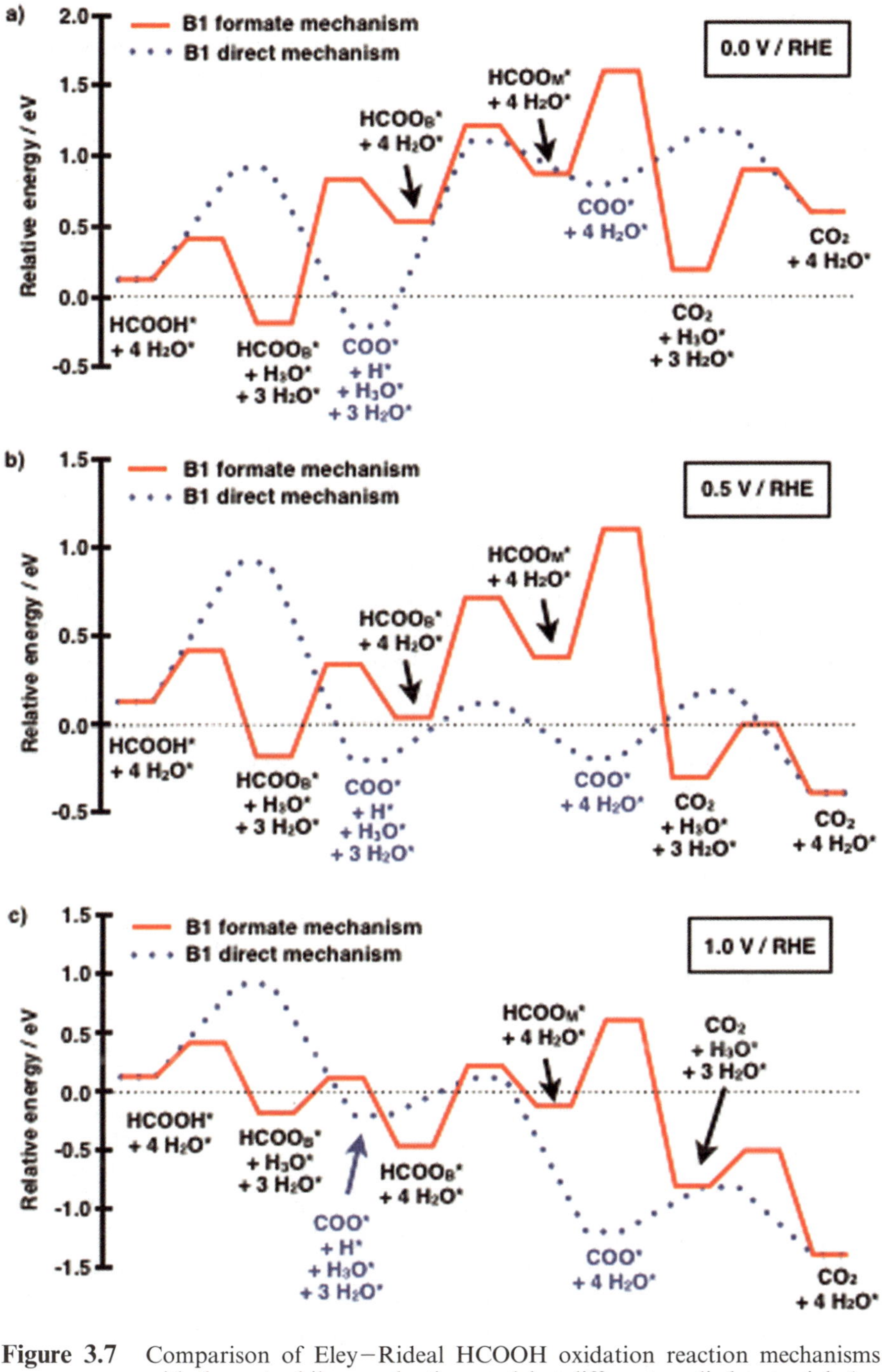

Figure 3.7 Comparison of Eley–Rideal HCOOH oxidation reaction mechanisms with the water bilayer solvation model at different applied potentials (*vs.* RHE): (a) U = 0.0 V; (b) U = 0.5 V; (c) U = 1.0 V.

of the CO_2 final intermediate has decreased to –0.4 eV with respect to the most stable HCOOH* structure found in model B2. Thus, the difference between the formate and the direct pathways has decreased substantially with increased applied potentials. Instead of the highest barriers being different from each other by ~0.3 eV at 0.0 V, the difference in barriers at 0.5 V is now only 0.06 eV.

At 1.0 V, the final product energies become even more stable due to the effect of the electrode potential on the final CO_2 product state (–1.4 eV overall). Nevertheless, at 0.5 V the highest barrier in both the formate and the direct pathways are potential-independent processes; this is still the case at 1.0 V. Thus, the difference in barriers for the formate and direct pathways are again <0.1 eV.

3.3.4 Kinetics Analysis

Naturally, formic acid oxidation is a highly complicated process. Undertaking a kinetics analysis using rate constants derived from first-principles calculations would be an ideal way to determine which pathways are relevant under different concentrations of intermediates. Such a kinetics analysis is unfortunately difficult since it would require coverage dependencies and reaction processes at step edges and defect sites to provide a realistic modeling of the real electrochemical system. However, our calculations can be incorporated into a reduced kinetic model that should report qualitative features of the reaction mechanisms at different applied potentials. We do this by comparing calculated potential-dependent rate constants obtained from Eyring's canonical transition state theory (TST),[48] as shown in eqn (3.3):

$$k(U) = \frac{k_B T}{h} \exp\left(\frac{-\Delta G_T(U)}{k_B T}\right) \quad (3.3)$$

where k_B is Boltzmann's constant, T is temperature, h is Planck's constant, and $G_T(U)$ is the highest barrier for a given mechanism at a given potential and temperature. Figure 3.8 shows a comparison between the potential-dependent rate constants for the formate and direct pathways in a potential range between 0.0−1.0 V.

As a first approximation, if one assumes that all intermediate concentrations are unity, Figure 3.8 would then correspond to the relative preference of the two HCOOH oxidation pathways at different potentials. At potentials <0.3 V, the formate pathway should be preferred over the direct pathway; however, it is not clear to which degree this would be expected, since this is the region where during the HCOOH oxidation Pt(111) is largely covered with CO*.

Recall that the highest barrier process in the formate pathway is first the CPET process after $HCOO_B$* formation, but with increasing potential the highest barrier process changes to the $HCOO_M \rightarrow CO_2$ process. Based on the relative energies of the possible intermediates, $HCOO_B$ should be observable under electrochemical conditions, and indeed it is.[7,8,11,12] Furthermore, at

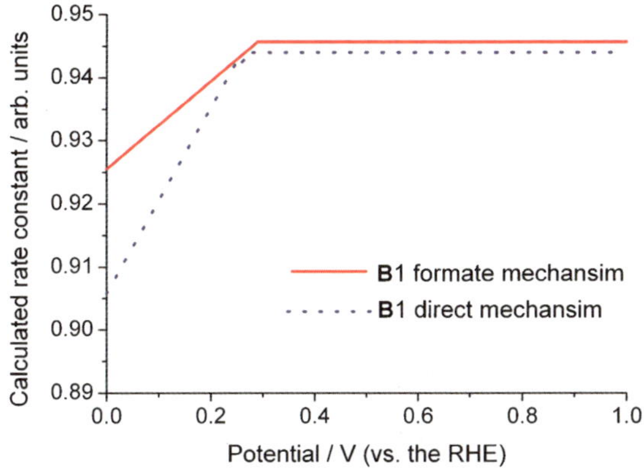

Figure 3.8 A comparison of potential-dependent rate constants as calculated from canonical TST. While the formate mechanism always has a larger calculated single rate constant than the direct pathway, both mechanisms appear to converge at potentials >0.2 V, implying that both mechanisms should be operational. At potentials <0.2 V, the formate pathway would ideally predominate on a Pt(111) terrace.

potentials where the $HCOO_M \rightarrow CO_2$ process has the highest barrier, formate intermediates should accumulate on the surface. Based on our results alone, one would expect the $HCOO_M \rightarrow CO_2$ process to be slow at potentials ~0.3 V and higher. Indeed, experiments have found that a formate adlayer starts to accumulate in the potential ranges >0.2 V.[12]

At potentials >0.3 V, our calculations show the two pathways converge towards very similar values, suggesting that both pathways should be largely indistinguishable at potentials higher than 0.3 V. As the highest reaction barriers decrease from 0.0 to 0.3 V, the rates for HCOOH oxidation should increase. Indeed, experimental faradaic currents increase with increasing potential in the range of 0.1–0.7 V.[7,8,12] It has been rationalized that at potentials >0.7 V the surface is poisoned by oxide species, and hence should result in a decrease in current activity. Since our calculations have not investigated HCOOH oxidation on Pt oxide surfaces, there is no reason to think that hypothesis is unreasonable.

In our calculations, we considered the influence of solvation by adding different amounts of explicit water molecules to the system. Although the water movements as well as entropic effects introduced by the solvent are not included, we believe the present representation already provides important insights on the influence of electrochemical conditions. Certainly, suitable molecular dynamics (MD) studies, that would allow for using thermodynamic integration techniques, metadynamics, *etc.*, would help understanding the role of these additional effects on reaction energies and activation barriers. Although in some cases these effects might become important, unfortunately widely useable MD techniques still have difficulties in separating electrostatic from non-electrostatic contributions to the free energy of solvation as well as reducing uncertainties in the molecular mechanics representation of the solvent.[48]

Importantly, we find that solvation changes the dominant pathway for HCOOH oxidation from strictly the formate pathway in the gas phase[21] *via* the $HCOO_B$ intermediate to a dual pathway mechanism involving both the same formate pathway as well as the direct pathway with a highly transient CO_2* intermediate in electrochemical conditions. The theoretically based observation in the gas phase is fully consistent with experimental results in ultrahigh vacuum,[3,4] while our results in electrochemical conditions provide more fundamental understanding of this process at electrochemical interfaces to better explain corresponding experiments.[7,8,11–13] However, it should be mentioned that most experiments use rather rough surfaces, which should be kept in mind when comparing theory and experiment.

In conclusion, so far we found that electrochemical oxidation of HCOOH on platinum proceeds *via* a dual-path mechanism involving a formate pathway (the $HCOO_B$* intermediate) and a direct pathway from HCOOH* (a highly transient CO_2* intermediate). When calculated on idealized terraced surfaces, these two pathways are nearly identical in energy; however, the contribution of the formate pathway may become larger with increasing potential because the coverage of formate increases and available sites for the direct pathway become reduced. In addition, the presence of other electrochemical species (CO*, OH* or acid radicals) may be also important for the above pathways. Since such co-adsorbates might additionally influence the reaction mechanism, in the following we will aim for a detailed and comprehensive study of these effects, for which the results obtained so far provide the basis.

3.3.5 Role of Co-adsorbed CO and OH

Now we will simulate the HCOOH electro-oxidation under the influence of the co-adsorbed CO* and/or OH* molecule, which is the next logical step to more realistically model this complicated electrochemical reaction. Experimental measurements of CO* on Pt(111) in the presence of HCOOH* revealed two adsorption geometries: bridge-bonded CO* (CO_B) and top-bonded (or linearly bonded) CO* (CO_L). The surface coverage of CO_B is low and nearly independent of the applied current, while the coverage of CO_L depends strongly on the applied current.[7] Thus, here we neglect CO_B and focus on the potential-dependent effects of co-adsorbed CO_L.

As shown in Figure 3.9(b) and Table 3.3, our discussions only focus on the more preferred structure of CO* (model C_1) at $^1/_9$ ML. The corresponding energy diagrams are given in Figure 3.10 for 0 V and in Figure 3.11 for 1.0 V. There are two possible structures for the higher coverage of CO* ($\theta_{CO} = {}^2/_9$ ML), depending on whether the second CO molecule adsorbs on T2 or T3 sites. The stabilities of these two structures are nearly the same (Table 3.3). However, because T3 is the more active site for HCOOH oxidation, we only consider the case where the second CO* adsorbs on T2 (model C_2), because we expect that the adsorption of CO* on the active site blocks HCOOH oxidation.

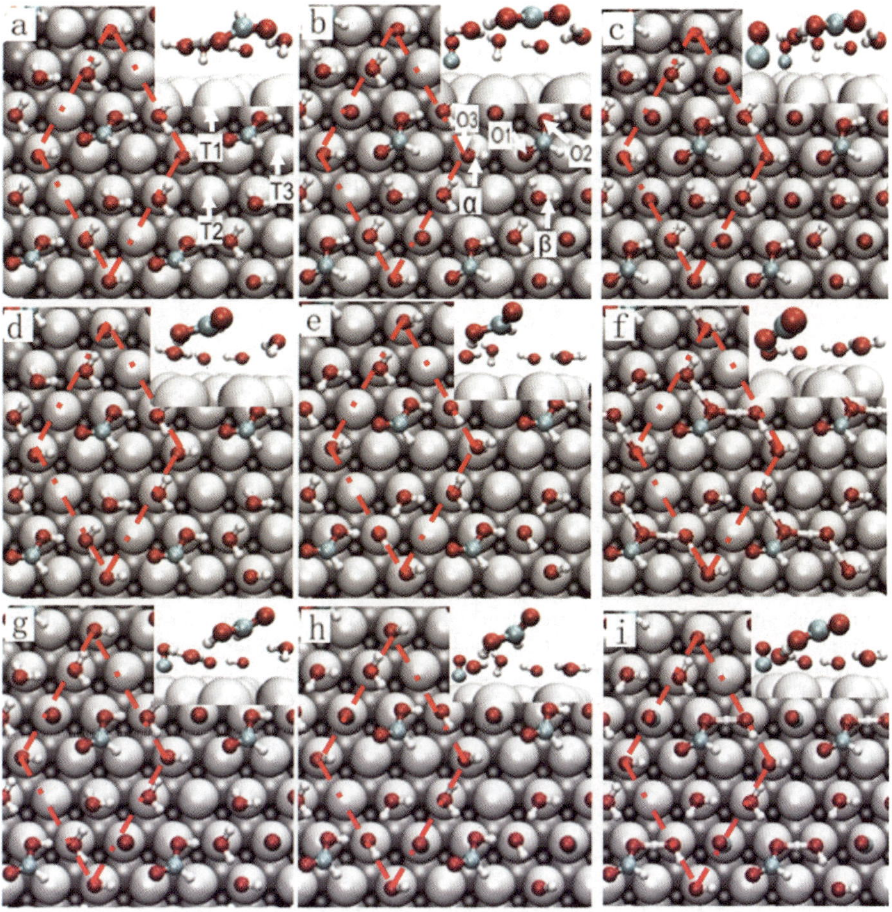

Figure 3.9 (a) Solvation model B_1 having HCOOH structures with four waters on Pt(111). (b, c) Single CO* models C_1 and C_2 originating from CO* adsorption in model B_1. (d–f) Single OH* models D_1, D_2 and D_3 originating from H_2O dissociation in model B_1. (g–i) Co-CO* and OH* models F_1, F_2 and F_3 based on single CO* and single OH* models. *Insets* show side views of the particular interface. T1, T2, T3 in (a) indicate the different top sites; O1, O2, O3 in (b) present the different O atoms; α and β in (b) denote the different water positions; these symbols are suitable for all of structures in this figure.

Table 3.4 summarizes the energies for HCOOH oxidation pathways investigated using models B−F. In model C_1, the formate pathway clearly dominates at low potentials, with a barrier of E_a = 1.19 eV at U = 0 V, which involves a CPET process as the rate-determining step. As the potential increases, the decomposition of monodentate formate ($HCOO_M$*) to CO_2 (E_a = 0.93 eV) in the formate pathway, and the decomposition of HCOOH* to CO_2* (E_a = 0.91 eV) in the direct pathway, become increasingly important.

Table 3.3 Surface binding energies (E_{bind}) calculated by subtracting the energy of the bare slab and all adsorbates infinitely separated from the slab and one another in the gas phase from the energy of the bound system. The adsorption energy per CO ($E_{CO\text{-}ad}$) is calculated by computing the adiabatic binding energy for the relevant system and dividing by the number of CO adsorbates.[a]

Species	Sites	E_{bind} (eV)	$E_{CO\text{-}ad}$ (eV)
CO	T1	3.74	1.26
	T2	3.64	1.16
	T3	3.68	1.20
2CO	T1 + T2	5.03	1.28
	T1 + T3	5.04	1.28
OH	W1	4.53	–
	W2	4.55	–
2OH	W1 + W2	6.66	–
CO + OH	T1 + W1	5.82	1.29
	T1 + W2	5.88	1.33
CO + 2OH	T1 + W1 + W2	7.41	0.75
2CO + OH	T1 + T2 + W1	6.80	1.13

[a]Surface adsorption sites for CO* are denoted T1, T2 and T3 for various top sites in model B. Which water is deprotonated to form OH* is indicated by W1 and W2 for the water containing the O3 oxygen and water β, respectively.

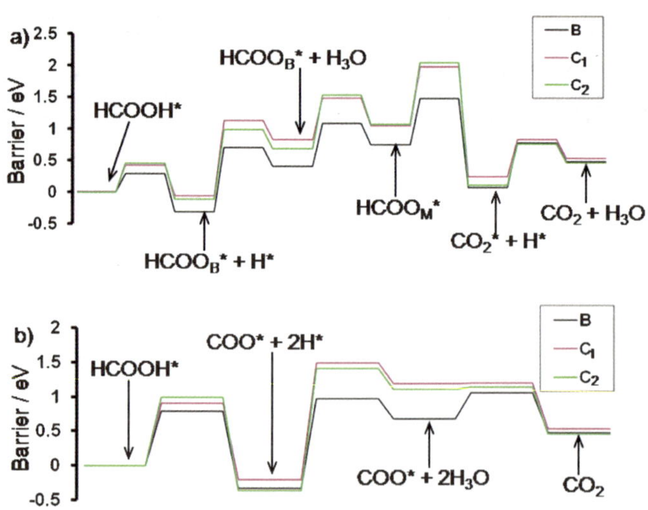

Figure 3.10 Potential energy surface for HCOOH oxidation in models B, C_1 and C_2 at 0.0 V. These surfaces demonstrate the influence of various CO* coverages, as C_1 has $^1/_9$ ML CO* and C_2 has $^2/_9$ ML CO*, while the solvation model B (no CO*) is shown for comparison.[21] Part (a) presents the formate pathway and part (b) the direct pathway.

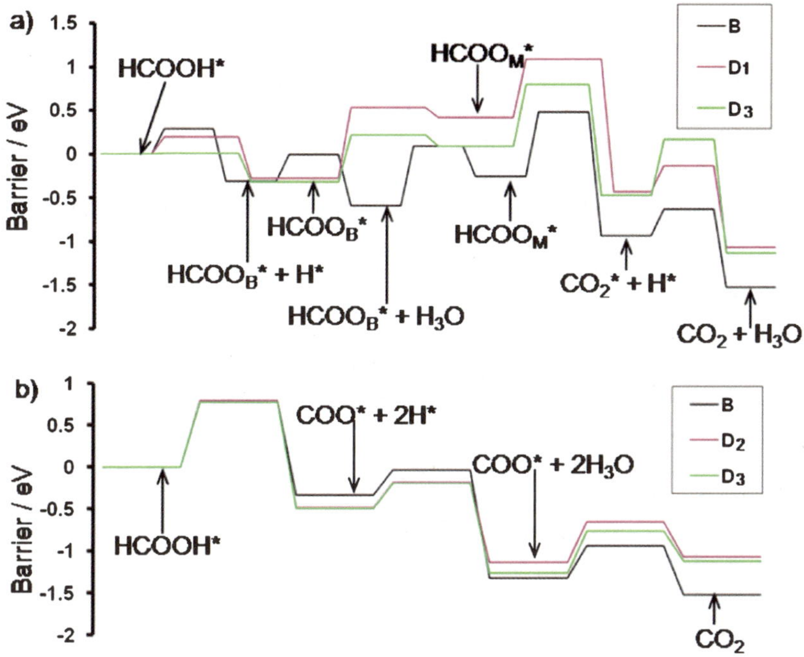

Figure 3.11 Potential energy surfaces for HCOOH oxidation in models B, D_1, D_2 and D_3 at a potential of 1.0 V. These surfaces reveal the role of various OH* coverages as D_1 and D_2 have $^1/_9$ ML OH* and D_3 has $^2/_9$ ML OH*. Solvation model B (no OH*) is shown for comparison.[21] Part (a) presents the formate pathway and part (b) the direct pathway.

Table 3.4 Energy barriers (eV) for the individual steps of HCOOH oxidation in the different models at U = 0 V.

	Formate pathway						Direct pathway		
	1^a	2^b	3^c	4^d	5^e	6^f	1^g	2^h	3^i
B_1	0.29	1.02	0.68	0.73	1.97	0.71	0.79	1.31	0.38
C_1	0.42	1.19	0.66	0.93	1.18	0.59	0.91	1.69	0.01
C_2	0.45	1.10	0.84	0.97	1.21	0.62	0.99	1.77	0.03
D_1	0.19	–	0.81	0.67	2.07	0.66	1.07	0.54	0.50
D_2	0.21	–	0.64	0.96	1.74	0.77	0.79	0.64	0.48
D_3	0.01	–	0.54	0.71	1.69	0.64	0.77	0.53	0.50
F_1	0.51	–	0.96	0.92	1.80	0.68	0.33	0.67	0.50
F_2	0.35	–	0.60	0.74	1.20	0.68	0.59	0.58	0.44
F_3	0	–	0.61	0.80	1.02	0.59	0.33	0.55	0.45
F_4	0.34	–	0.91	1.07	1.24	0.76	0.47	0.56	0.59

[a] $HCOOH^* \rightarrow HCOO_B^* + H^*$ or $HCOOH^* + OH^* \rightarrow HCOO_B^* + H_2O^*$. [b] $HCOO_B^* + H^* + H_2O \rightarrow HCOO_B^* + H_3O^+$. [c] $HCOO_B^* \rightarrow HCOO_M^*$. [d] $HCOO_M^* \rightarrow CO_2^* + H^*$. [e] $HCOO_B^* \rightarrow CO_2^* + H^*$. [f] $CO_2^* + H^* + H_2O \rightarrow CO_2 + H_3O^+$. [g] $HCOOH^* \rightarrow CO_2^* + 2H^*$ or $HCOOH^* + OH^* \rightarrow CO_2^* + H^* + H_2O^*$. [h] $CO_2^* + nH^* + nH_2O \rightarrow CO_2^* + nH_3O^+$. [i] $CO_2^* \rightarrow CO_2$.

This behavior is consistent with that of model B, indicating that the presence of up to $^1/_9$ ML of CO* does not substantially alter the HCOOH oxidation mechanism. Nevertheless, the increase of the reaction barriers in the presence of $^1/_9$ ML of CO* reveals that CO* evidently still suppresses HCOOH oxidation. Indeed, an experimental study has already revealed the suppressive influence of CO* on the formate pathway.[5]

The presence of additional CO* [up to $^2/_9$ ML as in model C_2 (Figure 3.9c)] does not significantly alter either the structure of the HCOOH/H_2O network (Table 3.4) or the HCOOH oxidation mechanism. Consequently, the barriers for the individual reaction processes in model C_2 are only slightly higher than the barriers for the corresponding processes in model C_1 (Table 3.4). Thus, low coverage of CO* ($^1/_9$–$^2/_9$ ML) suppresses HCOOH oxidation by increasing the reaction barriers, without substantially altering the basic mechanistic steps (see Figure 3.10).[5] Nevertheless, we expect that further increasing the coverage of CO* would eventually block the active sites for HCOOH oxidation, resulting in more drastic changes.

Models D_1–D_3 are obtained by removing H from either one or both of the two most acidic water molecules in model B. Models D_1 and D_2 (Figures 3.9d and 3.9e) each contain one OH molecule per unit cell and have similar energies. However, model D_1 favors the formate pathway (E_a = 0.81 eV), while model D_2 favors the direct pathway (E_a = 0.79 eV). Thus, at low coverage of OH* we expect HCOOH oxidation to proceed *via* a dual-path mechanism, involving both pathways depending on the local position of OH*. Moreover, the step with the highest barrier is no longer associated with a CEPT process, but rather with either the cleavage of a $HCOO_B$* surface bond to form $HCOO_M$* or the decomposition of HCOOH* to give CO_2*. In contrast to the mechanism observed in model B (no OH*), the steps with the highest barriers now are independent of the applied potential. This difference in behavior is primarily due to the fact that OH* eliminates a potential-dependent CPET step in the HCOOH oxidation process. Thus, OH* not only participates in the catalyzing hydrogen bond network in place of water, but also acts as a reducing agent in the oxidation of HCOOH.

When we oxidize both waters to form OH* (model D_3, $\theta_{OH} = {}^2/_9$ ML), we find that the steps with the highest barriers for each pathway are potential independent and have similar values (E_a = 0.71 and 0.77 eV for the formate and direct pathways, respectively). The small difference between these barriers suggests that HCOOH oxidation proceeds *via* a dual-path mechanism here also.

Surface OH* is only stable for electrode potentials above 0.5 V.[49] In this potential range the highest barriers in the minimum energy pathway without OH* (E_a = 0.73 and 0.79 eV for model B) are similar to the barriers with OH* (models D). Thus, OH* only has a minor effect on the reactivity of HCOOH oxidation. Nevertheless, its presence breaks the rate constant's dependence on the applied potential.

The oxidation of HCOOH under the simultaneous influence of CO* and OH* is modeled by adding CO* to models D_1–D_3 to produce models F_1–F_4.

For $U > 0.5$ V and low surface coverage of CO* and OH* ($^1/_9$ ML for each in models F_1 and F_2), the direct pathway dominates. The desorption of CO_2 (E_a = 0.50 eV) is the step with the highest barrier in the case of model F_1, while the double dehydrogenation of formic acid ($HCOOH^* \rightarrow CO_2^*$; $E_a = 0.59$ eV) has the highest barrier in the case of model F_2. As neither of these processes involves a CPET, the barriers for these processes are independent of the applied potential. They are as much as 0.29 eV lower in energy than in the absence of CO* and OH* ($E_a = 0.79$ eV in model B). Thus, together OH* and CO* promote the direct pathway, while suppressing the formate pathway by increasing its barriers by as much as 0.23 eV (0.96 eV and 0.74 eV for models F_1 and F_2, respectively, compared with 0.73 eV for model B).

Next, we consider the presence of additional OH* by deprotonating an additional water in model F_1 to construct model F_3, which contains $^2/_9$ ML OH* and $^1/_9$ ML CO*. The corresponding energy scheme is shown in Figure 3.12. This additional OH* does not change the overall mechanism of HCOOH oxidation, but reduces the direct pathway barrier to 0.45 eV, further promoting the reaction.

Finally, we model the effects of additional CO* by adding CO* to model F_1 to form model F_4, which contains $^2/_9$ ML CO* and $^1/_9$ ML OH*. While the additional CO* increases the barrier for the direct pathway by 0.09 eV, it is still preferred over the formate pathway by 0.48 eV. Thus, increasing the coverage of CO* beyond $^1/_9$ ML, in the presence of co-adsorbed OH*, adversely affects HCOOH oxidation.

The presence of neither CO* nor OH* alone is able to promote the activity of Pt(111) for HCOOH oxidation; however, surprisingly, the combination of CO* and OH* is able to do so. We attribute this synergy to a combination of changes in the hydrogen bond network and the induction of a surface charge.

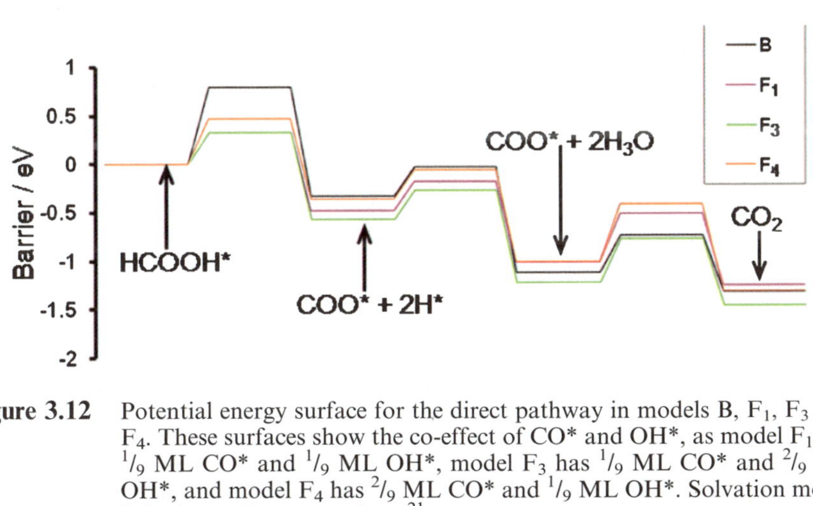

Figure 3.12 Potential energy surface for the direct pathway in models B, F_1, F_3 and F_4. These surfaces show the co-effect of CO* and OH*, as model F_1 has $^1/_9$ ML CO* and $^1/_9$ ML OH*, model F_3 has $^1/_9$ ML CO* and $^2/_9$ ML OH*, and model F_4 has $^2/_9$ ML CO* and $^1/_9$ ML OH*. Solvation model B is shown for comparison.[21]

Table 3.5 Geometric data (Å) for HCOOH structures in different models.

Model	d_{O1-Pt}	d_{O2-Pt}	d_{O3-Pt}	d_{O1-H1}[a]	d_{O1-H2}[b]
B_1	3.55	2.93	3.19	1.97	3.52
C_1	3.79	3.57	3.26	2.42	1.81
C_2	3.80	3.59	3.33	2.53	1.84
D_1	3.83	3.19	2.15	2.43	3.89
F_1	4.35	3.37	2.17	3.15	3.35

[a]The distance between O1 and the nearest hydrogen in α water. [b]The distance between O1 and the nearest hydrogen in the β water.

The addition of CO* to model B to form model C_1 results in a repulsive interaction between CO* and HCOOH (d_{O-O1} = 2.98 Å), which causes HCOOH to rotate and redistribute its hydrogen bonding with the surrounding water molecules. Most importantly, the O1 is rotated toward the β water to form a hydrogen bond (d_{O1-H2} = 3.52 Å for model B and d_{O1-H2} = 1.81 Å for model C_1 in Table 3.5, for which we estimate a hydrogen bond increase of 0.16 eV in Table 3.6). On the other hand, the hydrogen bond distance between the carbonyl O of HCOOH (O1) and the α water increases (compare d_{O1-H1} = 1.97 Å for model B and d_{O1-H1} = 2.42 Å for model C_1 in Table 3.5, for which we estimate a hydrogen bond energy decrease of 0.03 eV in Table 3.6). Thus, the net hydrogen bond energy increases by ~0.13 eV. Because both of these hydrogen bonds must be broken in the rate-limiting step of the direct pathway, we expect that an increase in their net strength will result in increases to both the reaction barrier and reaction energy for this step.

The barrier for the initial step in the formate pathway increases (0.29 eV to 0.41 eV) for the same reason. However, a later step, the deprotonation of $HCOO_M$, is rate limiting, namely cleaving the C−H bond. In the presence of

Table 3.6 Interaction energies (in eV) between HCOOH* (CO_2*) and neighboring water molecules.[a]

Model	E_1	E_2	E_3
B	0.18	0.02	0.09
C_1	0.15	0.18	0.08
D_1	0.25	0.01	0.12
D_2	0.33	0	0.23
F_1	0.07	0	0.10

[a]The interaction energies are calculated by fixing the atomic positions in the bound system and then removing the Pt(111) slab along with any adsorbates not directly involved in the interaction of interest. The energy of the remaining system (with fixed atomic positions) is then calculated. Then, the energy of a non-interacting system (interacting molecules are infinitely separated, but internal atomic positions remain fixed) is subtracted from this interacting system to compute the interaction energy. The interaction energy between the carbonyl O of HCOOH (O1) and the α water is E_1, the interaction energy between O1 and the β water is E_2, and E_3 is associated with the interaction of O2 in CO_2* and the nearest neighboring water.

CO*, $HCOO_M$ is repulsed away from the surface, increasing the distance between this C−H bond and the bridge site, at which H binds in the final product, by 0.05 Å (d_{H-Pt} = 2.46 Å for model B and d_{H-Pt} = 2.51 Å for model C_1). We associate this increased distance for H to traverse during the reaction with the 0.20 eV increase in the barrier (0.73 eV to 0.93 eV).[21,50,51]

Models D_1 and D_2 are formed by deprotonating either the W1 or the W2 water. The O atom in the newly formed OH is now able to form a proper covalent bond to the surface, as reflected in the marked reductions in the Pt−O bond distances. In the case of deprotonating the W2 water, the physisorption distance of d_{Pt-O} = 3.19 Å in model B is reduced to covalent bond distances of and d_{Pt-O1} = 2.25 Å in model D_2. This covalent bond is polarized and induces a positive charge on the Pt(111) surface, as can be seen in Figure 3.13(c). Interestingly, this surface charge appears to be spread across the whole surface, rather than being localized directly under the adsorbate, as is the case for CO* (compare Figures 3.13b and 3.13c). Theoretical work by Hartnig *et al.* suggests that inducing a positive charge on the Pt surface facilitates breaking the C−H bond.[18] This activation is reflected in a stronger interaction between H and the charged Pt surface, which is correlated with an increase in the C−H and Pt−H bond distances in the transition state (TS) in model D_2. We observe d_{C-H} = 1.55 and d_{Pt-H} = 1.82 Å in the TS in model D_2 compared to d_{C-H} = 1.46 and d_{Pt-H} = 1.75 Å in the TS in model B. We would expect the effects of the induced surface charge to correspond to a decrease in the reaction barrier; however, here we also find an off-setting counter effect.

The higher polarity of OH compared to H_2O results in a strengthening of the hydrogen bond network. Based on the changes in hydrogen bond distances, we estimate a hydrogen bond increase of 0.27 eV compared to model B. As in the

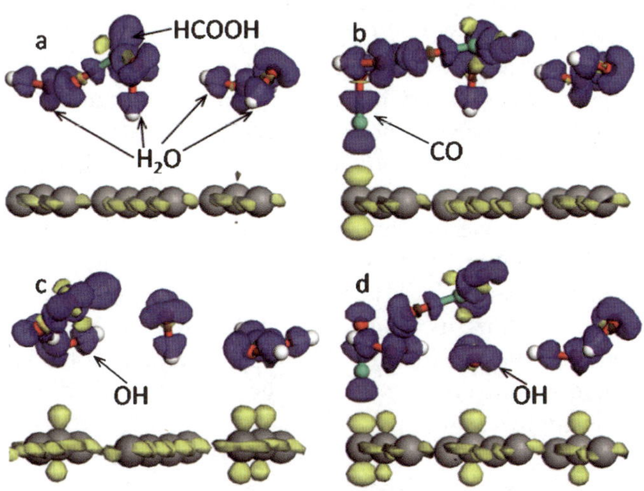

Figure 3.13 Charge density differences: (a) model B_1; (b) model C_1; (c) model D_1; and (d) model F_1. The *blue region* shows electron accumulation, while *yellow regions* show electron loss.

case of adsorbed CO, where these hydrogen bonds were also strengthened, we expect an increase in the barrier to result in this case. As we find little change in the barriers due to the conversion of one or both water molecules to OH, the effects of increased hydrogen bonding and the increased reactivity with the charged surface on the direct pathway appear to have cancelled each other out.

For the formate pathway of model D_1, the intermediate preceding the rate-limiting step is relatively unchanged compared to model B after the process of $HCOOH + OH/3H_2O \rightarrow HCOO_B + 4H_2O$; thus the barrier remains substantially unchanged.

In model F_1, which has $^1/_9$ ML CO* and $^1/_9$ ML OH*, we are only concerned with the direct pathway, because it is strongly preferred. The electron density differences illustrated in Figure 3.13(d) show clearly that OH* induces a positive charge on the Pt(111) surface, which we expect to decrease the reaction barrier as discussed above. Meanwhile, rather than enhancing (as was the case in model C), CO* disrupts the interaction between HCOOH and the neighboring molecules in the hydrogen bond network. In particular, the hydrogen bond involving the carbonyl O is broken (d_{O1-H1} = 1.97 Å for model B and d_{O1-H1} = 3.15 Å for model F_1, as in Table 3.5) at an estimated cost of 0.13 eV. As a result, the co-effects of CO* and OH* drastically promote HCOOH oxidation by reducing the barrier by 0.29 eV.

Comparing our results with experimental studies requires choosing the appropriate surface model to match the experimental surface coverage conditions. Thus, we must determine the relevant surface coverage of CO (θ_{CO}) and OH (θ_{OH}) on Pt(111) for electrode potentials of interest. Based on IR spectroscopy studies, Osawa *et al.* estimated that θ_{CO_L} is in the range of 0.1−0.3 ML below U = 0.85 V, after which it decreases until it nearly reaches zero at U = 1.0 V.[8] Thus, we assume $\theta_{CO} = {}^2/_9$ ML for the approximately constant coverage region[7] for $U < 0.85$ V, a $\theta_{CO} = 0$ for $U > 1.0$ V,[5,52] and $\theta_{CO} = {}^1/_9$ ML for the region between. Experiments first observe OH* at U = 0.5 V; however, its coverage is still very low at U = 0.6 V.[49] By the time U = 0.85 V, we assume the OH coverage has increased to $\theta_{OH} = {}^2/_9$ ML. Thus, for $0 < U <$ 0.6 V, we utilize model C_2 ($\theta_{CO} = {}^2/_9$ ML, $\theta_{OH} = 0$). We anticipate that a more complex model will be necessary to account for the water dissociation that takes place in the negative differential resistance (NDR) region ($0.6 < U <$ 0.85 V), and thus do not attempt to model this region in this work. For $0.85 <$ $U < 1.0$ V, model F_3 ($\theta_{CO} = {}^1/_9$ ML, $\theta_{OH} = {}^2/_9$ ML) is appropriate. Finally, for 1.0 V $< U$ 1.2 V, we utilize model D_3 ($\theta_{CO} = 0$, $\theta_{OH} = {}^2/_9$ ML), which ignores the possibility of surface oxidation. Thus, we obtain a potential-dependent picture of the overall rate constant (R) for the HCOOH oxidation process as a function of U in Figure 3.14.

Starting from U = 0.0 V, R increases as the potential increases to 0.4 V, at which point it reaches a plateau that lasts at least to the beginning of the NDR region (U = 0.6 V). On the other side of the NDR region (U > 0.85 V), it achieves its maximum value before entering the final region we consider beginning at U = 1.0 V, where we find a lower rate that lasts until at least U =

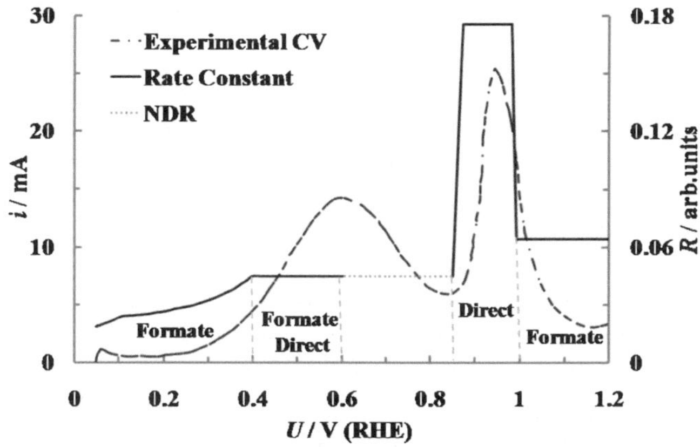

Figure 3.14 Comparison of potential-dependent rate constants (R) and experimental CVs.[7]

1.2 V. These trends are compatible with the experimental CV curve obtained over $0.0 < U < 1.2$ V. Moreover, our calculations further reveal the mechanistic processes underlying this rate behavior.

The initial oxidation, beginning at $U = 0.0$ V, is dominated by the formate mechanism, which has the rate-limiting step associated with the potential-dependent CEPT process. Because the $HCOO_B$ immediately precedes the rate-limiting CEPT process, we expect $HCOO_B$ to be observable at these low potentials. Experiments fulfill these expectations, finding that formate appears around $U = 0.2$ V in the spectra.[6,12] When the potential reaches $U = 0.4$ V, the dual-path mechanism, which includes both the formate pathway and the direct pathway, begins to dominate the HCOOH oxidation process. This domination continues at least to the beginning of the NDR region ($U = 0.6$ V). Because the experimentally observed current decay in the NDR is most likely due to the adsorption of a poisoning species, which would not be accounted for in our present models, we do not consider this region here.[7,9,52]

Moving beyond the NDR region, for $0.85 \text{ V} < U < 1.0$ V, HCOOH is oxidized almost exclusively through the direct pathway ($E_a = 0.45$ eV). We find that the co-effect of CO* and OH* here effectively promotes HCOOH oxidation, challenging the traditional conception of CO* as always acting as a poison in HCOOH oxidation. Unfortunately, the oxidative removal of this promoting CO* at higher potentials limits this efficient mechanism to a relatively narrow potential range. During this process, the formation of $HCOO_B$* (the first step in the formate pathway) is nearly barrierless ($E_a \approx 0$ eV). However, because the subsequent decomposition of $HCOO_M$* to form CO_2 has a high barrier ($E_a = 0.80$ eV), we expect HCOO* to rapidly accumulate on the surface. Indeed, experiments observe this accumulation beginning at $U \approx 0.9$ V and peaking around $U \approx 1.0$ V. We expect this excess HCOO* to suppress HCOOH oxidation by blocking surface sites, which would otherwise be available for the adsorption and oxidation of additional HCOOH through the direct pathway.

Beyond 1.0 V, HCOO now decomposes to CO_2 under the influence of OH* with a barrier of E_a = 0.71 eV, rapidly reducing the formate coverage as observed in experiments.[5,7] The experimentally observed decay of current for potentials above U = 1.0 V is usually ascribed to oxidation of the Pt surface.[5] Although our model does not include this surface oxidation, our results are consistent with this hypothesis, insofar as we are unable to reproduce this experimental decrease using a model that does not include Pt oxide.

The deposition of a monolayer of CO* on the Pt electrode prior to the oxidation of HCOOH results in the total suppression of the first current peak in the CV, and a positive shift of the second current peak ($\Delta U \approx$ 0.1V), coinciding with a shift in the growth of the HCOO coverage.[5] The total suppression of the first current peak is easily explained by the inaccessibility of surface active sites for HCOOH oxidation due to the high coverage of CO*. Furthermore, the oxidative removal of CO* requires OH*, which is also required for HCOOH oxidation. Thus the competition with excess CO* for surface sites and OH* molecules decreases the rate of HCOOH oxidation, as both the build-up of OH* and the decomposition of CO*, to reach the optimal coverages required to synergistically promote HCOOH oxidation, as our model predicts, are delayed, resulting in a positive shift in the second peak. The delayed growth of the HCOO coverage, observed in experiments, is also consistent with this interpretation.

A typical chronamperogram reveals that when the electrode potential is stepped up from 0.05 V to 0.9 V, CO* is completely oxidized within two seconds.[6] Within these 2 s, the current (I) quickly decays following an initial spike and the HCOO coverage drastically increases. The behavior of I corresponds to the rapid oxidation and disappearance of CO*. Thus, model F_3 ($^1/_9$ ML of CO* and $^2/_9$ ML of OH*) is initially relevant, followed by model D_3 (0 ML of CO* and $^2/_9$ ML of OH*) once CO* has been oxidized away. The low barrier ($E_a \approx$ 0 eV) to form HCOO*, coupled with the high barrier (E_a = 0.80 eV) to convert it to product, results in the initial build-up of HCOO*. We anticipate that the formation of surface oxide (experimentally identified as Pt_4OH near 0.9 V, but not included in our current model) hinders the rate of formic acid oxidation here.[49]

Stepping U to 0.6 V (following the 2 s at 0.9 V) results in an immediate increase in I followed by a gradual decrease as the HCOO coverage drops by a fraction of its initial value and quickly converges to a new constant. Under the reduced potential we anticipate that the Pt oxide is quickly reduced, returning the metal surface to its initial state. This frees up additional surface sites for HCOOH oxidation, resulting in an increased oxidation rate, which slightly reduces the concentration of HCOO. Because the surface is initially covered with HCOO, this must first decompose under surface conditions best represented by model D_3 (E_a = 0.71 eV). As HCOO is oxidized away, surface sites become available for the adsorption of HCOOH. The condition of the surface is probably now best represented by a combination of models B, D_1 and D_1 (E_a = 0.73, 0.79 and 0.81 eV), which all have a higher reaction barrier than model D_3 (E_a = 0.71 eV). Thus, the removal of OH* from the surface results in a slower HCOOH oxidation rate, corresponding to the gradual decrease in *I* that is observed.

In conclusion, we find that CO* (in the absence of OH*) adversely affects HCOOH oxidation, not only by blocking active sites, but also by retarding the rates of important reaction steps. On the other hand, OH* (in the absence of CO*) does not significantly alter the reactivity of HCOOH oxidation on a clean Pt surface. However, at high potentials it oxidizes the Pt surface, and thus poisons the reaction by blocking active sites. Nevertheless, we find that CO* and OH* have a synergistic co-effect, such that relatively low coverage of both effectively promotes HCOOH oxidation. These solitary and synergistic behaviors of CO* and OH* successfully explain the CV of the electro-oxidation of HCOOH in the potential range of 0.0–1.2 V.

3.4 Summary

In this chapter we have discussed how quantum-mechanical modeling can provide useful insights into electrocatalytic reactions as complex as the oxidation of formic acid on Pt(111). We discussed that HCOO is the active intermediate in HCOOH oxidation in the gas phase, which is consistent with the experimental studies under UHV conditions. With increasing water coverage, the dominant pathway of HCOOH oxidation does not change under the influence of the low water coverage (θ = $^1/_9$, $^2/_9$, $^1/_4$ ML), which is also agreement with the experiments. In contrast, both the HCOO pathway and the direct pathway are active under the influence of the bilayer water (θ = $^2/_3$ ML) at high electrode potentials, which succinctly rationalizes conflicting experimental observations in electrochemistry. Further, including the effects of co-adsorbed CO and/or OH, we found that changing the coverage of CO/OH drastically influences the mechanism of HCOOH oxidation under different electrode potentials. A kinetic analysis derived from these mechanistic studies elucidated the atomistic surface phenomena underlying the experimental CV measurements in the potential range between 0.0 and 1.2 V.

Acknowledgements

The authors gratefully acknowledge support by the Deutsche Forschungsgemeinschaft (DFG) and the Alexander von Humboldt foundation (AvH), as well as the bw-grid for computing resources [bwGRiD (http://www.bw-grid.de), a member of the German D-Grid initiative, is funded by the Ministry for Education and Research (Bundesministerium für Bildung und Forschung) and the Ministry for Science, Research and Arts Baden-Württemberg (Ministerium fuer Wissenschaft, Forschung und Kunst Baden-Württemberg)]. Further support by the European Union through the Marie-

Curie Initial Training Network ELCAT, proposal no. 214936-2, 2008–2012, is acknowledged.

References

1. (a) M. W. Breiter, *J. Electroanal. Chem.*, 1967, **14**, 407; (b) A. Capon and R. Parsons, *J. Electroanal. Chem.*, 1973, **44**, 1; (c) A. Capon and R. Parsons, *J. Electroanal. Chem.*, 1973, **45**, 205; (d) K. Kunimatsu and H. Kita, *J. Electroanal. Chem.*, 1987, **218**, 155; (e) D. S. Corrigan and M. J. Weaver, *J. Electroanal. Chem.*, 1988, **241**, 143; (f) S. Wilhelm, W. Vielstich, H. Buschmann and T. Iwasita, *J. Electroanal. Chem.*, 1987, **229**, 377; (g) R. Parsons and T. VanderNoot, *J. Electroanal. Chem.*, 1988, **257**, 9; (h) C. Lamy and J. M. Leger, *J. Chim. Phys. Phys.-Chim. Biol.*, 1991, **88**, 1649; (i) N. Kizhakevariam and M. J. Weaver, *Surf. Sci.*, 1994, **310**, 183.
2. (a) B. Beden, J. M. Leger and C. Lamy, in *Modern Aspects of Electrochemistry*, ed. J. O'M. Bockris, B. E. Conway and R. E. White, Plenum, New York, 1992, vol. 22, p. 97; (b) T. D. Jarvi and E. M. Stuve, in *Electrocatalysis*, ed. J. Lipkowski and P. Ross, Wiley-VCH, New York, 1998, pp. 75–153; (c) S.-G. Sun, in *Electrocatalysis*, ed. J. Lipkowski and P. N. Ross, Wiley-VCH, New York, 1998, p. 243.
3. M. R. Columbia, A. M. Crabtree and P. A. Thiel, *J. Am. Chem. Soc.*, 1992, **114**, 1231.
4. M. R. Columbia and P. A. Thiel, *Surf. Sci.*, 1990, **235**, 53.
5. G. Samjesk, A. Miki, S. Ye and M. Osawa, *J. Phys. Chem. B*, 2006, **110**, 16559.
6. M. Osawa, K. Komatsu, G. Samjeské, T. Uchida, T. Ikeshoji, A. Cuesta and C. Gutiérrez, *Angew. Chem. Int. Ed.*, 2011, **50**, 1159.
7. G. Samjeské and M. Osawa, *Angew. Chem. Int. Ed.*, 2005, **44**, 5694.
8. G. Samjeské, A. Miki, S. Ye, A. Yamakata, Y. Mukouyama, H. Okamoto and M. Osawa, *J. Phys. Chem. B*, 2005, **109**, 23509.
9. Y. Mukouyama, M. Kikuchi, G. Samjeské, M. Osawa and H. Okamoto, *J. Phys. Chem. B*, 2006, **110**, 11912.
10. V. Grozovski, F. J. Vidal-Iglesias, E. Herrero and J. M. Feliu, *ChemPhysChem*, 2011, **12**, 1641.
11. Y. X. Chen, M. Heinen, Z. Jusys and R. J. Behm, *Angew. Chem. Int. Ed.*, 2006, **45**, 981.
12. Y. X. Chen, M. Heinen, Z. Jusys and R. J. Behm, *Langmuir*, 2006, **22**, 10399.
13. Y. X. Chen, M. Heinen, Z. Jusys and R. J. Behm, *ChemPhysChem*, 2007, **8**, 380.
14. P. Strasser, M. Lübke, F. Raspel, M. Eiswirth and G. Ertl, *J. Chem. Phys.*, 1997, **107**, 979.
15. G. Lu, A. Crown and A. Wieckowski, *J. Phys. Chem. B*, 1999, **103**, 9700.
16. H. Okamoto, W. Kon and Y. Mukouyama, *J. Phys. Chem. B*, 2005, **109**, 15659.

17. I. Bak and G. Pálinkás, *Surf. Sci.*, 2006, **600**, 3809.
18. C. Hartnig, J. Grimminger and E. Spohr, *J. Electroanal. Chem.*, 2007, **607**, 133.
19. M. Neurock, M. Janik and A. Wieckowski, *Faraday Discuss.*, 2009, **140**, 363.
20. H. F. Wang and Zh. P. Liu, *J. Phys. Chem. C*, 2009, **113**, 17502.
21. W. Gao, J. A. Keith, J. Anton and T. Jacob, *Dalton Trans.*, 2010, **39**, 8450.
22. W. Gao, J. A. Keith, J. Anton and T. Jacob, *J. Am. Chem. Soc.*, 2010, **132**, 18377.
23. W. Gao, J. E. Mueller, Q. Jiang and T. Jacob, *Angew. Chem. Int. Ed.*, in press.
24. M. D. Segall, P. J. D. Lindan, M. J. Probert, C. J. Pickard, P. Hasnip, S. J. Clark and M. Payne, *J. Phys.: Condens. Matter*, 2002, **14**, 2717.
25. D. Vanderbilt, *Phys. Rev. B*, 1990, **41**, 7892.
26. J. P. Perdew, K. Burke and M. Ernzerhof, *Phys. Rev. Lett.*, 1996, **77**, 3865.
27. N. Govind, M. Petersen, G. Fitzgerald, D. King-Smith and J. Andzelm, *Comput. Mater. Sci.*, 2003, **28**, 250.
28. J. A. Keith, G. Jerkiewicz and T. Jacob, *ChemPhysChem*, 2010, **11**, 2779.
29. W. Gao, M. Zhao and Q. Jiang, *ChemPhysChem*, 2008, **9**, 2092.
30. X. Q. Gong, P. Hu and R. Raval, *J. Chem. Phys.*, 2003, **119**, 6324.
31. L. C. Grabow, A. A. Gokhale, S. T. Evans, J. A. Dumesić and M. Mavrikakis, *J. Phys. Chem. C*, 2008, **112**, 4608.
32. J. Chen, J. Kubota, A. Wada, J. N. Kondo and K. Domen, *J. Phys. Chem. C*, 2008, **112**, 12477.
33. S. E. Mason, I. Grinberg and A. M. Rappe, *Phys. Rev. B*, 2004, **69**, 161401.
34. W. Liu, Y. F. Zhu, J. S. Lian and Q. Jiang, *J. Phys. Chem. C*, 2007, **111**, 1005.
35. M. Gajdoš, A. Eichler and J. Hafner, *J. Phys.: Condens. Matter*, 2004, **16**, 1141.
36. L. Köhler and G. Kresse, *Phys. Rev. B*, 2004, **70**, 165405.
37. G. Kresse, A. Gil and P. Sautet, *Phys. Rev. B*, 2003, **68**, 073401.
38. B. Shan, Y. J. Zhao, J. Hyun, N. Kapur, J. B. Nicholas and K. Cho, *J. Phys. Chem. C*, 2009, **113**, 6088.
39. Y. Y. Yeo, L. Vattuone and D. A. King, *J. Chem. Phys.*, 1997, **106**, 392.
40. T. Jacob and W. A. Goddard *J. Am. Chem. Soc.*, 2004, **126**, 9360.
41. J. S. Spendelow, J. D. Goodpaster, P. J. A. Kenis and A. Wieckowski, *J. Phys. Chem. B*, 2006, **110**, 9545.
42. H. Kim, Y. Kim, M. Yoon, S. Lim, S. M. Park, G. Seo and K. Kim, *J. Am. Chem. Soc.*, 2010, **132**, 12200.
43. M. D. Maciá, E. Herrero and J. M. Feliu, *Electrochim. Acta*, 2002, **47**, 3653.
44. A. Alavi, P. Hu, T. Deutsch, P. L. Silvestrelli and J. Hutter, *Phys. Rev. Lett.*, 1998, **80**, 3650.
45. S. Desai and M. Neurock, *Electrochim. Acta*, 2003, **48**, 3759.

46. J. O'M. Bockris and A. K. Reddy, *Modern Electrochemistry*, Kluwer/Plenum, New York, 2001, vol. 2B, pp 1539–1550.
47. A. V. Marenich, C. J. Cramer and D. G. Truhlar, *J. Chem. Theory Comput.*, 2008, **4**, 877.
48. H. Eyring, *J. Chem. Phys.*, 1935, **3**, 107.
49. H. Angerstein-Kozlowska, B. E. Conway and W. B. A. Sharp, *J. Electroanal. Chem.*, 1973, **43**, 9.
50. H. Lesnard, M.-L. Bocquet and N. Lorente, *J. Am. Chem. Soc.*, 2007, **129**, 4298.
51. W. Gao, M. Zhao and Q. Jiang, *J. Chem. Phys.*, 2008, **129**, 164705.
52. T. Iwasita and S. H. Xia, *J. Electroanal. Chem.*, 1996, **411**, 95.

CHAPTER 4

Gold Leaf Based Electrocatalysts

RONGYUE WANG AND YI DING*

School of Chemistry and Chemical Engineering, Shandong University, Jinan 250100, China
*E-mail: yding@sdu.edu.cn

4.1 Introduction

Different from hydrogen-fuelled polymer electrolyte membrane fuel cells (PEMFCs), alcohol-fuelled direct oxidation fuel cells (DOFCs) could be used as portable power sources without worrying about the intractable hydrogen storage issues.[1,2] However, unlike the simple and fast oxidation of hydrogen on a platinum catalyst surface, the oxidation of small organic molecules such as methanol or formic acid is sluggish, and poisoning intermediates form almost spontaneously on the catalyst surface. These two effects result in the massive usage of precious metal catalysts and low fuel cell performance, both of which are unacceptable for ultimate commercialization.

In the last several decades, extensive research has been devoted to the development of highly active and efficient catalysts for small organic molecule oxidation. To improve the intrinsic catalytic activity, a variety of binary, ternary, and even quaternary alloys have been investigated and the fundamental mechanisms of catalytic activity enhancement have also been clarified to some extent, thanks to the rapid advance of surface science and *in situ* electrochemical spectroscopy.[3–5] For methanol electro-oxidation, the most active binary catalyst is commonly accepted to be PtRu nanoalloys in a close to 1:1 atomic composition, based on the bifunctional mechanism in which Pt atoms contribute to the dehydrogenation of methanol and the

RSC Energy and Environment Series No. 6
Catalysts for Alcohol-Fuelled Direct Oxidation Fuel Cells
Edited by Zhen-Xing Liang and Tim S. Zhao

Published by the Royal Society of Chemistry, www.rsc.org

neighboring Ru atoms facilitate the adsorption of oxygen-containing species and ease the poisoning effect.[6] For formic acid electro-oxidation, methods that allow isolation of small Pt ensembles, such as molecule adsorption, atom deposition, and alloying with other metals, have all been demonstrated to benefit the desired direct oxidation path.[7–9]

Boosted by the advance of nanotechnology, research in electrocatalysis during the last decade has been shifted to the controlled synthesis of various nanostructured catalysts with tailored sizes or dimensions, structural uniformity and dispersity, crystallographic morphology, and surface configuration, aiming at increasing the catalytic activity.[10,11] As comprehensively discussed in the previous chapters, core–shell nanostructured catalysts have been developed to improve the precious metal utilization,[12,13] while morphologically controlled nanocatalysts have received particular interest in the light of the possibility that different crystal surfaces have distinct activities.[14] In order to function in real fuel cells, these nanocrystal catalysts have to undergo further processing by adsorbing onto high surface area conducting substrates such as carbon black, to facilitate catalyst dispersion and fuel diffusion. As the catalyst nanoparticles are only physically adsorbed on the carbon support, they tend to aggregate to form larger nanoparticles or even lose contact with the support under the operational conditions.[15] Moreover, the large contact resistance between carbon particles aroused from the incorporation of poorly conductive polymer electrolytes such as Nafion, when preparing the membrane electrode assembly (MEA), leads to an increase of internal resistance in the catalytic layer and a decrease of overall cell performance. Although several other supports, such as ordered nanoporous carbon,[16] carbon nanotubes,[17] fullerene films,[18] and more recently graphene, have been investigated with an eye on improving the catalyst–support interaction, most research still focuses on carbon materials.[19,20]

In this chapter, we introduce the recent research on a new kind of metallic nanoporous catalyst made by dealloying, in particular electrocatalysts based on nanoporous gold (NPG) leaf. With a large surface area, a bicontinuously porous structure, high conductivity, and great corrosion resistance, ultrathin NPG leaf could be an exceptional catalyst or catalyst support for fuel cell applications. We will begin with a historical retrospect, and then provide a detailed description for the fabrication, structural properties, and evolution of NPG leaf. After discussing the catalytic properties of bare NPG leaf, we will focus on platinum-plated NPG leaf (Pt-NPG). Topics include the plating method, structure characterization, stability, electrocatalytic properties, and performance in hydrogen fuel cells. In the fourth section, we will give a particular example regarding the design and fabrication of highly efficient electrocatalysts for direct formic acid fuel cells. The chapter will end with a conclusion and our perspectives that may warrant future investigation.

4.2 Nanoporous Gold Leaf

4.2.1 History and Formation Mechanism of NPG

Dealloying, also known as selective dissolution, has long been used by ancient artisans to create artifacts with a protective surface of nearly pure gold from Cu/Au-based alloys, although the underlying science remained unknown for centuries.[21,22] During the last century, dealloying was primarily investigated in order to understand the corrosion behavior of technologically important alloys such as stainless steel and brass.[23–25] Corrosion has long been notorious for its harmfulness to alloys and engineering materials, with probably one exception when Raney showed that selectively leaching away aluminum and/or silicon from their alloys could generate very useful catalysts for the chemical industry, now known as Raney metals.[26] From the pioneering work of Pickering, Swann, and Forty, between the early 1960s and the late 1970s, we know that dealloying of gold-based alloys can result in nanoporous structures with ligament and pore sizes of the order of 10 nm (see Figures 4.1a and 4.1b).[21,27,28] Since the 1980s, Sieradzki and Newman initiated the systematic investigation of the corrosion process by using Au alloys as model systems.[22,29,30] With electrochemical scanning tunneling microscopy and other advanced X-ray techniques such as *in situ* X-ray diffraction with picometre-scale resolution, they revealed significant structural information about the initial dealloying stage and two key parameters that determine the corrosion process. One key parameter is called the parting limit, which is associated with the composition of the pristine alloy. The corrosion process can only proceed when the noble metal content is below the parting limit, which is determined by the alloy system. Another key parameter is the critical potential below which the surface of the alloy could be passivated by the formation of a noble metal enriched layer.

The progress in experiments was accompanied by some attempt to reveal the formation mechanism of nanoporous metals by dealloying. It was thought previously that the porous structure pre-existed in the alloy and dealloying simply excavated this structure from the precursor solid, but this mechanism was proved to be incorrect by diffraction experiments.[31,32] Later on, percolating clusters within the solid solution of the alloy were considered to

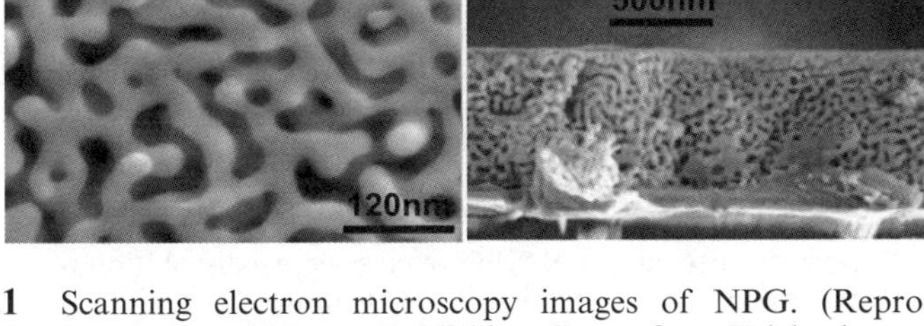

Figure 4.1 Scanning electron microscopy images of NPG. (Reproduced with permission of Nature Publishing Group from Erlebacher *et al.*[35])

influence the formation of porous structure.[33,34] However, this attempt could not explain all the experimental observations. In 2001, Erlebacher *et al.*[35,36] proposed a theoretical model which realized high consistency between simulation and experiments during alloy selective dissolution. As proposed in this model, the formation of the porous structure was controlled by an intrinsic dynamical pattern formation process which involved the competition of corrosion-induced surface roughening and diffusion-induced surface flattening. As shown in Figure 4.2(a), less noble metal atoms in the first layer of the alloy could be firstly dissolved upon contact with a corrosion regent, leaving the noble metal atoms adsorbed on the terrace. As the ad-atoms are supersaturated, they tend to aggregate into small clusters. At the same time, the second layer is attacked and more ad-atoms of the noble metal will be released and transported to the base of the just-formed clusters. As the corrosion process proceeds, the uncovered surface area becomes smaller and the clusters grow into small hills with an alloy core and a noble metal enriched surface (see Figure 4.2b). With the increase of the perimeters at the base of the hills (see Figure 4.2c), the noble metal atoms can no longer passivate the surface and the hills become undercut (see Figure 4.2d). Then the first layer of the porous structure forms and more pristine alloy surfaces are released for further dealloying. The newly formed nanoscale ligaments are actually

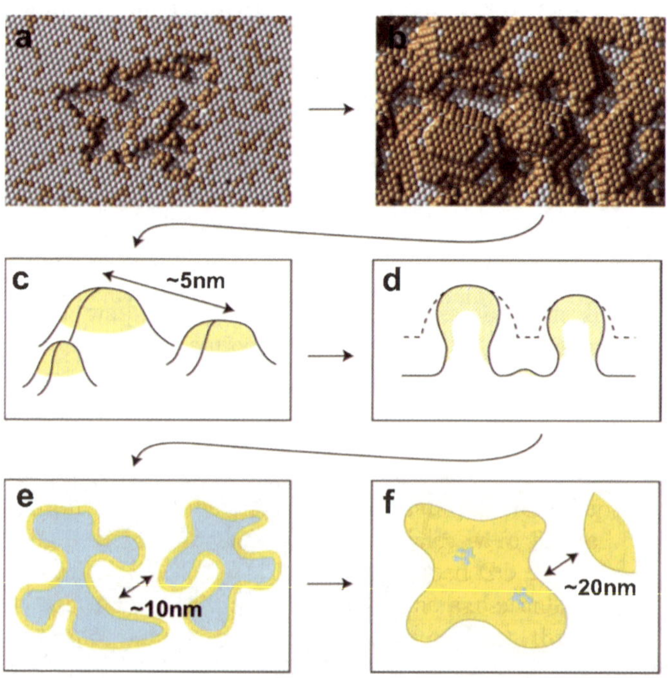

Figure 4.2 Formation mechanism of NPG from selective dissolution of Au–Ag alloy. (Reproduced with permission of Cambridge University Press from Erlebacher *et al.*[36])

composed of an alloy core with a noble metal enriched surface (see Figure 4.2e). The final formation of nanoporous metal with nearly pure noble metal is caused by the coarsening effect, which is driven by the surface energy induced atomic diffusion (see Figure 4.2f).

4.2.2 Structural Properties of NPG Leaf

In principle, nanoporous metals could be made into any kind of form, depending on the shape and geometry of the pristine alloy. Indeed, the applications of NPG in actuation,[37] sensing,[38] and catalysis[39,40] benefit greatly from the monolithic character of these unique nanostructured materials. However, for those pursued in clean energy technologies, such as fuel cells, a high surface area and ultra-thin membranes are highly desired as they will allow functional integration within a very narrow yet highly effective reaction/catalysis zone just next to the proton-transporting layer.

Inspired by the depletion gilding process used by ancient artisans, Ding *et al.* developed an ingenious method to make large-area, free-standing, ultra-thin NPG membranes by dealloying commercially available white gold leaf.[41] Like decorating pure gold leaf, white gold leaf is made of Ag–Au alloys and fabricated on an industrial scale by beating. This simple manufacturing process is remarkably effective to make large-area metal leaves with thicknesses down to about 70 nm, an amazing class of macroscopic nanostructured materials long before the term "nano" was widely used in the academic field. Taking 100 nm thick, 12 carat white gold leaf as an example, the material contains about 50 wt% silver, and upon dealloying the novel NPG leaf thus produced contains only about 0.1 mg of Au per centimeter square.

White gold leaf can be easily transferred from one solution to another by a glass slide or a graphite roller (see Figures 4.3a and 4.3b). This process eases the fabrication of high-quality large-area NPG leaves, which could be further surface functionalized for a wide variety of applications. After dealloying on the surface of concentrated nitric acid for just several minutes, white gold leaves turn into free-standing copper-colored and translucent thin membranes made of almost pure gold. Compared with the pristine alloy leaf, the NPG leaf is more brittle, but still can be transferred from one solution to another for rinsing or decoration upon gentle operation.

As mentioned in Section 4.2.1, the formation of a nanoporous structure is associated with atomic corrosion and diffusion at the interface between the alloy and the electrolyte. The grain structure of NPG leaf is primarily determined by the pristine white gold leaf, which has a large lateral grain dimension on the order of 10 μm. However, considering that the pore and ligament size of NPG can be made down to several nanometers, almost three magnitudes smaller than the lateral grain dimension, one would expect an unusual single-crystalline yet porous grain structure for NPG leaf. This unique single-crystal feature can be demonstrated by electron diffraction and high-resolution transition electron microscopy (HRTEM) observations. As shown in Figure 4.3(c), the structure is composed of interconnected ligaments with sizes of about 15 nm. Selected-area electron diffraction often gives sharp single-crystal diffraction patterns composed of spots with square symmetry, which indicates the entire area of observation is from one grain with excellent crystallinity. The (100) texture of NPG leaf is also demonstrated by X-ray diffraction (XRD) results, which often show sharp (200) diffraction peaks. From the HRTEM image shown in Figure 4.3(d), {200} lattice fringes are found to extend continuously from one ligament across the pore to another. This unique single-crystal feature of NPG leaf is markedly different from traditional porous metal structures based on the controlled aggregation of nanoparticles, which renders the material with superior electrical conductivity[42] that is highly desired for electrocatalysis and fuel cell applications.

The reaction of silver with nitric acid is rather dramatic and under free corrosion conditions the white gold leaf can be etched away and form the nanoporous structure in seconds, although at this stage there is still quite an amount of silver atoms left within the structure. After about 10 minutes, more than 99% of the silver component can be dissolved and leave behind a

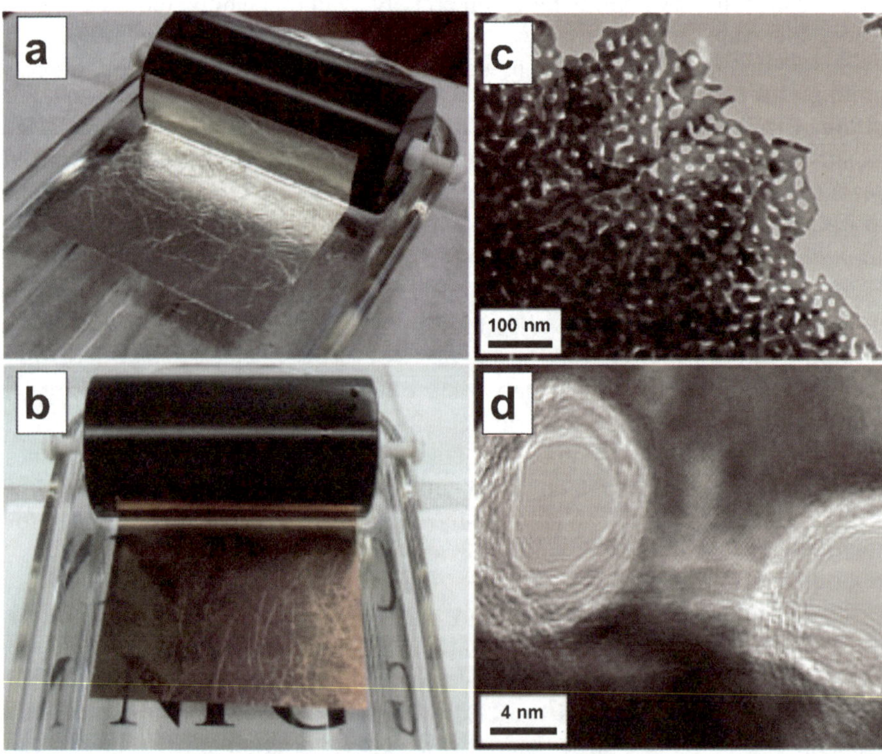

Figure 4.3 (a, b) Digital photographs of a whole page of white gold leaf before and after dealloying in concentrated nitric acid for 15 min. (c, d) TEM and HRTEM images of NPG leaf. (Reproduced with permission of Wiley from Ding *et al.*[41])

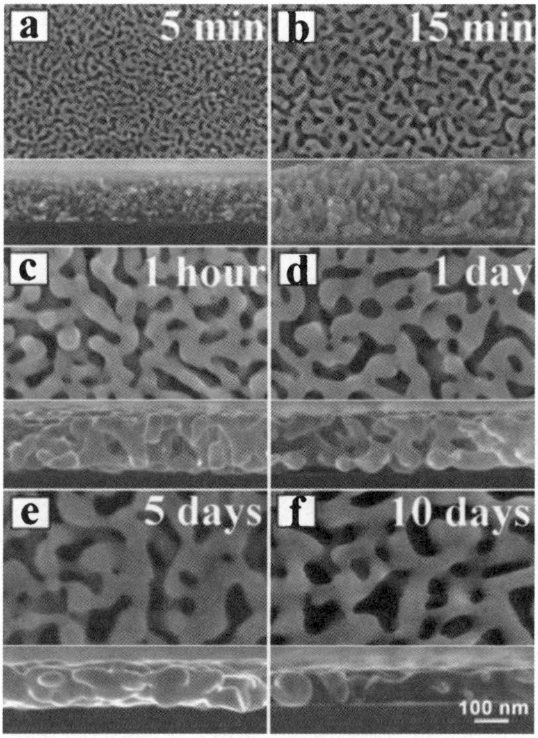

Figure 4.4 Plan-view and cross-sectional SEM images of NPG leaves upon immersion in nitric acid for an extended time. (Reproduced with permission of Wiley from Ding *et al.*[41])

nanoporous material made of almost pure gold. Interestingly, the porous structure would undergo a further coarsening process if the sample is left in the concentrated nitric acid for a prolonged time. Figure 4.4 depicts the structural evolution of NPG leaf at different stages of dealloying time. This is driven by the surface energy reduction effect and facilitated by the enhanced surface diffusion rate of gold atoms in the electrolyte, a phenomenon similar to the annealing process at elevated temperatures.[43] As the total thickness of the NPG leaf is only 100 nm, the porous structure would eventually evolve into a two-dimensional network after dealloying for several days. It should be pointed out that the coarsening process could be easily halted by removing the sample from the dealloying medium. This simple technique was used and modified to fabricate nanoporous metals with similar ligament sizes but different compositions.[44]

4.2.3 Electrocatalysis of NPG Leaf

NPG leaf possesses several distinct structure features which make it unique compared with traditional nanoparticle-based catalysts. Firstly, as the NPG leaf is dealloyed from a highly corrosive regent, the ligament surface is free of contaminants such as surfactants and organic surface passivating species that are commonly seen in traditional nanoparticles made *via* wet chemical methods. This structural advantage is of particular importance for catalysis. Secondly, the whole structure is interconnected with highly conductive ligaments and open pore channels, which is greatly desired for electrocatalysis as electrons can be easily transported from any reaction sites on the ligament surface to the current collector. This feature is also markedly different from that in nanoparticle-based catalysts, where functioning nanoparticles have to be isolated from each other to avoid aggregation and then be dispersed on conducting substrates such as carbon black. It can be imagined that a significant portion of the nanoparticles would be under non-optimized conditions and eventually become inoperative. Finally, as the grain size in NPG leaf is typically about three orders of magnitude larger than the length scale of the ligaments (~10 μm *vs.* ~10 nm), this 3D nanoporous structure should consist of different kinds of crystal facets exposed on the ligament surfaces, and most of them are highly active for electrocatalysis. Indeed, computer simulation has revealed that the dealloying process proceeds on the atomic terrace layer-by-layer, which leaves a large density of steps on the ligament surface.[35] These low-coordinated step atoms are often favorable reaction sites for catalysis and electrocatalysis. In contrast, crystal facets exposed on traditional metal nanoparticles are often low indexed, such as {111} and {100}, which have relatively lower surface energy and reactivity.

With these unique structural properties, NPG itself can be an interesting electrocatalyst used for potential energy saving and electrochemical sensing technologies. Methanol electro-oxidation in alkaline solution was the first reaction examined using NPG electrodes.[45] With stronger chemisorption of OH^- ions on the NPG surface than on the bulk gold surface, NPG shows much enhanced catalytic activity toward methanol electro-oxidation. Although this reaction was performed on thick NPG foils, these results have been reproduced on NPG leaves with different ligament sizes. The ultra-thin yet robust enough NPG in leaf form makes it very easy to be assembled into an MEA. Zeis *et al.*[46] have examined the fuel cell performance using NPG leaf as the cathode (see Figure 4.5a), and observed a surprisingly high open circuit of ~850 mV. Without any hydrogen peroxide detected in the output of the fuel cell, it was suggested that the most possible reaction mechanism was a two-step four-electron reduction process in which oxygen was first reduced to hydrogen peroxide, and then further reduced to water. Further experiments on rotating ring disk electrochemistry (RRDE) demonstrated that the reduction of oxygen to hydrogen peroxide on NPG leaf was similar to results on a planar gold electrode. However, the reduction of hydrogen peroxide to water was quite different (see Figure 4.5b). In contrast to the small reduction currents on a planar gold electrode, the reduction current on NPG was much enhanced. As the reduction of hydrogen peroxide was highly catalyzed by NPG, the reduction of oxygen became the rate-determining step. Under the fuel cell

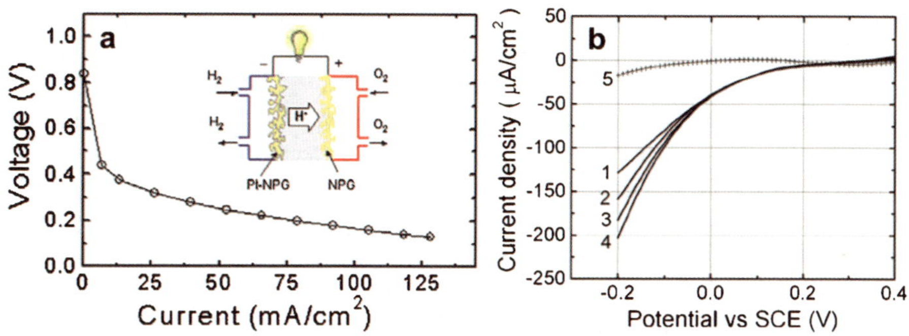

Figure 4.5 (a) Voltage–current polarization of an NPG-based MEA (cathode: NPG leaf; anode: Pt-NPG leaf) at room temperature. (b) ECSA-specific disk current densities during peroxide reduction on NPG at various rotation rates: (1) 400; (2) 900; (3) 1600; and (4) 2500 rpm; and on planar gold (5) at ω = 1600 rpm. (Reproduced with permission of Elsevier from Zeis *et al.*[46])

operating conditions, the hydrogen peroxide once formed would then be further reduced to water, mediated by the NPG leaf catalyst. This is why the open circuit potential could reach as high as 850 mV. It was suggested that the high density of steps on the ligament surface of NPG leaf contributed to the high catalytic activity toward hydrogen peroxide reduction. Very recently it was shown that NPG had high catalytic activity toward the oxidation of borohydride, which suggests its potential in direct borohydride fuel cells.[47]

With a bicontinuous porous structure and high catalytic activity toward several important reactions, NPG leaf was proposed to be a low-cost substrate material for electrochemical sensing. It was shown by Meng *et al.*[48] and Ge *et al.*[49] that NPG leaves could be efficient and highly selective sensor materials for the electrochemical detection of hydrogen peroxide and nitrite over wide concentration ranges. NPG was also proved to be active for the electro-oxidation of D-glucose in alkaline aqueous solutions and may be used for amperometric glucose detection in biological samples.[50,51] Another example associated with NPG leaf is the electrochemical detection of *p*-nitrophenol.[52] Different from the voltammetric behavior on polycrystalline gold electrodes, the CV profiles associated with the 4-(hydroxyamino)phenol/4-nitrosophenol reaction couple on NPG leaf are highly symmetric and hardly affected by its isomers. This means the detection selectivity of NPG toward *p*-nitrophenol could be very high. These characteristics make NPG a promising electrochemical sensor substrate for the detection of trace *p*-nitrophenol in wastewaters.

4.3 Platinum-Plated Nanoporous Gold Leaf

Although NPG leaf has shown high catalytic activity toward several electrochemical reactions (mostly in alkaline solution), compared with Pt, gold is much less active in most cases, especially in acidic solution. On the other hand, several key issues remain for further improvement in the traditionally used Pt/C-like catalysts, including the accessibility of Pt atoms and stability. As NPG leaf is highly conductive and corrosion resistant, decoration of NPG surfaces with catalytically active metals such as Pt could be an alternative strategy to develop highly effective electrocatalysts; these could bypass the intrinsic weaknesses of traditional nanoparticle-based catalysts such as carbon corrosion at high potentials and the weak interaction between nanoparticle catalysts and carbon substrates.

Metal plating can be classified into two main categories: electroless plating and electrochemical plating, both of which involve electrochemical mechanisms in which metal ions obtain electrons and are deposited onto the desired substrate. Unlike planar substrates, it is intrinsically difficult to modify a highly porous surface at a nanometer scale. Inspired by the phenomenon that NPG leaf tends to float on the surface of solutions, Ding *et al.* developed a general method to plate foreign metals on NPG leaf with very high precision.[53,54] Later, Liu *et al.* reported a new method to plate NPG leaf.[55] Based on a combination of Cu under potential deposition and an *in situ* redox replacement reaction, Pt could be deposited onto the ligament surface of NPG leaf sequentially with atomic layer control. This new electrochemical method provides better control than the electroless one and opens up new opportunities for the design and fabrication of advanced electrocatalysts with integrated functions.

4.3.1 Plating Methods

4.3.1.1 Electroless Method to Plate NPG Leaf

The electroless method was first developed to plate NPG leaf with silver to make nanoporous gold with hierarchically porous nano-architecture.[53] Thereafter, it was demonstrated that Pt could also be plated using this method and the resulting nanocomposites showed great potential for electrocatalysis and fuel cell applications. The electroless plating method was actually a gas/electrolyte/substrate interfacial plating technique. The plating procedure can be illustrated in Figure 4.6(a). NPG leaf is placed on the surface of plating solution and a reducing agent (typically N_2H_4) is introduced from air above. When the reducing gas approaches and dissolves into the thin wetting layer of the plating solution within the porous leaf, the redox reaction takes place. As the reducing agent vapor and metal ions are separately supplied from each side of the leaf, the redox reaction is essentially confined to the very thin porous layer of the nanostructure.

With this method, Pt could be deposited onto the whole ligament surfaces of NPG leaf in minutes at room temperature in the absence of any structural directing agents or surfactants (Figure 4.6b). The Pt loading (deposition amount) could be easily monitored by controlling the time of reaction, which could be terminated by removing the reducing agent vapor. Typically, the

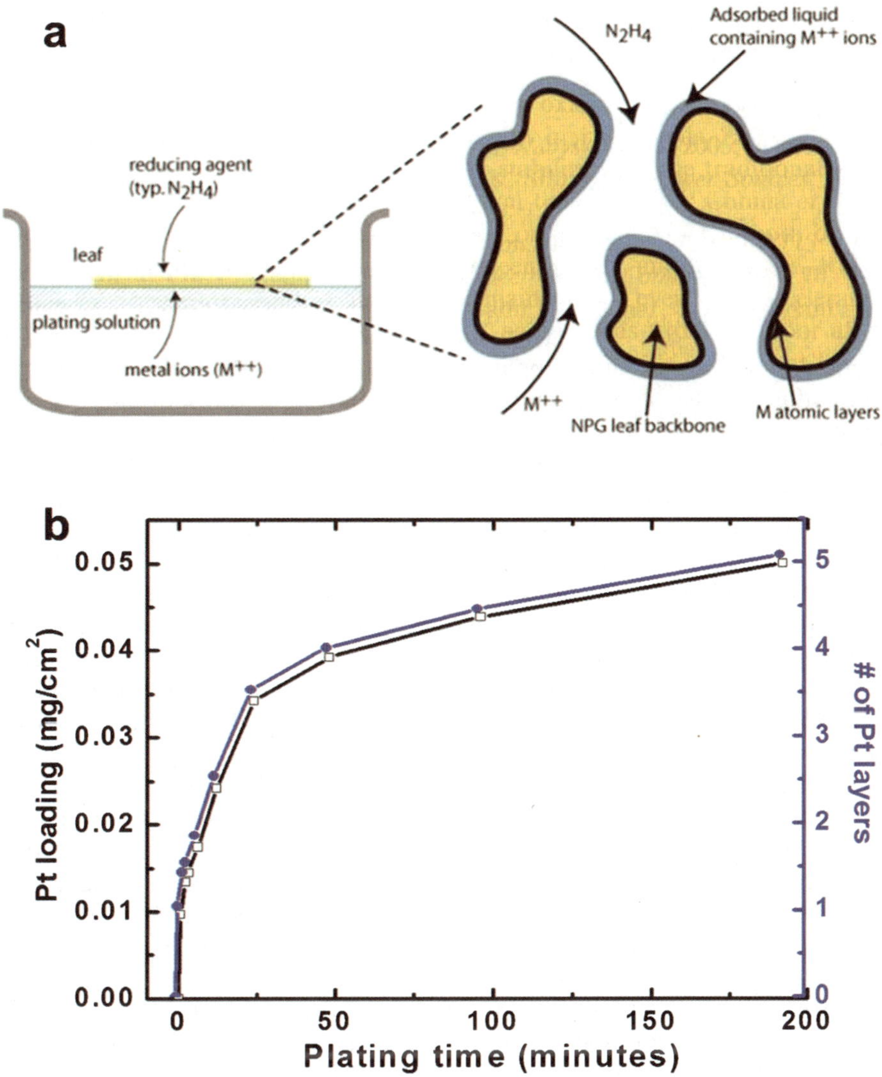

Figure 4.6 (a) Schematic illustration of the electroless plating method. (b) Kinetics for Pt plating on NPG leaf.

calculated Pt layer (based on the loading and surface area of the NPG leaf) reaches one monolayer after several minutes. The plating is self-limiting and saturates at a loading of approximately 0.05 mg cm^{-2} after about 200 min.

4.3.1.2 Underpotential Deposition Mediated Method to Plate NPG Leaf

The electroless plating process is governed by the redox reaction of metal ions and reducing reagents, which depends significantly on reaction parameters such as temperature, concentration, and the combination of metal ions and reducing reagents of choice. It is also nearly inevitable that some reducing gas will be dissolved into the solution and react with the metal ions, leading to the deterioration of the whole plating system. In contrast, electrochemical methods allows precise control of the reducing process by finely tuning the potential applied to the electrodes. However, traditional electrochemical deposition methods could not be easily applied to the plating of NPG leaf, as the crystal nucleation on the high curvature surfaces and mass transportation in the porous structure would greatly affect the plating process.

In 2001, Brankovic *et al.*[56] developed a new metal deposition method which was a combination of Cu underpotential deposition (UPD) and a galvanic replacement reaction. UPD is a phenomenon where the foreign metal could deposit on the noble metal substrate in a monatomic layer configuration just above the equilibrium potential.[57] This monolayer of metal atoms could then be displaced by targeted more noble metal atoms (*t*-metals) *via* a galvanic replacement reaction and result in the deposition of *t*-metal on the substrate metal surface at atomic precision. This method has been demonstrated to be a versatile procedure to make highly efficient nanocatalysts with core–shell structures.[12,13]

In 2009, Liu *et al.*[55] demonstrated that the UPD-mediated deposition method could be used to plate noble metals such as Pt on NPG leaf with remarkable structural uniformity and precision (Figure 4.7). It was found that the deposition potential and time are two key parameters that influence the completeness of the UPD layer and consequently the state of the Pt layer. Typically, the optimized deposition condition in 0.5 M H_2SO_4 solution containing 0.5 mM Cu^{2+} ions is to hold the potential at 0.24 V *vs.* RHE for 120 s, when a complete monolayer of Cu can be deposited on NPG leaf. The galvanic replacement of the monolayer Cu with a solution containing Pt^{2+} ions results in the monolayer deposition of Pt on NPG. As Cu could also form a

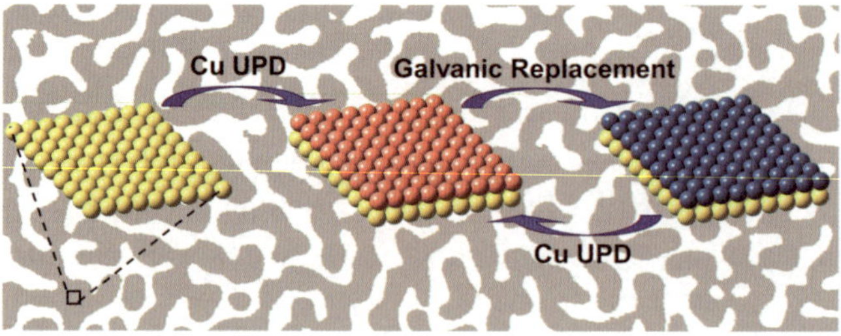

Figure 4.7 Schematic illustration of the fabrication procedure for a UPD-mediated method to plate NPG leaf.

UPD layer on Pt, this process could then be repeated to deposit more Pt on the NPG in a layer-by-layer mode. With this method, one can not only investigate the structure–property relationships of NPG-supported Pt catalysts, but also design and fabricate other artificial nanocatalysts with tailored functionalities.

4.3.2 Structure and Stability of Pt-NPG Leaf

The catalytic activity and stability of electrocatalysts are often influenced by the catalyst surface and the interfacial structure between catalyst and support. The surface structure of Pt-plated NPG leaf has been investigated by scanning electron microscopy (SEM).[58] As shown in Figure 4.8, the ligaments of the porous metal grow a little thicker after 2 min Pt plating, but the surface is still quite smooth. In contrast, the sample with 128 min Pt plating has much thicker ligaments and lots of small clusters are observed on the surface. These results indicate that Pt has been successfully plated onto the ligament surfaces of NPG leaf and the decoration is very uniform. The presence of Pt has been demonstrated by energy dispersion X-ray spectroscopy (EDS) and inductively coupled plasma atomic emission spectroscopy (ICP-AES) analyses.

The detailed structure has been investigated by transmission electron microscopy (TEM).[54] As shown in Figure 4.9(a), after plating for 24 min the whole surface of the NPG is covered by a thin layer of Pt clusters. The uniform Pt deposition reveals that the plating process is not controlled by ion diffusion or reducing agent transportation, which would otherwise lead to gradient Pt deposition across the thickness of the leaf. HRTEM observations and corresponding Fourier transform (FT) patterns indicate that the Pt clusters are grown epitaxially on the Au substrate (Figure 4.9b). Despite the different deposition method, the structures of Pt-NPG leaves using the Cu-UPD-

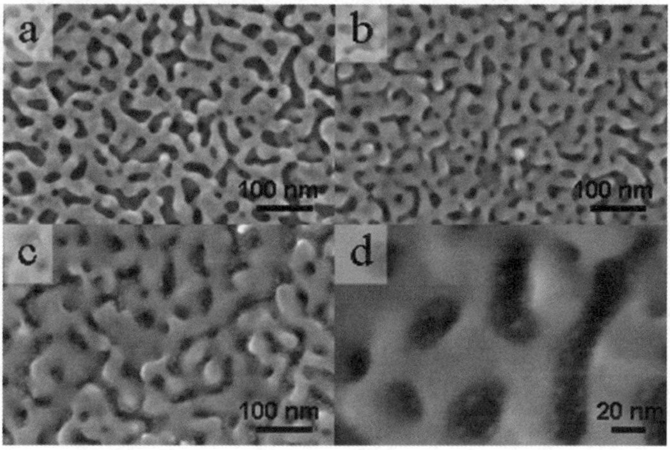

Figure 4.8 SEM images of (a) pristine NPG leaf and (b–d) Pt-NPG leaf for (b) 2 and (c, d) 128 min. (Reproduced with permission of the American Chemical Society from Ge *et al.*[58])

Figure 4.9 (a) TEM and (b) HRTEM images of Pt-NPG leaf. *Inserts* are the electron diffraction pattern, the FT pattern and the Fourier filtered image, which provide direct evidence for the epitaxial relationship between plated Pt and the NPG leaf substrate. (Reproduced with permission of the American Chemical Society from Ding *et al.*[54])

mediated method are quite similar (Figure 4.10).[55] During the first two cycles of Pt deposition, Pt appears to cover the NPG surface in a layer-by-layer mode, and further deposition leads to the formation of Pt nano-clusters of size 2–3 nm.

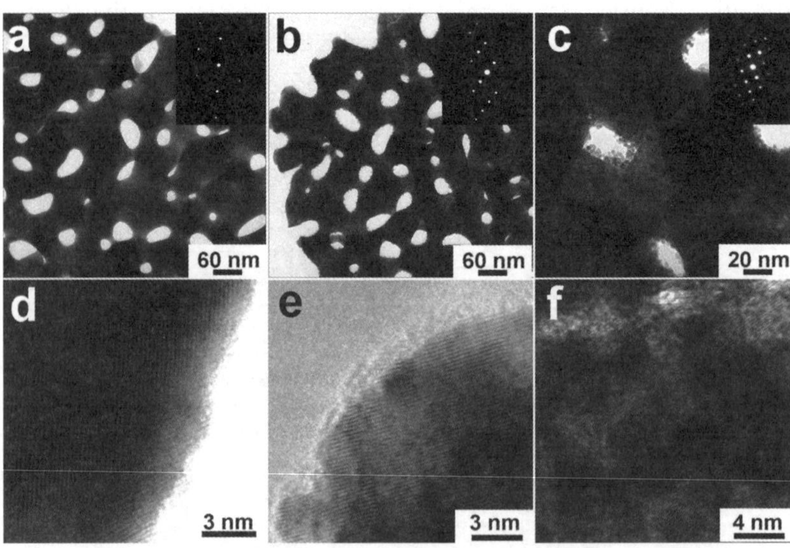

Figure 4.10 TEM and HRTEM images for (a, d) NPG-Pt2, (b, e) NPG-Pt4, and (c, f) NPG-Pt6. *Insets* are the respective electron diffraction patterns for these samples. (Reproduced with permission of the American Chemical Society from Liu *et al.*[55])

Platinum-plated NPG leaf possesses several structural advantages compared with the traditional Pt/C catalyst. (1) The utilization of Pt is significantly improved, especially for low Pt loading samples, where almost all the Pt atoms are accessible by reactants and function in the catalytic reaction. (2) The highly conductive NPG substrate greatly facilitates the electron transportation from surface Pt through the interconnected porous structure to the current collector. (3) The NPG substrate is more corrosion resistant compared with carbon, which would contribute to the electrochemical stability of the composite catalyst under the harsh operating conditions. (4) Unlike the weak interaction between Pt nanoparticles and a carbon support, the strong metallic bond between Pt and the NPG substrate could stabilize the Pt nanoparticles from detaching from the substrate.

Despite the above advantages, the stability of Pt-NPG nanocomposites should be addressed before discussing their potential application in fuel cells. The structural evolution of Pt-NPG leaf has been investigated at different elevated temperatures.[59] It has been found that the porous structure can be retained after annealing at 300 °C for 48 h (Figures 4.11a and 4.11d), which is markedly different from that of bare NPG which would undergo significant coarsening at this temperature even for a couple of minutes. The stabilization effect of Pt on NPG leaf should be ascribed to the relatively slow diffusion rate of Pt compared with Au. Interestingly, although the porous structure maintains very well upon low-temperature (below 300 °C) annealing, careful structural analyses with HRTEM still reveal quite dramatic surface structure evolution details. After annealing at 150 °C, the growth of Pt islands was

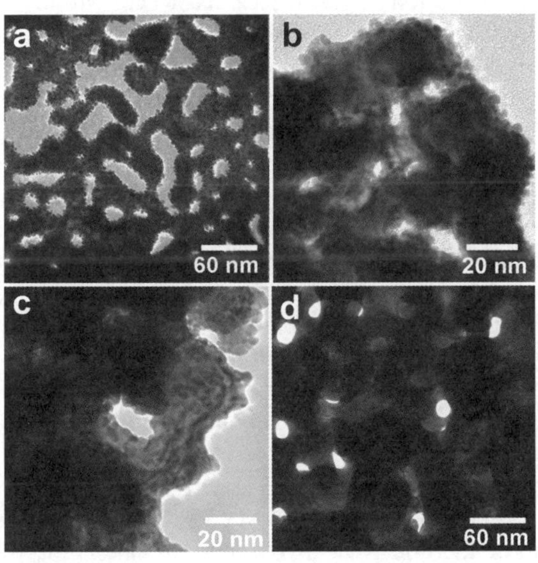

Figure 4.11 TEM images of Pt-NPG leaves (a) before and after annealing at (b) 100 °C for 5 h, (c) 150 °C for 2 h, and (d) 300 °C for 48 h. (Reproduced with permission of the American Chemical Society from Ge *et al.*[59])

already seen. The situation was much more apparent if the temperature was raised even higher to 300 °C, where all the Pt nano-islands were found flattened out on the surface of the NPG substrate. Coupled with surface-sensitive electrochemical characterization techniques, this surface structure evolution was ascribed to the atomic inter-diffusion between Pt islands and the Au substrate. This inter-diffusion process changed the originally core–shell type of the Au–Pt nanostructure into the eventual Au–Pt surface alloys at temperatures down to about 150 °C.

As the operation of fuel cells is commonly under much lower temperatures (<80 °C), the demonstrated thermal stability of Pt-NPG leaf paves the way for most PEMFC applications. However, as the fabrication of a MEA is often conducted at about 140 °C, caution should be made when processing these new types of low-Pt electrocatalysts into an MEA. Indeed, MEAs made by hot pressing at about 110 °C often exhibit a better performance in fuel cells than those treated at higher temperatures.

4.3.3 Electrocatalysis of Pt-NPG Leaf

With the above-mentioned structural advantages, Pt-NPG leaf should be very suitable for electrocatalytic applications. The surface structures of Pt-NPG leaves made by different methods[55,58] have been investigated by cyclic voltammetry (CV) in 0.5 M H_2SO_4 and the results are shown in Figure 4.12. Both plating methods result in complete coverage of Au by a Pt overlayer, which was proved by the disappearance of a gold oxide reduction peak especially for higher Pt loading samples. However, at low Pt loading the difference between the two fabrication methods is quite obvious. The Cu-UPD-mediated method shows better control of the plating layer, which is demonstrated by nearly complete inhibition of a gold oxide reduction signal even after one layer of Pt plating. The sample made by the electroless method shows a significant gold reduction peak even after 2 min plating and complete coverage of Au by Pt could be achieved after 8 min plating, which equals more than one layer of Pt. From the CV curves one can see that the surface area of Pt increases with the increase of Pt loading, although the utilization of Pt shows an opposite trend. In addition, the platinum oxide reduction peaks demonstrate a positive shift with the increase of Pt loading. This phenomenon should be caused by the electronic interaction between Au and Pt, which induces improved adsorption of oxygen on the Pt surface, as demonstrated by both density functional theory (DFT) simulations[60] and electrochemical experimental results.[61]

The electrocatalytic activity of Pt-NPG leaves toward methanol oxidation was investigated by CV in 0.5 M H_2SO_4 + 1.0 M CH_3OH mixed solution. Despite the different fabrication methods, the two types of nanocomposite show rather similar electrocatalytic performance; Figure 4.12 depicts the CV profiles for samples made by the Cu-UPD-mediated method The catalytic activities have been normalized to ECSA (electrochemically active surface area) and Pt

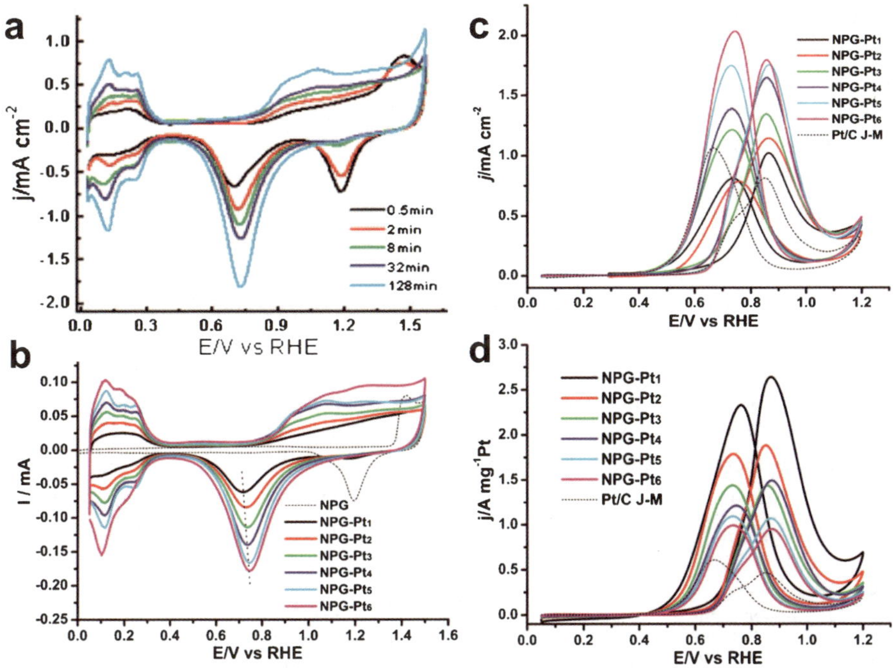

Figure 4.12 Cyclic voltammograms (CVs) of Pt-NPG leaves made by (a) electroless and (b) Cu-UPD-mediated methods with different Pt loadings. Curves (c) and (d) are ESCA- and Pt mass-specific catalytic activities of different Pt-NPG leaves made by the Cu-UPD method in 0.5 M H_2SO_4 + 1.0 M CH_3OH. (Reproduced with permission of the American Chemical Society from Liu *et al.*[55] and Ge *et al.*[58])

loadings, which are critical for the understanding of catalytic performance from fundamental and technological points of view, respectively. As shown in Figure 4.12(c), the ECSA-specific activities of Pt-NPG leaves increase with the increase of Pt loading and all are superior to commercial Pt/C catalysts, as revealed by the enhanced current densities. The increased catalytic activities with the increase of Pt loading should be caused by a decrease of the electronic perturbation of the Pt surface by the underlayer of the gold substrate that has been discussed above. The improved catalytic activities of samples with a higher Pt loading compared with the Pt/C catalyst must be caused by their different surface configurations. In Pt-NPG, epitaxial Pt nano-islands are essentially linked together and form an extended overlayer structure with comparatively less compressive stress, as often seen in spherical Pt nanoparticles of similar size. This is somewhat consistent with the observation by Frelink and co-workers that larger Pt nanoparticles showed improved catalytic activity toward methanol electro-oxidation compared with smaller ones.[62]

The Pt mass-specific catalytic activities of Pt-NPG leaf catalysts are shown in Figure 4.12(d). With much improved Pt utilization (close to 100% for the sample with one deposition layer), all Pt-NPG leaf samples show significantly enhanced mass-specific activity compared with the Pt/C catalyst. The highest enhancement factor is about 7 for the sample with one deposition layer. With the increase of the Pt layer, the Pt mass-specific activities decreased, which is reasonable considering the decreased Pt utilization value.

With higher energy density and much lower toxicity, ethanol was also believed to be an attractive fuel option.[63] The catalytic activities of Pt-NPG leaves toward ethanol electro-oxidation were investigated by CV in 0.5 M H_2SO_4 + 1.0 M C_2H_5OH mixed solution. The ECSA- and Pt mass-specific activities are shown in Figure 4.13 with those of Pt/C catalyst for comparison. It is expected that the Pt mass-specific activities are all much higher than the commercial catalyst. However, different from methanol electro-oxidation, the Pt-NPG leaf sample with one Pt deposition layer showed much enhanced catalytic activity even compared with samples with more Pt deposition layers. The decrease of ECSA-specific catalytic activities with the increase of Pt loading implied that the electronic perturbation of Pt by the underlayer of gold substrate should facilitate ethanol electro-oxidation. The different response of the oxidation kinetics between methanol and ethanol to the electronic

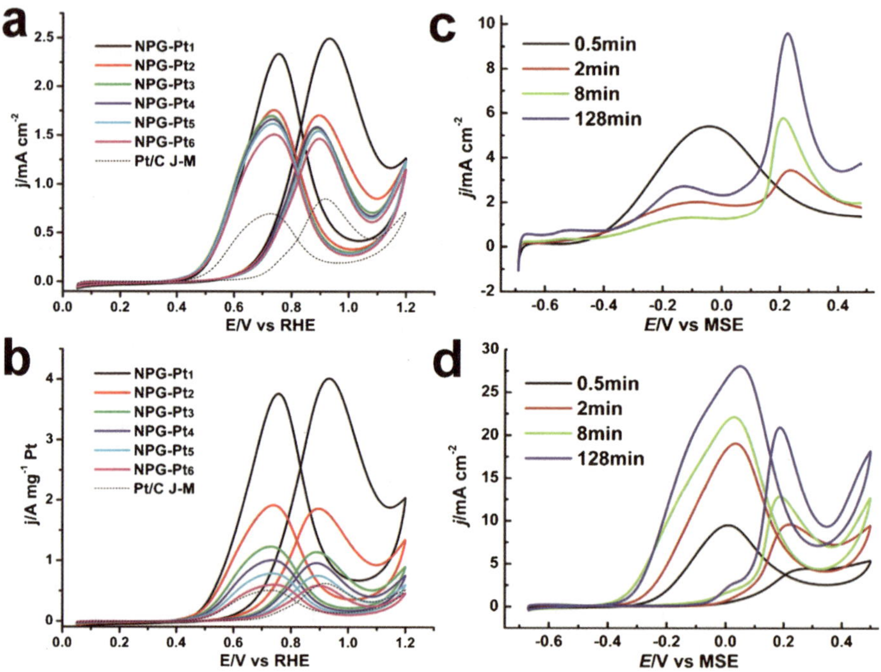

Figure 4.13 (a, b) ECSA- and Pt mass-specific catalytic activities of Pt-NPG leaves with different Pt loadings made by the Cu-UPD method in 0.5 M H_2SO_4 + 1.0 M C_2H_5OH. (c, d) Electrochemical catalytic activities of Pt-NPG leaves made by the electroless method with different Pt loadings in 0.1 M $HClO_4$ + 0.1 M HCOOH and 0.1 M $HClO_4$ + 0.1 M HCHO mixed solutions. (Reproduced with permission of the American Chemical Society from Liu *et al.*[55] and of Elsevier from Ge *et al.*[64])

perturbation of Pt should be caused by the quite different reaction paths and this phenomenon deserves further investigation.

The electro-oxidation reactions of formic acid and formaldehyde are often investigated, on the one hand to understand the oxidation mechanism of small organic molecules fundamentally and on the other hand to develop new types of fuel cells with alternative fuel options. The catalytic activities of Pt-NPG leaf toward formic acid and formaldehyde electro-oxidations were investigated by CV in 0.1 M $HClO_4$ + 0.1 M HCOOH and 0.1 M $HClO_4$ + 0.1 M HCHO mixed solutions.[64] As expected, the formaldehyde oxidation currents increased with the increase of Pt loading, which is caused by the improved electrochemical surface area of the Pt. The electro-oxidation of formic acid was quite different. At higher Pt loading the oxidation currents increased with the increase of Pt loading. However, for the low Pt loading sample, only one oxidation peak appears at lower potential, which is totally different from that for the high Pt loading samples, where there are two broad anodic peaks. The distinct voltammetric behavior for low Pt loading samples was caused by the switch of the reaction path of formic acid electro-oxidation on small Pt ensembles and will be discussed in detail in Section 4.4.

4.3.4 Fuel Cell Performance of Pt-NPG Leaf

To function in fuel cells, a traditional Pt/C catalyst is processed by first mixing with an ionomer solution such as Nafion, which acts as binder and proton carrier to form a catalyst ink. The catalyst suspension is then brushed or sprayed onto a gas diffusion layer (GDL) to form a catalyst layer which, in turn, is hot pressed onto each side of a solid electrolyte membrane to form a MEA. Different from traditional catalysts, Pt-NPG leaf has a monolithic structure which makes it much easier to prepare an MEA. Moreover, characterized with some attractive features such as high Pt utilization, high conductivity and stability, and low overall precious metal loading, Pt-NPG leaf-type electrocatalysts hold great potential as a new generation of catalysts for fuel cell applications.[54,65] Zeis *et al.* demonstrated a neat process to make such an MEA.[65] As shown in Figure 4.14, Pt-NPG leaf floating on a water surface was first caught by a mica sheet from underneath. After drying in air, the samples on the mica sheets were hot pressed onto each side of a moisturized Nafion film. It was reported that MEAs prepared at 70 °C demonstrated the highest and reproducible performance.

The prepared Pt-NPG leaf-based MEAs were in turn sandwiched between two pieces of Teflon-treated carbon cloth and then fixed between two stainless steel current collection plates for test and characterization in actual fuel cells. The MEAs were firstly activated by CV cycling between 0.03 and 1.5 V at a sweep rate of 550 mV s^{-1} and the electrochemical surface areas were investigated using the hydrogen under potential adsorption/desorption (H-UPD) method. The charge densities associated with H-UPD on Pt-plated NPG leaf MEAs with various Pt loadings are shown in Figure 4.15(a). It is clearly

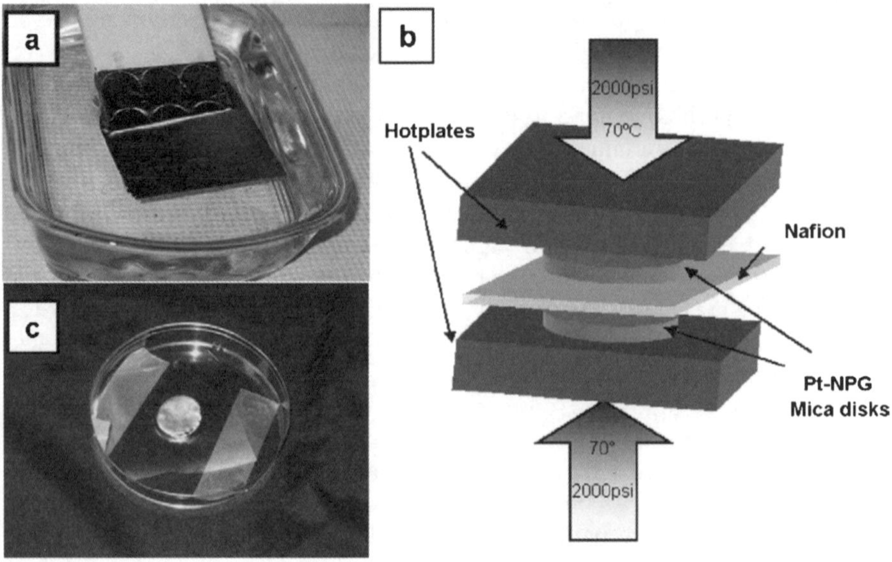

Figure 4.14 Process of MEA fabrication with NPG leaves: (a) transfer of Pt-NPG leaf with mica sheets; (b) stamping procedure to make the MEA; and (c) the final MEA. (Reproduced with permission of Elsevier from Zeis *et al.*[65])

seen that the H-UPD charges increased with the Pt loading for samples with a loading below 0.03 mg cm^{-2}. However, the H-UPD charges saturate at around 3.6 mC cm^{-2} regardless of further increase of the Pt loadings, indicating the surface area had reached the highest value at this point of Pt loading. As discussed above, the Pt layer grew from a sub-monolayer into monolayer and then nano-islands formed. During this stage the surface area of the Pt undergoes a straight increase. After the formation of Pt islands, further increasing the Pt loading results in coverage of the pre-formed Pt surface but the total surface area of the Pt remains the same. The specific electrochemical surface area of the Pt could be calculated by dividing the real surface area with the Pt loading and the results are shown in Figure 4.15(b). It was expected that the specific ECSA value would decrease when the Pt loading reached 0.3 mg cm^{-2}. However, the relative smaller ECSA values at low Pt loadings seem contradictory with the high Pt utilization results obtained in half-cell tests. This may be caused by the different measurement methods. In the gas-phase measurement in fuel cells, the protons from the H-UPD layer on Pt-NPG leaf catalysts are transported from the anode side. As there is no ionomer nearby, the transport of protons should be *via* a surface diffusion mechanism which is similar to that on a Pt mesh cathode.[66] However, in a half-cell test the protons could be easily transported through the solution and the H-UPD layers could be formed on all surface Pt atoms, which results in the high apparent ECSA values for low Pt loading samples.

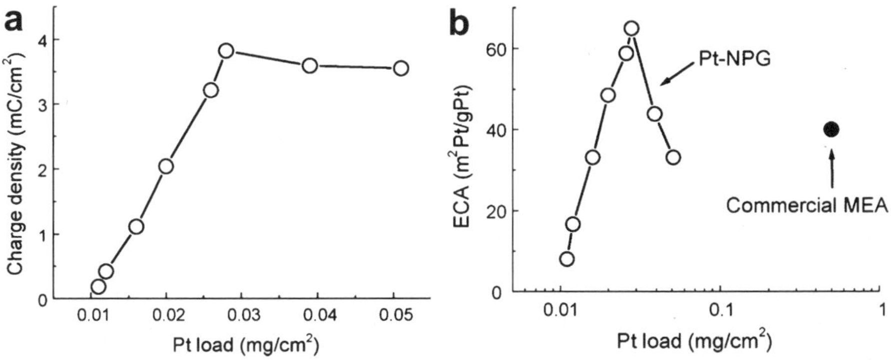

Figure 4.15 (a) Charge density corresponding to adsorbed hydrogen on Pt-NPG leaf MEAs *vs.* Pt loadings. (b) Electrochemically active surface area *vs.* Pt loadings. ECSA of a commercial MEA (0.5 mg cm^{-2}) is also shown for comparison. (Reproduced with permission of Elsevier from Zeis *et al.*[65])

The fuel cell performance with Pt-NPG leaves is shown in Figure 4.16. It is obvious that the open circuit potentials increase with the Pt loading and the highest value approaches ~0.9 V. Compared with the commercial Pt/C catalyst, this value is about 0.1 V lower, which may be caused by the kinetic losses at the cathode side at such low Pt loadings. The maximum power densities as a function of Pt loadings are shown in Figure 4.16(b). The curve follows the trend of electrochemical surface areas, which increase at low Pt loading and remain unchanged upon further increasing the loading. Further increasing the electrochemical surface area may be a potential direction to increase the fuel cell performance. Although the fuel cell performances are inferior to those of commercial Pt/C catalysts, the Pt specific activity [2.5–4.5 kW $(g\text{-}Pt)^{-1}$] is typically within a factor of two of what is reported for many state-of-the-art fuel cells. Further research should be conducted to reveal the

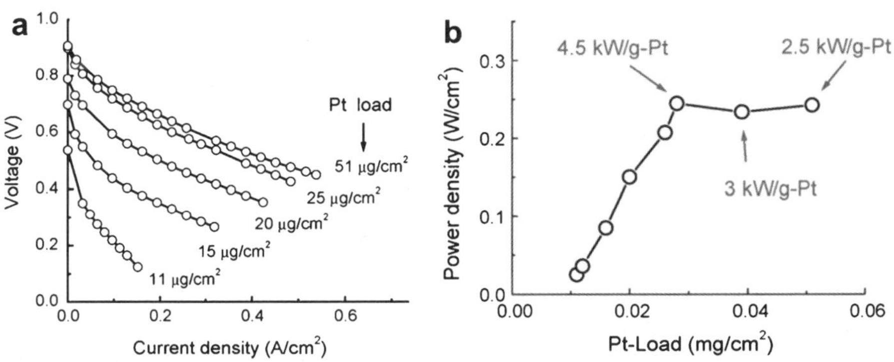

Figure 4.16 (a) Voltage–current polarization curves and (b) maximum power densities of Pt-NPG leaf MEAs with different Pt loadings. (Reproduced with permission of Elsevier from Zeis *et al.*[65])

fundamental issues governing the performance of this kind of material to guide catalyst design and fabrication.

4.4 NPG-Based Electrocatalysts for Formic Acid Oxidation

For an ideal fuel cell catalyst, at least three parameters should be achieved simultaneously. The first and foremost is that the catalyst should be highly active for the electrocatalytic reactions. Secondly, the activity should be highly durable against poisoning intermediates and for long-term operation. The last one is that the loading of precious metals such as Pt should reach an affordable low level. To rationally design such a catalyst, one should first be aware of the fundamental mechanisms of electrode reactions on distinct catalyst surfaces. More importantly, this design has to be achieved by current fabrication techniques or tools for potential scale-up. In this section we will discuss how to prepare highly active and stable electrocatalysts for formic acid electro-oxidation on the surface of NPG leaf at ultra-low Pt loadings.

It is commonly accepted that formic acid electro-oxidation takes place on Pt *via* a dual-path mechanism.[67] One path is to form a CO intermediate by dehydration and then oxidation to CO_2. As the oxidation of CO takes place at a higher potential, it thus functions as a poisoning species and passivates the Pt catalyst surfaces at low potentials. This mechanism has been proved by direct spectroscopic observation.[5] The other path is to form CO_2 *via* an active intermediate which is believed to be formate and has also been demonstrated by time-resolved surface enhanced infrared absorption (SEIRA) spectroscopy.[5] However, the dual-path mechanism was recently challenged by Chen *et al.*, who believed there should be at least another path to explain the apparent difference in the ratio between formate band intensity and faradic current at different potentials, although they have no direct spectroscopic evidence.[68] Thus they proposed a triple-path mechanism, which is shown in Figure 4.17(a). Although the exact mechanism is still under debate,[69] it is quite clear and accepted that the desired direct reaction path requires small Pt ensembles,[70] which is also consistent with reported experimental results that adsorbed atoms on Pt could facilitate the reaction[7–9] and simulation results that at least three adjacent Pt atoms are needed in order to proceed *via* the unfavorable dehydration path to form CO (see Figure 4.17b).

Guided by the understanding of the desired structure, Wang *et al.* proposed a novel strategy to prepare highly active and stable catalysts for formic acid electro-oxidation.[71] The preparation procedure can be illustrated by the scheme in Figure 4.18. Monolayer Pt is firstly deposited onto the ligament surface of NPG leaf using the Cu-UPD mediated method to achieve very high Pt utilization. Then sub-monolayer Au clusters are deposited further onto the monolayer Pt-covered NPG leaf in order to divide the continuous surface into surface ensembles that are small enough to facilitate the desired direct oxidation path. With a monolayer structure, the Pt loading in this kind of

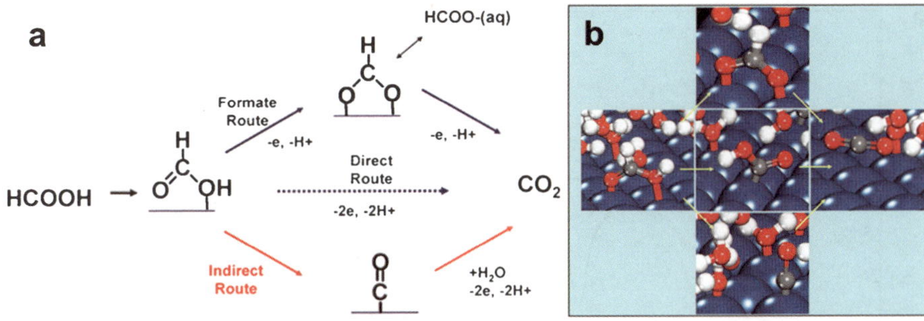

Figure 4.17 (a) Reaction paths involved in formic acid electro-oxidation. (b) DFT-calculated surface ensembles required for the direct oxidation path (*center*), the indirect path through CO (*bottom*), and the formate path (*top*) for the oxidation of formic acid over Pt(111) to CO_2 at a constant potential of 0.5 V.[70]

material is only 2 μg cm^{-2} of geometric area. As the monolayer Pt has been divided into small ensembles, it should be very active for formic acid electro-oxidation. Also, because of the protection of the outer-layer Au, the monolayer Pt should be stabilized against dissolution, which would guarantee the durability of the low Pt loading catalyst.

The Au-covered monolayer Pt catalysts supported on NPG leaf are denoted as NPG-Pt_1-Au_x where x means the Cu-UPD layer. Specifically, NPG-Pt_1-$Au_{0.5}$ and NPG-Pt_1-Au_1 indicate samples made by a replacement reaction between Au and Cu at half-monolayer and monolayer thicknesses, respectively. NPG-Pt_1-Au_2 describes a sample prepared by a Cu-UPD-mediated method for two times each with monolayer Cu. The electrochemical responses of a series of catalysts are shown in Figure 4.19. It is clear that the surface of NPG is almost totally covered by Pt after monolayer deposition. After gold

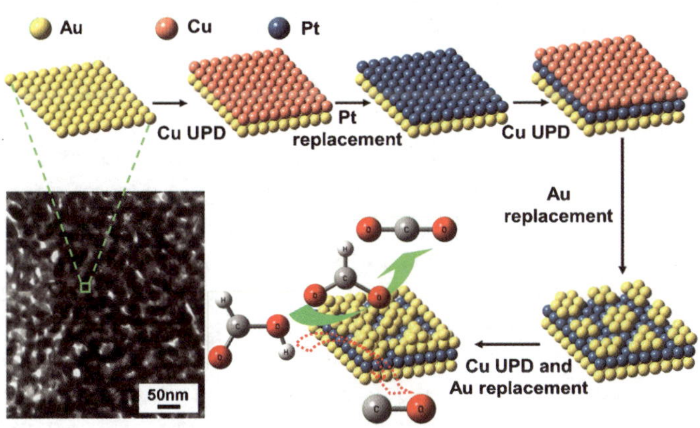

Figure 4.18 Schematic illustration of the fabrication procedure of a gold-covered monolayer Pt catalyst on NPG leaf.

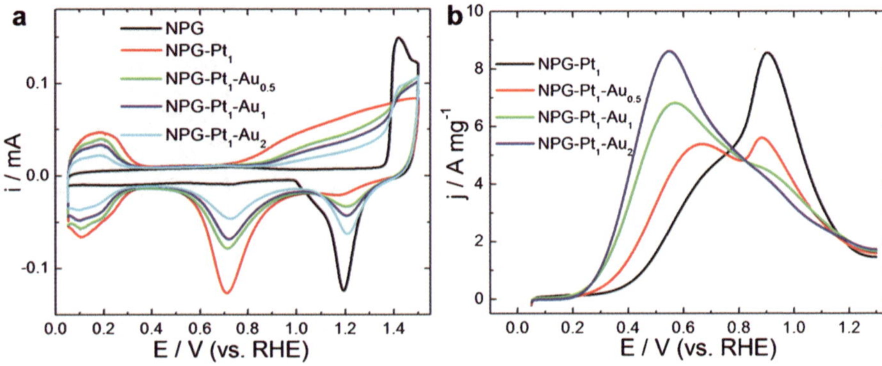

Figure 4.19 (a) Electrochemical CV curves for NPG-Pt_x-Au_y catalysts in 0.5 M H_2SO_4. (b) Pt mass-specific CV curves for NPG-Pt_x-Au_y catalysts in 0.1 M $HClO_4$ and 0.05 M HCOOH (only forward segments are shown for clarity). (Reproduced with permission of Wiley from Wang *et al.*[71])

deposition, the characteristic profile of Pt is reduced and the peaks of gold oxidation and reduction re-emerge. These results demonstrate the precise control of metal deposition, which forms a sound basis for the final catalyst performance.

The results of the formic acid electro-oxidation using these catalysts are shown in Figure 4.19(b). The CV profile of the NPG-Pt_1 sample shows a peak at 0.9 V, with a shoulder at lower potentials between 0.4 and 0.7 V. The result is similar to that of the bulk Pt surface, which means the whole Pt surface is severely poisoned by the CO intermediate.[5] After a little amount of gold deposition, the peak current density at 0.9 V decreases and the shoulder evolves into a new peak at about 0.6 V. Further deposition of gold on the surface results in a much enhanced current density at a lower potential of 0.5 V and nearly complete suppression of the peak at a higher potential of ~0.9 V. Because CO could not be oxidized at such a low potential, the much enhanced current density can only be ascribed to the oxidation of formic acid molecules at CO-free sites. These results provide clear evidence that the catalytic reaction path of formic acid electro-oxidation has been changed from an indirect path on a monolayer Pt surface to a direct one on a gold-covered monolayer Pt surface.

To understand the detailed mechanism of the catalytic activity improvement, time-resolved SEIRA spectroscopy was used to monitor the intermediate on the electrode surface during the potential cycling; the results are shown in Figure 4.20. From the spectra, we can see there is a little CO formation on the gold-covered Pt surface, but is far less compared with the bulk Pt surface. To determine the coverage of CO during the formic acid electro-oxidation, the spectra were compared with that of the monolayer CO-covered surface, which provides a CO coverage of about 16% for the new catalyst, indicating a

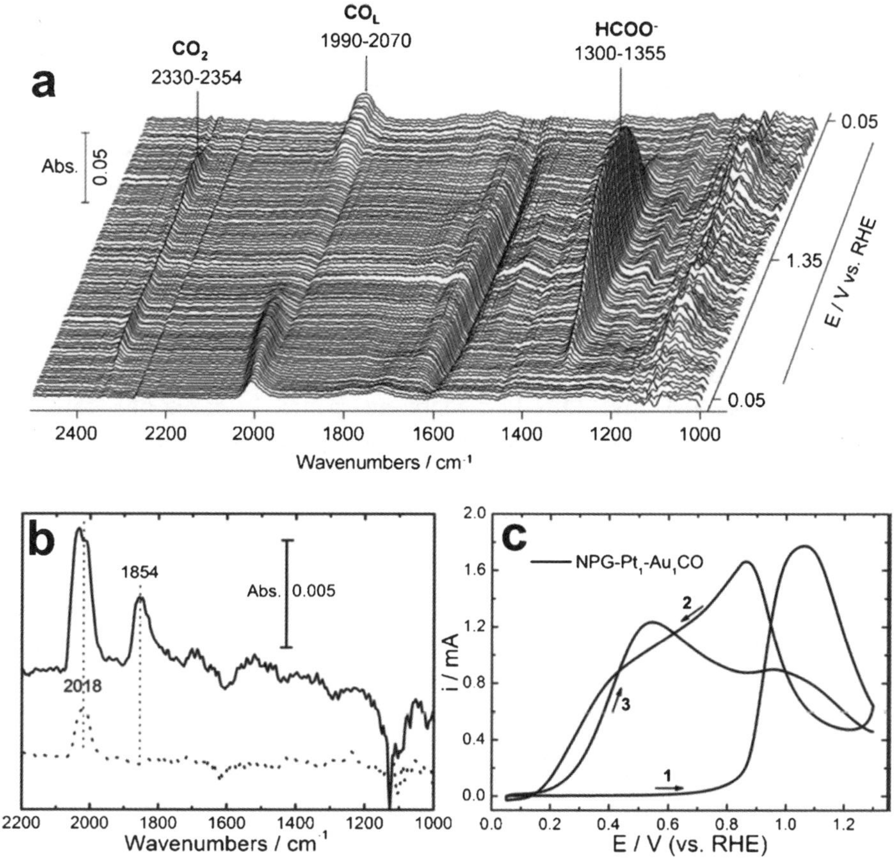

Figure 4.20 (a) Series of SEIRA spectra for the Si-Au-Pt_1-Au_2 electrode (made *via* a similar fabrication procedure with the NPG-Pt_1-Au_2 catalyst) acquired simultaneously with the CV cycling between 0.05 and 1.35 V. (b) Comparison of SEIRA spectra for a Si-Au-Pt_1-Au_2 electrode in 0.1 M $HClO_4$ + 0.05 M HCOOH (*dotted line*) and pre-dosed with monolayer CO (*solid line*) in 0.1 M $HClO_4$. (c) Electrochemical response of the NPG-Pt_1-Au_1 electrode (pre-adsorbed with monolayer CO) in 0.1 M $HClO_4$ + 0.05 M HCOOH. (Reproduced with permission of Wiley from Wang *et al.*[71]).

structural perfectness of better than 80% on this high surface area electrode. These results demonstrate that after Au deposition the monolayer Pt surface has been separated into small ensembles, most of which are small enough to suppress the formation of the CO poisoning intermediate but large and active enough to facilitate the direct reaction path. To provide a clear picture for the reaction path changing, the NPG-Pt_1-Au_1 electrode was firstly covered by monolayer CO and then tested for formic acid electro-oxidation. As shown in Figure 4.20(c), there is no anodic current below 0.8 V during the first CV segment because the surface was totally poisoned by CO. After CO oxidation, there is a tremendous anodic current onset around 0.2 V during the third CV segment, which can be ascribed to formic acid electro-oxidation at a gold-covered Pt surface *via* the direct pathway.

The stability of this kind of catalyst was demonstrated by continuous potential cycling. After the same CV excursion, the NPG-Pt_1-Au_2 sample showed even better stability than the commercial Pt/C catalyst. The enhanced stability was attributed to the protection effect of out-layer Au clusters, which could suppress the oxidation of Pt.[72] The implementation of these new catalysts into real fuel cells is currently underway, and preliminary results show orders of magnitude improvement in Pt efficiency in both single cells and stacks, with remarkable catalyst durability. The successful fabrication of high-performance catalysts for DFAFCs represents a nice example of a rational design strategy to make new generation advanced materials that allows realistic functional integration at the molecular level.

4.5 Summary

This chapter aimed to introduce NPG leaf as a new type of electrocatalyst. The formation and structural properties of NPG leaf were described in view of their possible applications in electrocatalysis and fuel cells. The structural evolution and surface modification were discussed in great detail to show their versatility in wide range of applications with tailored surface properties. We have provided one particular example to demonstrate the feasibility of the design and fabrication of highly effective electro-catalysts on the basis of NPG leaf material, which could exhibit intriguing properties and performance absolutely not possible for traditional materials.

Featured by a flexible porous membrane structure, NPG leaf can be used as a high-performance electrode or electrode substrate in many advanced energy technologies. For example, metal oxide decorated NPG leaf was demonstrated to be highly efficient supercapacitor electrodes.[73] The excellent interfacial structure allows the performance of thus-made nanocomposites to approach the theoretical value. Conductive polymers, such as polypyrrole, were also deposited on the surface of NPG leaf as a very thin overlayer and the composite materials could be used to fabricate flexible, all-solid-state supercapacitors with high power and energy densities.[74] As another example, tin nanocrystalline decorated NPG leaf could be used as an alternative anode for rechargeable lithium ion batteries, with much improved electrochemical performance.[75] These recent examples not only expand the application of NPG leaf but also open new opportunities for important energy storage and conversion technologies. It can be expected that with deeper understanding of the fundamental mechanisms of electrode reactions and more advanced nanofabrication tools and techniques, NPG leaf and related nanoporous metal nanocomposites will find great opportunities in the new energy era.

Acknowledgements

This work was sponsored by the National Basic Research Program of China (2012CB932800) and the National Science Foundation of China (90923011, 51171092). Y.D. is a Tai-Shan Scholar supported by the Independent Innovation Foundation of Shandong University (IIFSDU), the Research Fund for the Doctoral Program of Higher Education of China (20090131110019), and the Shandong Natural Science Fund for Distinguished Young Scholars.

References

1. R. F. Service, *Science*, 2002, **296**, 1222.
2. X. W. Yu and P. G. Pickup, *J. Power Sources*, 2008, **182**, 124.
3. E. Reddington, A. Sapienza, B. Gurau, R. Viswanathan, S. Sarangapani, E. S. Smotkin and T. E. Mallouk, *Science*, 1998, **280**, 1735.
4. N. M. Markovic and P. N. Ross, *Surf. Sci. Rep.*, 2002, **45**, 121–229.
5. G. Samjeske, A. Miki, S. Ye and M. Osawa, *J. Phys. Chem. B*, 2006, **110**, 16559.
6. L. Zhuang, J. Jin and H. D. Abruna, *J. Am. Chem. Soc.*, 2007, **129**, 11033.
7. J. Clavilier, A. Fernandez-Vega, J. M. Feliu and A. Aldaz, *J. Electroanal. Chem.*, 1989, **258**, 89.
8. A. Lopez-Cudero, F. J. Vidal-Iglesias, J. Solla-Gullon, E. Herrero, A. Aldaz and J. M. Feliu, *Phys. Chem. Chem. Phys.*, 2009, **11**, 416.
9. E. Leiva, T. Iwasita, E. Herrero and J. M. Feliu, *Langmuir*, 1997, **13**, 6287.
10. Y. N. Xia, Y. J. Xiong, B. Lim and S. E. Skrabalak, *Angew. Chem. Int. Ed.*, 2009, **48**, 60.
11. A. R. Tao, S. Habas and P. D. Yang, *Small*, 2008, **4**, 310.
12. K. Sasaki, H. Naohara, Y. Cai, Y. M. Choi, P. Liu, M. B. Vukmirovic, J. X. Wang and R. R. Adzic, *Angew. Chem. Int. Ed.*, 2010, **49**, 8602.
13. M. H. Shao, K. Shoemaker, A. Peles, K. Kaneko and L. Protsailo, *J. Am. Chem. Soc.*, 2010, **132**, 9253.
14. C. Wang, H. Daimon, T. Onodera, T. Koda and S. H. Sun, *Angew. Chem. Int. Ed.*, 2008, **47**, 3588.
15. R. Borup, J. Meyers, B. Pivovar, Y. S. Kim, R. Mukundan, N. Garland, D. Myers, M. Wilson, F. Garzon, D. Wood, P. Zelenay, K. More, K. Stroh, T. Zawodzinski, J. Boncella, J. E. McGrath, M. Inaba, K. Miyatake, M. Hori, K. Ota, Z. Ogumi, S. Miyata, A. Nishikata, Z. Siroma, Y. Uchimoto, K. Yasuda, K. I. Kimijima and N. Iwashita, *Chem. Rev.*, 2007, **107**, 3904.
16. S. H. Joo, S. J. Choi, I. Oh, J. Kwak, Z. Liu, O. Terasaki and R. Ryoo, *Nature*, 2001, **412**, 169.
17. C. Wang, M. Waje, X. Wang, J. M. Tang, R. C. Haddon and Y. S. Yan, *Nano Lett.*, 2004, **4**, 345.
18. K. Vinodgopal, M. Haria, D. Meisel and P. Kamat, *Nano Lett.*, 2004, **4**, 415.

19. W. M. Zhang, P. Sherrell, A. I. Minett, J. M. Razal and J. Chen, *Energy Environ. Sci.*, 2010, **3**, 1286.
20. Y. Y. Shao, J. Liu, Y. Wang and Y. H. Lin, *J. Mater. Chem.*, 2009, **19**, 46.
21. A. J. Forty, *Nature*, 1979, **282**, 597.
22. R. C. Newman and K. Sieradzki, *MRS Bull.*, 1999, **24**, 12.
23. H. W. Pickering, *Corros. Sci.*, 1983, **23**, 1107.
24. D. E. Williams, R. C. Newman, Q. Song and R. G. Kelly, *Nature*, 1991, **350**, 216.
25. R. C. Newman and K. Sieradzki, *Science*, 1994, **263**, 1708.
26. M. Raney, *U.S. Pat.*, 1 628 190, 1927.
27. H. W. Pickering and P. R. Swann, *Corrosion*, 1963, **19**, 373.
28. P. R. Swann, *Corrosion*, 1969, **25**, 147.
29. K. Sieradzki and R. C. Newman, *U.S. Pat.*, 4 977 038, 1990.
30. I. C. Oppenheim, D. J. Trevor, C. E. D. Chidsey, P. L. Trevor and K. Sieradzki, *Science*, 1991, **254**, 687.
31. S. Corcoran, in *Critical Factors in Localized Corrosion III*, ed. R. G. Kelly, G. S. Frankel, P. M. Natishan and R. C. Newman, Electrochemical Society, Pennington, NJ, 2000, pp. 500–507.
32. H. W. Pickering and C. Wagner, *J. Electrochem. Soc.*, 1967, **114**, 698.
33. K. Sieradzki, R. R. Corderman, K. Shukla and R. C. Newman, *Philos. Mag. A*, 1989, **59**, 713.
34. K. Sieradzki, *J. Electrochem. Soc.*, 1993, **140**, 2868.
35. J. Erlebacher, M. J. Aziz, A. Karma, N. Dimitrov and K. Sieradzki, *Nature*, 2001, **410**, 450.
36. J. Erlebacher and R. Seshadri, *MRS Bull.*, 2009, **34**, 561.
37. J. Biener, A. Wittstock, L. A. Zepeda-Ruiz, M. M. Biener, V. Zielasek, D. Kramer, R. N. Viswanath, J. Weissmuller, M. Baumer and A. V. Hamza, *Nat. Mater.*, 2009, **8**, 47.
38. Z. Liu and P. C. Searson, *J. Phys. Chem. B*, 2006, **110**, 4318.
39. C. X. Xu, J. X. Su, X. H. Xu, P. P. Liu, H. J. Zhao, F. Tian and Y. Ding, *J. Am. Chem. Soc.*, 2007, **129**, 42.
40. V. Zielasek, B. Jurgens, C. Schulz, J. Biener, M. M. Biener, A. V. Hamza and M. Baumer, *Angew. Chem. Int. Ed.*, 2006, **45**, 8241.
41. Y. Ding, Y. J. Kim and J. Erlebacher, *Adv. Mater.*, 2004, **16**, 1897.
42. T. Fujita, H. Okada, K. Koyama, K. Watanabe, S. Maekawa and M. W. Chen, *Phys. Rev. Lett.*, 2008, **101**, 166601.
43. R. Li and K. Sieradzki, *Phys. Rev. Lett.*, 1992, **68**, 1168.
44. L. Zhang, L. Y. Chen, H. W. Liu, Y. Hou, A. Hirata, T. Fujita and M. W. Chen, *J. Phys. Chem. C*, 2011, **115**, 19583.
45. J. T. Zhang, P. P. Liu, H. Y. Ma and Y. Ding, *J. Phys. Chem. C*, 2007, **111**, 10382.
46. R. Zeis, T. Lei, K. Sieradzki, J. Snyder and J. Erlebacher, *J. Catal.*, 2008, **253**, 132.
47. L. C. Nagle and J. F. Rohan, *Int. J. Hydrogen Energy*, 2011, **36**, 10319.

48. F. H. Meng, X. L. Yan, J. G. Liu, J. Gu and Z. G. Zou, *Electrochim. Acta*, 2011, **56**, 4657.
49. X. B. Ge, L. Q. Wang, Z. N. Liu and Y. Ding, *Electroanalysis*, 2011, **23**, 381.
50. Z. N. Liu, L. H. Huang, L. L. Zhang, H. Y. Ma and Y. Ding, *Electrochim. Acta*, 2009, **54**, 7286.
51. B. Seo and J. Kim, *Electroanalysis*, 2010, **22**, 939.
52. Z. N. Liu, J. G. Du, C. C. Qiu, L. H. Huang, H. Y. Ma, D. Z. Shen and Y. Ding, *Electrochem. Commun.*, 2009, **11**, 1365.
53. Y. Ding and J. Erlebacher, *J. Am. Chem. Soc.*, 2003, **125**, 7772.
54. Y. Ding, M. W. Chen and J. Erlebacher, *J. Am. Chem. Soc.*, 2004, **126**, 6876.
55. P. P. Liu, X. B. Ge, R. Y. Wang, H. Y. Ma and Y. Ding, *Langmuir*, 2009, **25**, 561.
56. S. R. Brankovic, J. X. Wang and R. R. Adzic, *Surf. Sci.*, 2001, **474**, L173.
57. E. Herrero, L. J. Buller and H. D. Abruna, *Chem. Rev.*, 2001, **101**, 1897.
58. X. B. Ge, R. Y. Wang, P. P. Liu and Y. Ding, *Chem. Mater.*, 2007, **19**, 5827.
59. X. B. Ge, X. L. Yan, R. Y. Wang, F. Tian and Y. Ding, *J. Phys. Chem. C*, 2009, **113**, 7379.
60. M. O. Pedersen, S. Helveg, A. Ruban, I. Stensgaard, E. Laegsgaard, J. K. Norskov and F. Besenbacher, *Surf. Sci.*, 1999, **426**, 395.
61. B. C. Du and Y. Y. Tong, *J. Phys. Chem. B*, 2005, **109**, 17775.
62. T. Frelink, W. Visscher and J. A. R. van Veen, *J. Electroanal. Chem.*, 1995, **382**, 65.
63. A. Kowal, M. Li, M. Shao, K. Sasaki, M. B. Vukmirovic, J. Zhang, N. S. Marinkovic, P. Liu, A. I. Frenkel and R. R. Adzic, *Nat. Mater.*, 2009, **8**, 325.
64. X. B. Ge, R. Y. Wang, S. Z. Cui, F. Tian, L. Q. Xu and Y. Ding, *Electrochem. Commun.*, 2008, **10**, 1494.
65. R. Zeis, A. Mathur, G. Fritz, J. Lee and J. Erlebacher, *J. Power Sources*, 2007, **165**, 65.
66. J. Snyder, P. Asanithi, A. B. Dalton and J. Erlebacher, *Adv. Mater.*, 2008, **20**, 4883.
67. A. Capon and R. Parsons, *J. Electroanal. Chem.* 1973, **45**, 205.
68. Y. X. Chen, M. Heinen, Z. Jusys and R. J. Behm, *Angew. Chem. Int. Ed.*, 2006, **45**, 981.
69. M. Osawa, K. Komatsu, G. Samjeske, T. Uchida, T. Ikeshoji, A. Cuesta and C. Gutierrez, *Angew. Chem. Int. Ed.*, 2011, **50**, 1159.
70. M. Neurock, M. Janik and A. Wieckowski, *Faraday Discuss.*, 2008, **140**, 363.
71. R. Y. Wang, C. Wang, W. B. Cai and Y. Ding, *Adv. Mater.*, 2010, **22**, 1845.
72. J. Zhang, K. Sasaki, E. Sutter and R. R. Adzic, *Science*, 2007, **315**, 220.
73. X. Y. Lang, A. Hirata, T. Fujita and M. W. Chen, *Nat. Nanotech.*, 2011, **6**, 232.
74. F. H. Meng and Y. Ding, *Adv. Mater.*, 2011, **23**, 4098.
75. Y. Yu, L. Gu, X. Y. Lang, C. B. Zhu, T. Fujita, M. W. Chen and J. Maier, *Adv. Mater.*, 2011, **23**, 2443.

CHAPTER 5

Nanocatalysts for Direct Borohydride Oxidation in Alkaline Media

CHRISTOPHE COUTANCEAU*, STÈVE BARANTON AND MÁRIO SIMÕES

Université de Poitiers, IC2MP, UMR 7285 CNRS, 4 rue Michel Brunet, F-86022 Poitiers Cedex, France
*E-mail: christophe.coutanceau@univ-poitiers.fr

5.1 Introduction

Amongst fuel cells working at low temperature, alkaline fuel cells present great interest as an alternative to proton exchange membrane fuel cells (PEMFCs) for energy generation. The number of studies focusing on solid alkaline membrane fuel cells (SAMFCs) has indeed increased for the last few years. Recent progress in electrical characteristics and chemical stability of anion conducting membranes allowed for several advances in the development of such technology, considering different fuels like hydrogen,[1] alcohols[2,3] and borohydrides.[4] Moreover, alkaline membranes are now commercially available.[5–9]

Working in alkaline media leads to higher reaction kinetics than in acid electrolytes, so that catalytic materials other than platinum can be used, while keeping high activity and stability.[10–12] Alkaline media not only allow using less noble metal based catalysts, they also broaden the choice of molecules to be considered as fuels. Therefore, recent studies presented the feasibility of alkaline fuel cells fuelled by hydrogen, alcohols and borohydrides. Although these

RSC Energy and Environment Series No. 6
Catalysts for Alcohol-Fuelled Direct Oxidation Fuel Cells
Edited by Zhen-Xing Liang and Tim S. Zhao

Published by the Royal Society of Chemistry, www.rsc.org

researches have overcome several limitations of alkaline fuel cell systems, like stability of the materials, durability of the cells, cathode carbonation and global efficiency, more studies are still needed to understand the oxidation mechanisms of several of those molecules in alkaline media on different catalysts.

Alkali metal borohydrides, like $LiBH_4$, $NaBH_4$ and KBH_4, have been considered for fuel cell applications. Amongst these possible compounds as fuels, sodium borohydride possesses the best compromise between the specific energy density and the relative abundance of the elements that compose it. As stated by Demirci and Miele,[13] sodium borohydride has been used in chemistry for different purposes since it was discovered in 1940s. For example, it has been used for decades as a powerful reducing agent in organic chemistry, so that the idea of its application as a fuel to be oxidized at the anode of a fuel cell could be considered. A new research field was opened in the end of the 1990s, when Amendola *et al.*[14] presented the concept of an alkaline fuel cell fuelled by sodium borohydride.

A few years after the discovery by Professor H. C. Brown of sodium borohydride, Schlesinger *et al.*[15] proposed several ways to synthesize this compound. They first developed a rapid method to produce $NaBH_4$ of 90–96% purity with a 94% yield, by reacting sodium hydride with methyl borate at a temperature around 225–275 °C. The reaction was proposed to occur according to eqn (5.1):

$$4NaH + B(OCH_3)_3 \rightarrow NaBH_4 + 3NaOCH_3 \tag{5.1}$$

This group also performed the synthesis of $NaBH_4$ with a 60% yield at 330–350 °C using the reaction between sodium hydride and boron oxide. Several years later, Schubert *et al.*[16] prepared $NaBH_4$ by performing the reaction between dehydrated borax, quartz and sodium under a hydrogen atmosphere at high temperatures (450–500 °C). The same group also succeeded in synthesizing $NaBH_4$ from calcium hydride and sodium borate at high temperatures. The stoichiometry of the reactions based on sodium hydride shows that one molecule of $NaBH_4$ is produced when four molecules of NaH are consumed. However, all these reactions are highly energy consuming. Such methods for the synthesis of borohydride are thus rather expensive and difficult to transfer for industrial development.

More recently, Li *et al.*[17] proposed the synthesis of $NaBH_4$ at ambient temperature with a 43% yield, by carrying out the reaction between MgH_2 and $Na_2B_4O_7$ by ball milling. The reaction was proposed to take place according to eqn (5.2):

$$4MgH_2 + Na_2B_4O_7 \rightarrow 2NaBH_4 + 4MgO + B_2O_3 \tag{5.2}$$

The authors also evaluated the influence of adding a sodium-based compound to balance the sodium ratio between the reactants and reaction products in order to increase the reaction yield. The addition of sodium hydroxide to the reaction mixture allowed increasing the yield of the reaction to give $NaBH_4$ by up to 64%, whereas the addition of carbonate led to a 78%

yield. These results showed that sodium borohydride could be produced using cheaper compounds and at a low cost.

However, the complete oxidation of borohydride in a direct borohydride fuel cell (DBFC) is expected to produce sodium borate ($NaBO_2$). Some research groups then focused on the development of sodium borate recycling processes for $NaBH_4$. The challenges are to develop adapted processes for the low-cost production of $NaBH_4$ and the efficient recycling of sodium borate. In 2003, Kojima and Haga claimed 97–98% conversion yields of borate into borohydride by using MgH_2 or Mg_2Si compounds as reducing agent.[18] Kong *et al.*[19] recently obtained 71% conversion yield by performing the reaction between MgH_2 and $NaBO_2$ by ball milling under an inert atmosphere at 200 kPa. The chemical reaction was proposed to occur according to eqn (5.3):

$$2MgH_2 + NaBO_2 \rightarrow NaBH_4 + 2MgO \tag{5.3}$$

Li *et al.*[20] also obtained conversion yields of borate into borohydride higher than 70% by catalytically reducing sodium borate with hydrogen over magnesium particles at 400 °C.

All these results and developments are very encouraging and are also good indicators of the possibility of developing a viable market and economy based on borohydride conversion in an alkaline fuel cell. An energy cycle could then be developed considering that, in a first phase, sodium borohydride can be produced in large quantities at low cost, and assuming that, in a second phase, sodium borate could be efficiently recycled into $NaBH_4$. A sustainable energy cycle could then be implemented.

However, as it was stated earlier, more studies are still needed for the better understanding of the oxidation reaction mechanisms of sodium borohydride in alkaline media. It is also important to develop nanostructured catalysts in order to improve the cell efficiency in terms of power density (activity) and energy density (reaction pathway) and to make the technology economically viable. This chapter is devoted to the nanocatalysts used for the electro-oxidation of sodium borohydride dissolved in an aqueous phase and will first present the problems of the borohydride oxidation reaction (BOR) in alkaline media and the main results in the literature concerning nanostructured catalysts for borohydride oxidation on a DBFC anode. Then, results of the reaction kinetics on different catalysts, monometallic and multimetallic, involving different reaction mechanisms, namely the direct oxidation reaction of borohydride into borate and the indirect oxidation reaction involving the borohydride hydrolysis reaction into hydrogen and its oxidation, will be discussed.

5.2 Thermodynamics and Mechanism of Sodium Borohydride Oxidation

The oxidation of the borohydride ion can occur in alkaline media following two main reaction pathways. An electrochemical reaction involving eight electrons, usually named the "direct reaction pathway" (eqn 5.4):

$$BH_4^- + 8OH^- \rightarrow BO_2^- + 6H_2O + 8e^- \quad E^\circ_{BO_2^-/BH_4^-} = -1.24 \text{ V } vs. \text{ SHE} \tag{5.4}$$

and a complex chemical/electrochemical reaction pathway, involving the borohydride hydrolysis reaction and a subsequent hydrogen oxidation reaction, the so-called "indirect reaction pathway" (eqns 5.5 and 5.6):

$$BH_4^- + 2H_2O \rightarrow BO_2^- + 4H_2 \tag{5.5}$$

$$4H_2 + 8OH^- \rightarrow 8H_2O + 8e^- \quad E^\circ_{H_2O/H_2} = -0.83 \text{ V } vs. \text{ SHE} \tag{5.6}$$

In a DBFC the anodic reaction of borohydride oxidation is completed by the cathodic reaction of the oxygen reaction according to eqn (5.7):

$$O_2 + 2H_2O + 4e^- \rightarrow 4OH^- \quad E^\circ_{O_2/H_2O} = 0.4 \text{ V } vs. \text{ SHE} \tag{5.7}$$

In the case of the anodic indirect oxidation reaction pathway, the same faradic efficiency as in the case of the direct reaction pathway can be obtained by oxidizing the whole hydrogen molecule produced by the hydrolysis reaction. Even considering such a case, the theoretical efficiency of the energy conversion remains 25% higher when $NaBH_4$ is electrochemically oxidized following the reaction represented in eqn (5.4). Indeed, the standard equilibrium potential associated with the direct combustion reaction of $NaBH_4$ into $NaBO_2^-$ is determined from the standard free energy at the equilibrium in the cell, ΔG°_{eq}, According to eqn (5.8):

$$\Delta G^\circ_{eq} = -nFE^\circ_{eq} \tag{5.8}$$

where n is the number of exchanged electrons per oxidized BH_4^- molecule and F is the Faraday constant (96 485 C mol^{-1}).

Figure 5.1 shows the working principle of a DBFC for which the direct reaction pathway is considered. Considering eight exchanged electrons for the direct reaction pathway, $\Delta G^\circ_{eq} = -1266$ kJ mol^{-1}, against $\Delta G^\circ_{eq} = -949.5$ kJ mol^{-1} considering eight exchanged electrons for the indirect reaction pathway (consumption of the whole hydrogen formed by the hydrolysis reaction is assumed). From these ΔG°_{eq} values, the theoretical energy density W_e (expressed in kWh kg^{-1}) can be calculated from eqn (5.9):

$$W_e = \frac{\Delta G^\circ_{eq}}{3600 \times M} \tag{5.9}$$

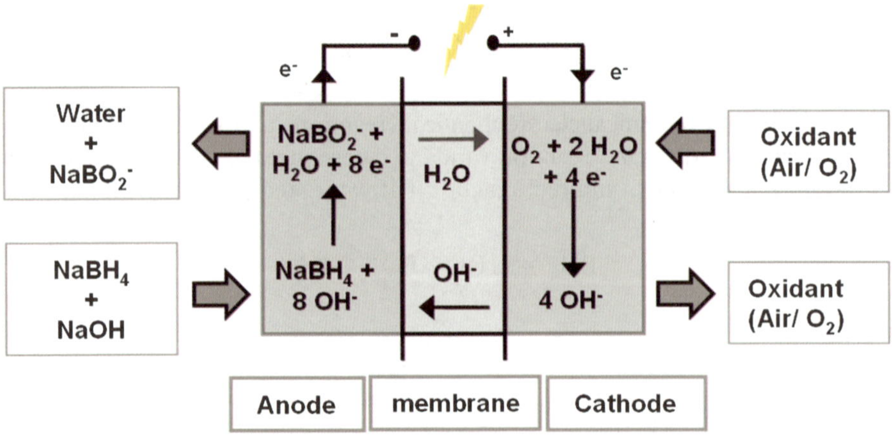

Figure 5.1 Schematic principle of a direct borohydride/oxygen fuel cell.

where *M* is the molecular weight of sodium borohydride in kg mol^{-1} and 3600 is expressed in seconds.

A theoretical energy density for sodium borohydride of *ca.* 9.3 kWh kg^{-1} is calculated for the direct reaction pathway and 7.0 kWh kg^{-1} is determined for the indirect reaction pathway involving the total oxidation of the produced hydrogen. However, because sodium borohydride contains 10.7 wt% hydrogen, some groups focused on the study of the indirect reaction pathway *via* $NaBH_4$ hydrolysis.[21–24] In this case, the BH_4^- ion is considered as a hydrogen carrier and, as a result, fuel cells can operate in the gas phase by recovering pure hydrogen from the $NaBH_4$ hydrolyzer. However, the theoretical energetic efficiency of such systems is lower than that of the direct oxidation of $NaBH_4$.

It was recently confirmed by several research groups[25–27] that the reaction pathways described in eqns (5.4)–(5.6) are the main ones, but the oxidation mechanism is actually more complex, involving several intermediates that can react differently, depending on the catalytic surface and on the electrode potential. Catalysis by the direct oxidation pathway is difficult, the borohydride oxidation being a complex anodic reaction that can be described by eqns (5.10)–(5.13):

(1) Reaction pathway involving the non-faradic hydrolysis reaction:

$$BH_4^- + xH_2O \rightarrow BH_{4-x}(OH)_x^- + xH_2 \qquad \text{with } x \leq 4 \tag{5.10}$$

followed either by the electrochemical reaction (if $x < 4$):

$$BH_{4-x}(OH)_x^- + (4-x)OH^- \rightarrow BO_2^- + 2H_2O + \frac{(4-x)}{2}H_2 + (4-x)e^- \tag{5.11}$$

or by the chemical reaction for $x = 4$:

$$B(OH)_4^- \rightarrow BO_2^- + 2H_2O \tag{5.11}$$

(2) Reaction pathway involving the faradic borohydride hydrolysis and oxidation reactions, *i.e.* either:

$$BH_4^- + xOH^- \rightarrow BH_{4-x}(OH)_x^- + \frac{x}{2}H_2 + xe^- \quad \text{with } x \leq 4 \tag{5.12}$$

followed by either reaction in eqns (5.11) or (5.11′), or the electrochemical oxidation reaction:

$$BH_4^- + xOH^- \rightarrow BO_2^- + (x-2)H_2O + (4-\frac{x}{2})H_2 + xe^- \quad \text{if } 5 \leq x \leq 8 \tag{5.13}$$

Different reaction pathways for the oxidation of borohydride are then possible. The number of exchanged electrons per oxidized molecule lies from four to eight, depending on the electrocatalyst nature and structure, on the reaction medium pH, on the electrode morphology and potential. Moreover, the chemical hydrolysis reaction of borohydride leads to the production of hydrogen, without any exchange of electrons, which can further be oxidized or not. The theoretical efficiency of the energy conversion will then be decreased.

Much progress has been made on the comprehension of reaction mechanisms on different catalytic metals in recent years, although these mechanisms are not yet totally elucidated. Numerous studies have been performed on several materials, namely platinum,[26,28–30] palladium,[25,31] gold,[25,26,32,33] nickel,[34,35] silver,[32,36] osmium,[37] ruthenium[38] or cobalt,[39] for example, to verify their activity towards the BOR and also to understand the reaction mechanism involved.

Several studies devoted to the catalytic activity of AB_5[40–42] or AB_2[43] compounds allowing hydrogen storage are also available in the literature.[44] Such structures have shown high activities for the BOR. Nevertheless, even if such structures allow hydrogen storage, the H_2 evolution kinetics can be higher than the H_2 oxidation reaction kinetics, leading to hydrogen losses decreasing the utilization rate of the fuel. However, several authors[40,43,45–54] have performed borohydride oxidation in an alkaline fuel cell fitted with AB_2 or AB_5 catalysts at the anode. Table 5.1 gives a non-exhaustive overview of the obtained DBFC performances.

Gold, which is often considered as a model catalytic surface for the direct oxidation of borohydride, has been extensively studied.[26,27,32,33,55–58] Initially, most of the basic researches on the determination of the reaction mechanism on massive gold stated that BH_4^- oxidation occurred *via* the direct reaction pathway with *ca.* eight electrons exchanged. Recently, however, different groups showed that BH_4^- hydrolysis could occur on gold, even at low potentials.[32,55,57] Finkelstein *et al.*[26] analyzed the BOR on a gold rotating disk electrode and on a ring disk rotating electrode. These authors proposed that the first oxidation step of the BH_4^- ion involved the presence of the BH_3OH^-

Table 5.1 Non-exhaustive review of performances of DBFCs fitted with anodic AB_2 and AB_5 catalysts under different operating conditions.

Anode catalyst[a]	Electrolyte	Cathode catalyst[b]	Temp./°C	Power density/mW cm^{-2}	Ref.
$Zr_{0.9}Ti_{0.1}Mn_{0.6}V_{0.2}Co_{0.1}Ni_{1.1}$	N117	Pt/C	50	100	43
$Zr_{0.9}Ti_{0.1}Mn_{0.6}V_{0.2}Co_{0.1}Ni_{1.1}$	N117	Pt/C	70	140	43
$Zr_{0.9}Ti_{0.1}Mn_{0.6}V_{0.2}Co_{0.1}Ni_{1.1}$	N117	Pt/C	85	190	43
$M_mNi_{3.55}Al_{0.3}Mn_{0.4}Co_{0.71}$	N117	Pt/C	40	146	45
$M_mNi_{3.55}Al_{0.3}Mn_{0.4}Co_{0.71}$	N117	Pt/C	70	352	45
$M_mNi_{3.55}Al_{0.3}Mn_{0.4}Co_{0.75}$	N117	Pt/C	30	35	46
$M_mNi_{3.55}Al_{0.3}Mn_{0.4}Co_{0.75}$	N117	Pt/C	70	150	46
$M_mNi_{4.5}Al_{0.5}$	N117	Pt/C	70	130	46
$M_mNi_{3.2}Al_{0.2}Mn_{0.6}B_{0.03}Co_{1.0}$	N117	Pt/C	70	125	46
$M_mNi_{3.2}Al_{0.2}Mn_{0.6}Co_{1.0}$	N117	Pt/C	70	100	46
$Zr_{0.9}Ti_{0.1}Mn_{0.6}V_{0.2}Cr_{0.5}Co_{0.75}Ni_{1.2}$	N117	Pt/C	70	70	46
$Zr_{0.9}Ti_{0.1}Mn_{0.6}V_{0.2}Co_{0.1}Ni_{1.1}$ + Pd/C (10%)	NRE211	Pt/C	60	290	47
$M_mNi_{3.55}Al_{0.3}Mn_{0.4}Co_{0.71}$	N961	Au-coated stainless-steel gauze	25	50	48
$M_mNi_{3.35}Al_{0.3}Mn_{0.4}Co_{0.75}$	6 M NaOH	MnO_2/C	25	70	49
$M_mNi_{3.55}Co_{0.75}Mn_{0.4}Al_{0.3}$	6 M NaOH	FePc/C	$T_{amb.}$	92	40
$M_mNi_{3.55}Co_{0.75}Mn_{0.4}Al_{0.3}$	6 M NaOH	CoPc/C	$T_{amb.}$	90	50
$M_mNi_{3.55}Al_{0.3}Mn_{0.4}Co_{0.75}$	N117	Prussian blue	$T_{amb.}$	70	51
Zr-based AB_2-type	N115	Pt/C	76	218	52
Corrugated anode	CEM	FeTMPP/C	–	200	53
$M_mNi_{3.55}Al_{0.3}Mn_{0.4}Co_{0.75}$	N117	$PbSO_4$/C	30	18	54
$M_mNi_{3.55}Al_{0.3}Mn_{0.4}Co_{0.75}$	N117	$PbSO_4$/C	70	82	54
$M_mNi_{3.55}Al_{0.3}Mn_{0.4}Co_{0.75}$	N117	$PbSO_4$/C	30	32	54
$M_mNi_{3.55}Al_{0.3}Mn_{0.4}Co_{0.75}$	N117	$PbSO_4$/C	70	120	54

[a]M_m = Ce-rich mixed mischmetal composed of 50 wt%Ce, 30 wt%La, 5 wt%Pr and 15 wt%Nd. [b]Pc = phthalocyanine; TMPP = tetramethoxyphenylporphyrin.

intermediate on such an electrode, the formation of which was accompanied by the production of H_2. In such a case, only four to five electrons are exchanged per BH_4^- ion. On the basis of differential electrochemical mass spectrometry measurements, Chatenet *et al.*[59] showed the formation of hydrogen in the course of the electro-oxidation of borohydride on gold. The extent of this reaction on gold is sensitive to the $[OH^-/BH_4^-]$ ratio. Molina Concha *et al.*[58] presented the first results obtained by *in situ* IR measurement performed during the oxidation of borohydride on a gold electrode. The results obtained by the different research groups studying the BOR on gold seem in agreement concerning the first reaction step of the oxidation. For a high $[OH^-/BH_4^-]$ ratio, the electro-oxidation of borohydride involves the formation at low potentials of BH_3OH^- as a reaction intermediate, accompanied with hydrogen evolution. The reaction is oriented towards the direct reaction pathway at high potentials with *ca.* eight electrons exchanged. Measurements carried out in a fuel cell fitted with an anode based on gold nanoparticles dispersed on a carbon support led to poor performances in terms of achieved power densities,[33,60] owing to the high borohydride oxidation overpotential on such materials.

Platinum is known to be an active material for the hydrogen evolution reaction from borohydride hydrolysis and for the hydrogen oxidation reaction. This metal is thus expected to favour the indirect reaction pathway. Although there are numerous studies carried out on platinum-based catalysts,[26,28–30] the understanding of the BOR mechanism on this material remains incomplete. Discrepancies in the number of electrons exchanged and on the reaction mechanism arise in the literature. The analysis of the results issued from electrochemical characterization methods of the BOR led to different values of the number of electrons exchanged per oxidized BH_4^- ion: Finkelstein *et al.*[26] found *ca.* seven electrons, Molina-Concha *et al.*[61,62] proposed two to eight electrons, depending on the catalytic layer thickness, on the $NaBH_4$ concentration and on the electrode structure (nanostructured catalyst or massive electrode), and Gyenge[56] claimed that four electrons were exchanged for the oxidation of one borohydride molecule on platinum. Although the reaction mechanism on platinum is very complex, the ability of platinum to catalyze the hydrogen evolution reaction is recognized by all groups. The influence of thiourea for the BOR on platinum electrodes was also studied.[30,56] It was concluded that the hydrogen evolution reaction was limited in the presence of thiourea. Nonetheless, the onset potential of the $NaBH_4$ oxidation wave is shifted by 0.4 V towards higher potentials. Other investigations have recently been performed on the BOR in order to better understand the reaction mechanism. Rostamikia and Janik[29] used density functional theory (DFT) to study the competition between the direct and indirect oxidation reactions on a Pt(111) oriented surface. They concluded that the BH_4^- ion adsorbs dissociatively on such a surface, leading to a high surface coverage by H^+ ions, which could facilitate the hydrogen evolution reaction.

Catalysts based on platinum alloys have also been evaluated for the $NaBH_4$ oxidation reaction. For example, Gyenge *et al.*[28] have studied the electrocatalysis of borohydride oxidation on colloidal Pt and Pt alloys (PtIr, PtNi and PtAu). They found that a PtAu catalyst led to an eight-electron process of borohydride oxidation, whereas hydrogen evolution and oxidation took place on a PtIr catalyst at low potentials. However, PtIr and PtNi were the most active catalysts in a direct borohydride fuel cell fed with a 2 M $NaBH_4$ + 2 M NaOH solution.

Several studies were performed to evaluate the activity of silver as a catalyst for the BOR. Chatenet and co-workers showed that the kinetics of the BOR on massive silver and on nano-dispersed silver on a carbon support was lower than that determined on gold[32] or platinum.[61,62] The Ag/C catalyst involves the exchange of *ca.* four electrons. They proposed that the direct reaction pathway be favoured on this material, silver displaying no activity towards the oxidation of hydrogen formed during the reaction. With a massive silver electrode, between two and six electrons are exchanged, depending on the pH of the electrolyte.

The activity towards the BOR of nanocatalysts based on silver and its alloys with Pt, Ir, Au or Pd has been evaluated by Atwan *et al.*[36] These authors confirmed the low activity of silver towards the BOR. On the other hand, the modification of silver by Ir or Pt led to dramatically increase the catalytic activity of the material.

Lam *et al.*[37] have shown that Os/C catalysts could perform the BOR with the exchange of seven electrons per borohydride molecule. Lam *et al.* and Atwan *et al.*[37,63] analyzed the BOR on osmium catalysts and its alloys such as OsSn, OsMo and OsV. These authors proposed that the direct BOR was favoured on Os-based catalysts with *ca.* seven electrons exchanged. According to these authors, Os-based alloys are not interesting for a SAMFC application because these materials present a higher BOR overpotential than that observed on the pure Os/C catalyst synthesized by a colloidal route. Moreover, these alloys deactivate more rapidly than the Os/C catalyst.

The BOR mechanism on nickel seems to be accepted by the scientific community. Results obtained by Indig and Snyder[34] in 1962 were generally confirmed by recent publications.[35,64] On a nickel surface the BOR occurs *via* the exchange of *ca.* four electrons at a lower potential than that of hydrogen oxidation in alkaline medium. The BOR is accompanied by hydrogen evolution, but nickel does not show any activity for its oxidation.[34]

Concerning the use of palladium as an anodic catalyst for borohydride electro-oxidation, relatively few studies are available in the literature. Most authors refer to the work of Kobukawa *et al.*,[65] published in 1968, where it was proposed that the BOR occurred *via* the exchange of six electrons at high potentials on palladium. More recently, Yang *et al.*[66] studied the activity of carbon-supported Pd nanocatalysts towards the electro-oxidation of borohydride. These authors found that the BOR occurred *via* the exchange of four electrons and was accompanied by hydrogen evolution. They also verified that the hydrogen production rate was dependent on the NaOH and $NaBH_4$ concentrations, particularly on the $[OH^-/BH_4^-]$ ratio. Li *et al.*[67] and Liu *et al.*[68] succeeded to limit the hydrogen evolution rate by optimizing the palladium and Nafion content in the formulation of a nickel-modified catalyst. Palladium is an interesting material for the study of the BOR because, according to the results obtained by Simoes *et al.* at a nanostructure Pd/C catalyst using the rotating disk electrode method,[25] different mechanisms are involved depending on the potential range. At low potentials the indirect mechanism is involved, but at high potentials the direct reaction pathway is favoured; in an intermediate potential range, both mechanisms coexist.

A non-exhaustive review of the electrical performance of DBFC under different operating conditions, classified according to the nature of the anodic catalyst, is presented in Table 5.2. Platinum-based catalysts (Pt/C[28,69–72] or PtX alloys[14,28,33,72–74]) are expensive but active for the $NaBH_4$ hydrolysis reaction. Catalysts based on Au,[33,67,75–78] Ni[79,80] or Pd[68,75,79,81,82] limit the borohydride hydrolysis reaction. These catalysts are very interesting in order to increase the DBFC efficiency, *i.e.* the number of exchanged electrons for the BOR.[14,43] Comparison of the electrical performance obtained in all these experiments is difficult, however, owing to the very different experimental conditions: BH_4^- concentration, electrolytes (Na^+ conductive membranes, OH^- conductive membranes or alkaline solutions), cathode catalysts, oxidant used (air, pure O_2 or H_2O_2), and fuel cell working conditions (humidification, pressure, reactants flow rates, temperature, *etc.*).

From the previous literature overview, it appears that few catalysts are available which can perform the $NaBH_4$ electro-oxidation from highly concentrated solutions at low potentials through the direct reaction pathway, without hydrogen evolution. The present chapter aims at presenting results on the BOR obtained with different nanostructured metallic catalysts. First the activity and mechanism on monometallic Au/C, Pd/C and Pt/C catalysts will be presented; then, the effect of their modification by foreign metals on the selectivity for the BOR will be presented and discussed. The influence of the catalyst nature and structure on the activity is also discussed. The kinetic parameters of the reaction determined for the different catalysts were drawn from the Koutecky–Levich and Butler–Volmer equations. Analysis of the catalyst activities is based on surface structure, onset potential, number of exchanged electrons and hydrogen evolution rate observations.

5.3 Experimental Details

5.3.1 Materials

Ultra-pure water (MilliQ, Millipore, 18.2 MΩ cm) was used for all experiments (synthesis, characterization and electrochemical measurements). For the synthesis of the catalysts, K_2PdCl_4, $H_2PtCl_6{\cdot}6H_2O$, $HAuCl_4{\cdot}3H_2O$, $NiCl_2$ and $BiCl_3$ (99.9% in purity, Alfa) were used. $NaBH_4$ wax (99% purity, Acros

Table 5.2 Non-exhaustive review of the performance of DBFCs fitted with anodic metallic-based catalysts under different operating conditions.

Anode catalyst[a]	Electrolyte	Cathode catalyst	Temp./°C	Power density/ $mW\ cm^{-2}$	Ref.
Pt/C	N117	Pt/C	25	20–40	28
PtAu/C, PtNi/C, PtIr/C	N117	Pt/C	60	40–70	28
Pt black	6 M KOH	Pt black	25	44.2	69
Pt/C	6 M KOH	Pt black	25	42	69
Pt/C	N115	Pt/C	70	120–160	70
Pt/Ni	KOH	MnO_2	25	19	71
Pt/C	ADP-Moragne (Solvay)	Air cathode	70	70	72
PtNi/C	ADP-Moragne (Solvay)	Air cathode	70	120	72
PtRu/C	Membrane?	Pt/C	60	200	73
MWCNT-Pt	AEM	Air cathode	$T_{amb.}$	16	74
F-MWCNT-Pt	AEM	Air cathode	70	44	74
AuPt/C	AEM	MnO_2/C	70	20	74
Au/Pt	AEM	Air cathode	25	20	14
Au/Pt	N117 or 115	Air cathode	70	63	14
Ni	NRE211	Pt/C	25	22	79
Ni	N112	Pt/C	25	25	79
Ni	N112	Pt/C	25	40	79
Ni	N117	Pt/C	25	35	79
Ni	N117	Ag/C	25	33	79
Ni	N112	Pt/C	25	38	80
Ni	N112	Ag/C	25	32	80
Ni	N112	Pt/C	60	180	67
Ni + Pd/C	N112	Pt/C	60	250	67
Ni + Au/C	N112	Pt/C	60	195	67
Ni + Ag/C	N112	Pt/C	60	210	67
Pt/C	N112	Ir black	25	~80	82
Pt/C	N112	Pd black	25	~75	82
Pt/C	N112	Au/C	25	~50	82
Ni/alumina silica	N112	Au/C	25	~12	82
Raney NiAl	N112	Au/C	25	~50	82
Nb	N112	Au/C	25	>10	82
Ir black	N112	Au/C	25	~55	82
Pd black	Cationic membrane	Au/C	25	~50	82
Pd electrodep.	Cationic membrane	Pd	$T_{amb.}$	~180	81
Pd electrodep.	Cationic membrane	Os	$T_{amb.}$	~160	81
Pd electrodep.	Cationic membrane	Au	$T_{amb.}$	~155	81
Pd electrodep.	Cationic membrane	Ir	$T_{amb.}$	~130	81
Pd electrodep.	Cationic membrane	Pd, Rh or Ru	$T_{amb.}$	~110	81

Table 5.2 (*Continued*)

Anode catalyst[a]	Electrolyte	Cathode catalyst	Temp./°C	Power density/ $mW\ cm^{-2}$	Ref.
Pd electrodep.	Cationic membrane	Ag	$T_{amb.}$	~90	81
Pd electrodep.	Cationic membrane	Au electrodep.	26	~270	81
Pd electrodep.	Cationic membrane	Au electrodep.	60	~680	81
Pd electrodep.	N112	Sputtered Au	26	~240	81
NiPd/C	N117	Sputtered Au	60	~600	81
Au	N117	Pt/C	25	80	68
AuPd	N117	Pt/C	60	22	33
$AuPt_{0.75}$	N117	Pt/C	60	33	33
Au/C	N117	Pt/C	60	47	33
Au/C	N117	Pt/C	20	34	75
Au/Ti	N117	Pt/C	85	79	76
Au/Ti	N117	Pt/C	25	27	77
Au/C	N117	Pt/C	85	80	77
Ag/Ti	N117	Pt/C	85	70	77
Au/Ti	N117	Pt/C	85	55	77
Au/C	N117	Pt/C	85	75	78
Au/C	N117	FeTMPP/C	85	60	78
Au/C	N117	Ag/C	85	50	78
Au/C	N117	Ni/C	85	40	78

[a]MWCNT = multi-walled carbon nanotube.

Organics), poly(ethylene glycol dodecyl ether) (BRIJ30 from Fluka) and *n*-heptane (99% purity, Acros Organics) were used as received. Chromasolv absolute ethanol (≤99.8% purity, VWR) and technical acetone (VWR) were used for the cleaning steps. For the electrode preparation, carbon powder (Vulcan XC72, Cabot), was previously treated under N_2 (U-purity, L'air Liquide) at 400 °C for 4 h. A Nafion solution (5 wt%, Aldrich) was added in the ink to serve as a binder. Electrochemical measurements were conducted in NaOH (semiconductor grade 99.99% purity, Sigma-Aldrich) solution.

5.3.2 Synthesis of Catalysts by the "Water-in-Oil" Microemulsion Method

By this method,[25,83,84] catalysts were prepared by mixing $NaBH_4$, as reducing agent, with a microemulsion carrying the specific reactants dissolved in an aqueous phase. BRIJ30 was chosen as surfactant and the organic phase was *n*-heptane. The desired amount of the metal salts was dissolved in ultra-pure water in order to obtain metallic nanoparticles with controlled compositions after reduction with $NaBH_4$. Carbon powder was directly added in the colloidal solution to obtain the desired metal loading and the mixture was kept under stirring for 2 h. In the present work, all the catalysts were synthesized in order to obtain 40 wt% metal loading. The mixture was filtered on a 0.22 μm Durapore membrane filter (Millipore). The resulting powder was abundantly rinsed with ethanol, acetone and ultra-pure water. The carbon-supported catalysts were dried overnight in an oven at 75 °C.

5.3.3 Electrochemical Measurements

Catalytic powders were deposited on a glassy carbon substrate according to the method proposed by Gloaguen *et al.*[85] The catalytic powder (25 mg) was added to a mixture of 0.5 mL Nafion solution in ultra-pure water. After ultrasonic homogenization of the catalyst/XC72-Nafion ink, a given volume (3 μL) was deposited from a syringe onto a freshly polished glassy carbon substrate, yielding a catalytic powder loading of 354 μg cm^{-2} with 21.85 μg of Nafion (6 wt%), leading to a catalytic layer thickness close to 1–1.5 μm.[62,86] The solvent was then evaporated in a stream of pure N_2 at room temperature. The electrochemical setup consisted of a computer-controlled Voltalab PGZ 402 potentiostat, a CTV 101 Radiometer Speed Control Unit and a Radiometer BM-EDI 101 rotating disk electrode (RDE). The solutions were prepared from NaOH, $NaBH_4$ and ultra-pure water. The electrochemical experiments were carried out at 20 °C in an N_2-purged supporting electrolyte, using a thermostated standard three-electrode electrochemical cell. The working electrode was a glassy carbon disk (0.071 cm^2 geometric surface area), the counter electrode was a glassy carbon plate (8 cm^2 geometric surface area) and the reference electrode was a reversible hydrogen electrode (RHE). All potentials are referred to the RHE.

5.3.4 TEM, XRD and XPS Characterization Methods

Catalysts were characterized by transmission electron microscopy (TEM) using a JEOL 2100 UHR microscope (200 kV) equipped with a LaB_6 filament. Images were taken with a Gatan Ultrascan 2k × 2k camera. The mean particle size and size distribution were determined by measuring the diameter of isolated particles using ImageJ free software,[87] although particle agglomeration is present in all catalysts. Between 200 and 300 particles were considered for each catalyst in order to have an acceptable statistical sample.

The microstructure of the catalytic powders was evaluated by X-ray diffraction (XRD). The powder diffraction patterns were recorded on a Bruker D5005 Bragg-Brentano (θ–θ) diffractometer operated with a copper tube powered at 40 kV and 40 mA ($Cu_{K\alpha1}$ = 1.54060 Å and $Cu_{K\alpha2}$ = 1.54443 Å). Measurements were taken from $2\theta = 15°$ to $2\theta = 90°$ in step mode, with steps of 0.06° and a fixed acquisition time of 10 s $step^{-1}$.

Oxidation states and possible electronic interactions between metals were evaluated by X-ray photoelectron spectroscopy (XPS). Spectra were collected using a VG ESCALAB 3 MKII spectrometer using a monochromatized Mg

K_α radiation (1253.6 eV). The source was operated at 300 W (15 kV and 20 mA). Powder analysis covered a surface of 2 mm × 3 mm.

5.4 Characterization of the Nanocatalysts

The complex reaction mechanism of borohydride oxidation described in Section 5.2 is very dependent on the catalyst nature and structure, which can orientate the reaction towards either the direct or the indirect pathway. Therefore, the catalysts have to be characterized as well as possible by DTA-TGA, TEM, XRD, ICP-OES and XPS in order to determine the relationship nature/structure/electrochemical behaviour. Extensive studies of the catalyst structures and of the catalytic surface considered in this chapter can be found in the literature.[25,83–88]

The results obtained for several catalysts by the different characterization techniques are summarized in Table 5.3. The metal loading of all carbon-supported catalysts measured by DTA-TGA was found to be between 36 wt% and 43 wt% on carbon. These results are in agreement with the nominal 40 wt% value.

ICP-OES analyses were conducted on the multimetallic catalysts to determine their bulk composition. The results given in Table 5.3 also confirmed that the synthesized materials possessed the expected atomic ratios. Both thermal and ICP-OES analyses evidenced that the metal salts were completely reduced during the synthesis procedure of the catalysts.

To further analyze the structural composition of the nanoparticles, TEM/EDX, XRD, XPS and electrochemical methods were used. The mean size of

Table 5.3 DTA-TGA, XRD, TEM and ICP-OES characterizations of monometallic and multimetallic catalysts.[a]

	XRD			TEM	ICP-OES Atomic composition / %					TDA-TGA
	$a/Å$	$a_b/Å$	L_v/nm	d/nm	Pd	Au	Ni	Bi	Pt	*wt%*
Au	4.067	4.069	8.0	7.4	–	100	–	–	–	39
Pd	3.938	3.889	4.1	4.0	100	–	–	–	–	36
Pt	3.924	3.924	5.5	5.3	–	–	–	–	100	37
$Pd_{0.9}Au_{0.1}$	3.950	–	–	4.7	90	10	–	–	–	35
$Pd_{0.7}Au_{0.3}$	3.986	–	–	4.9	69	31	–	–	–	37
$Pd_{0.5}Au_{0.5}$	4.001	–	3.9	5.0	50	50	–	–	–	37
$Pd_{0.7}Au_{0.3}$	4.018	–	–	5.1	31	69	–	–	–	40
$Pd_{0.9}Au_{0.1}$	4.048	–	–	5.3	11	89	–	–	–	38
$Pd_{0.5}Ni_{0.5}$	3.942	–	3.4	5.5	54	–	46	–	–	43
$Pd_{0.9}Bi_{0.1}$	3.938	–	4.5	5.2	92	–	–	8	–	38
$Pt_{0.9}Bi_{0.1}$	3.943	–	5.7	4.7		–	–	10	90	36
$Pd_{0.45}Pt_{0.45}Bi_{0.1}$	3.943	–	4.7	4.5	43	–	–	12	45	38

[a] *a* is the cell parameter of the nanoparticles, a_b is the cell parameter of the bulk metal, L_v is the Scherer length (diameter of crystallites) and *d* is the diameter of the particles.

the nanoparticles and their dispersion on the support were evaluated from TEM pictures. Figure 5.2 shows representative TEM pictures of different mono- and bimetallic nanomaterials dispersed on a carbon Vulcan XC72 support as well as their related particle size distributions. The results[25,83,84,88] showed that in the case of Pd_xAu_{1-x} materials the mean particle size increased monotonously with the gold content from 4.0 nm (Pd/C) to 7.4 nm for the Au/C catalyst. The average particle size was found close to 5 nm for the other catalysts (Pt, Pd_xBi_{1-x}, Pt_xBi_{1-x}). It can be seen that the macroscopic structures of all the catalysts are identical, with nanoparticles homogeneously dispersed on the carbon support despite the presence of agglomerates.

In Figure 5.3, a TEM picture of an isolated nanoparticle comprising the $Pd_{0.45}Pt_{0.45}Bi_{0.1}$/C catalyst as well as the corresponding EDX spectrum allowed us to verify that this nanoparticle is composed of Pd, Pt and Bi elements. Other isolated particles were also analyzed by EDX, giving identical results.

XRD measurements were performed in order to determine the microstructure of the catalysts (Figure 5.4). Considering the position of the diffraction peaks relative to the different crystallographic planes, all catalysts presented a face-centred cubic (fcc) crystalline structure. The lattice parameter of each catalyst was calculated considering the mean value of the four lattice parameters obtained for (111), (200), (220) and (311) planes using the Bragg eqn (5.14):[89]

$$\frac{4\sin^2\theta}{\lambda^2} = \frac{1}{d^2} = \frac{h^2+k^2+l^2}{a^2} \Leftrightarrow a = \frac{\lambda\sqrt{(h^2+k^2+l^2)}}{2\sin\theta} \tag{5.14}$$

For all materials, the crystallite size of the nanoparticles was estimated using the simple and quick Scherrer formula given in eqn (5.15):[90]

$$L = \frac{\lambda}{\text{FWHM} \times \cos\theta} \tag{5.15}$$

where L is the crystallite size, λ is the wavelength of the XRD beam and FWHM is the full with at half maximum of the diffraction peak corresponding to the (111) plane at the Bragg angle θ (half deviation angle).

The monometallic catalysts exhibited the crystalline structure of the pure metals. Even so, the Pd/C catalyst presented a small shift of the diffraction peaks towards lower 2θ values. As a consequence, the calculated lattice parameter of the Pd/C catalyst (a = 3.938 Å) was found to be higher than that for bulk palladium (a = 3.891 Å). For instance, the diffraction peak related to the Pd(111) crystallographic plane is shifted to the 2θ position close to the Pt(111) plane. Such a high lattice parameter value rather corresponds to the Pd–H system, for which enlargement of the palladium lattice parameter is expected.[91] The formation of palladium hydride can occur during the Pd(II) reduction step of the colloidal synthesis, where hydrogen evolution coming from hydrolysis of BH_4^- is involved; a part of this hydrogen can be absorbed by palladium metal, inducing a relaxation of the Pd crystalline network.

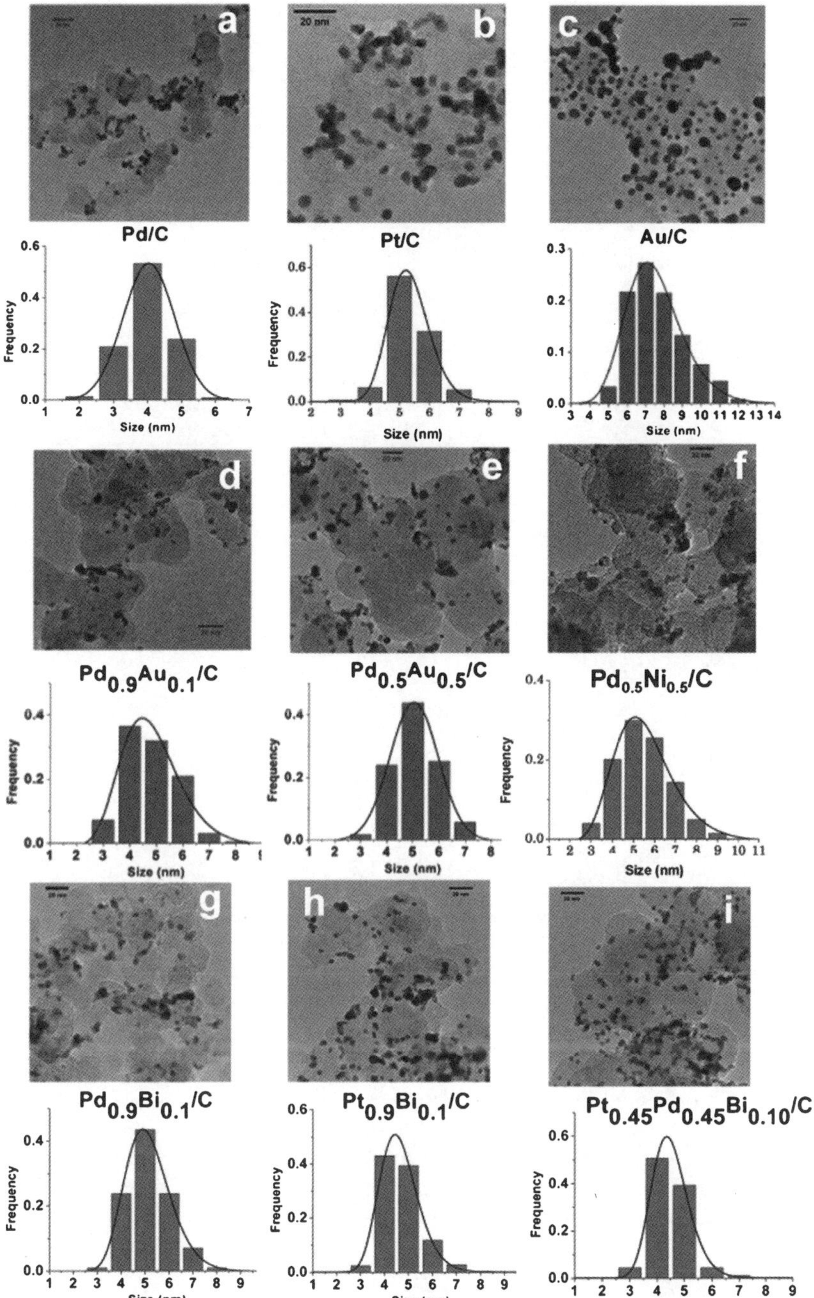

Figure 5.2 TEM picture and related particle size distribution of different mono-, bi- and trimetallic nanocatalysts (40 wt% metal loading on carbon): (a) Pd/C; (b) Pt/C; (c) Au/C; (d) $Pd_{0.9}Au_{0.1}/C$; (e) $Pd_{0.5}Au_{0.5}/C$; (f) $Pd_{0.5}Ni_{0.5}/C$; (g) $Pd_{0.9}Bi_{0.1}/C$; (h) $Pt_{0.9}Bi_{0.1}/C$; and (i) $Pd_{0.45}Pt_{0.45}Bi_{0.1}/C$.

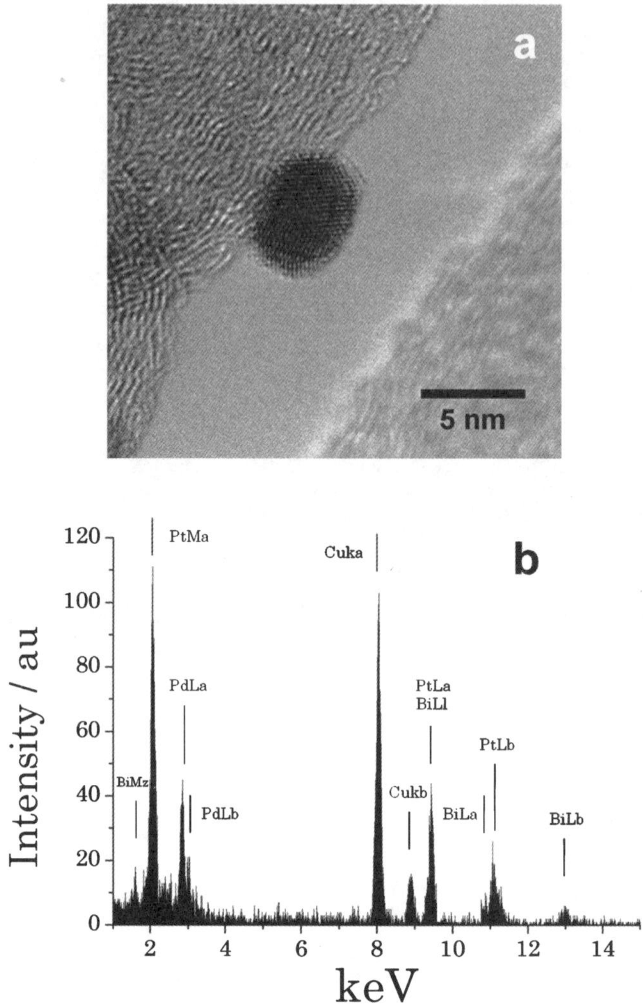

Figure 5.3 (a) TEM picture of an isolated nanoparticle comprising the $Pd_{0.45}Pt_{0.45}Bi_{0.1}/C$ catalyst. (b) EDX spectra of the nanoparticle presented in the TEM picture.

The $Pd_{1-x}Au_x/C$ catalysts were found to be composed of ordered alloys, formed between gold and palladium. This fact was verified using Vegard's law,[25] as presented in Figure 5.5. The results are again presented in Table 5.3.

It can be seen in Figure 5.4 that the crystallographic structures of the $Pd_{0.5}Ni_{0.5}/C$ and $Pd_{0.9}Bi_{0.1}/C$ catalysts are identical to that of the Pd/C monometallic catalyst. Nickel and bismuth clusters may then present an amorphous structure. The comparison of the mean particle size obtained by TEM for the Pd/C catalyst and the calculated crystallite size for the same material indicated that the nanoparticles that comprise the Pd/C catalyst were monocrystalline. On the $Pd_{0.5}Ni_{0.5}/C$ material, an important deviation of those

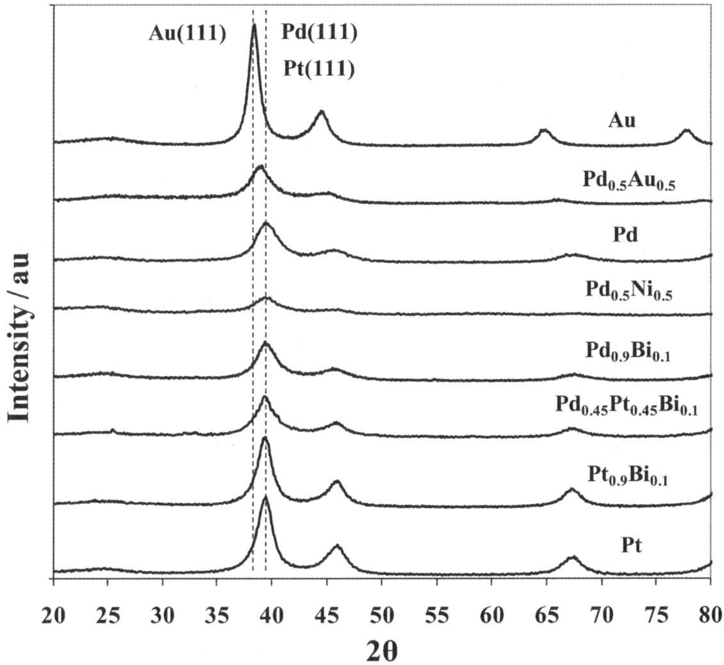

Figure 5.4 XRD diffraction patterns for mono-, bi- and trimetallic nanocatalysts.

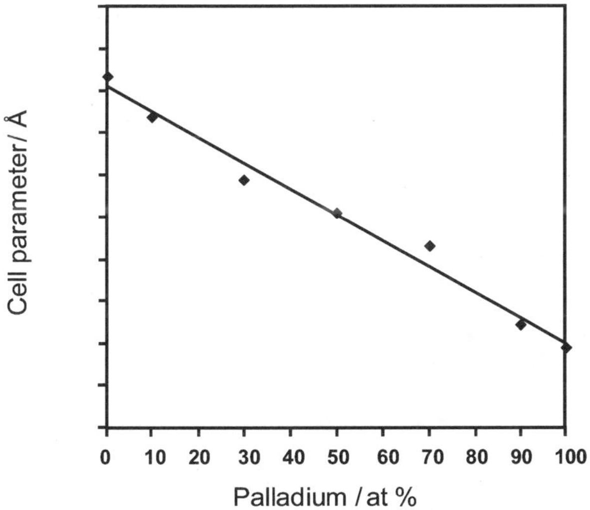

Figure 5.5 Plot of the lattice parameter of Pd_xAu_{1-x} (determined from the XRD patterns) as a function of the palladium atomic content in the catalyst (determined by ICP-OES).

values was verified, indicating that the structure of the $Pd_{0.5}Ni_{0.5}$ nanoparticles is complex. It was proposed that nickel atoms could form amorphous clusters interacting with palladium; an electrochemical effect related to the interaction between both metals was also verified by Simoes *et al.*[83]

The influence of the bismuth interaction with palladium and platinum nanoparticles, obtained from a co-reduction by the water-in-oil microemulsion method, was recently proposed for the electro-oxidation of sodium borohydride.[84] XPS measurements confirmed the interaction between bismuth and palladium and/or platinum. XRD and TEM/EDX analyses coupled to electrochemical experiments highlighted the presence of bismuth agglomerates directly supported on the carbon substrate when the bismuth atomic concentration was increased. A detailed structural analysis of a trimetallic $Pd_{0.45}Pt_{0.45}Bi_{0.1}/C$ catalyst is presented as an example. XRD measurements confirmed that this material displayed a face centred cubic (fcc) crystalline structure with a lattice parameter close to that calculated for the Pt/C and $Pt_{0.9}Bi_{0.1}/C$ materials. The XPS spectra of $Pd_{0.45}Pt_{0.45}Bi_{0.1}/C$ catalyst are presented in Figure 5.6.

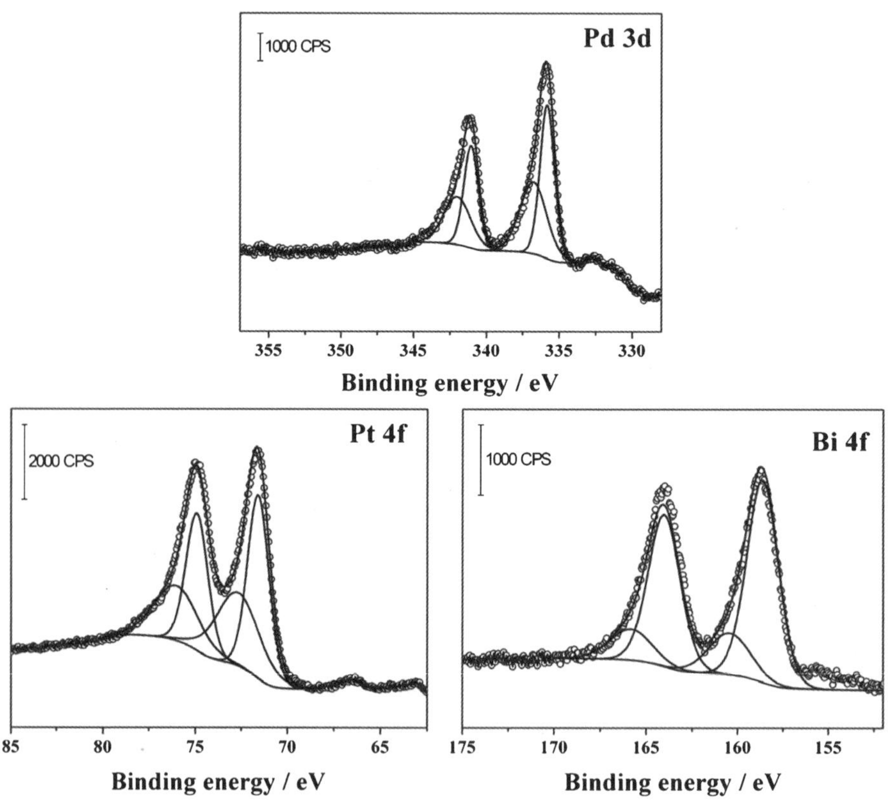

Figure 5.6 Detailed XPS spectra for Pd 3d, Pt 4f and Bi 4f orbitals on the $Pd_{0.45}Pt_{0.45}Bi_{0.1}/C$ catalyst.

The analysis was based on the *Handbook of X-Ray Photoelectron Spectroscopy* data;[92] the results are presented in Table 5.4. The Pd 3d core level spectrum presented identical signals as the bimetallic $Pd_{0.9}Bi_{0.1}/C$ material.[84] The Pd 3d core level spectrum presented two asymmetrical peaks related to Pd $3d_{5/2}$ and Pd $3d_{3/2}$, with spin–orbit splitting of *ca.* 5.3 eV. The signal obtained for Pd $3d_{5/2}$ could be fitted by two peaks, located at 335.7 eV and 336.8 eV. The first one could be assigned to metallic Pd and the second one to PdO, meaning that monometallic nanoparticles presented a degree of palladium oxidation other than the zero valence state. In agreement with the analysis conducted on $Pd_{0.9}Bi_{0.1}/C$, the presence of bismuth on the surface of the trimetallic catalyst inhibited the palladium oxidation. Nonetheless, the presence of bismuth did not influence platinum in the same manner as palladium. Two asymmetrical signals corresponding to Pt $4f_{7/2}$ and Pt $4f_{5/2}$ spin–orbits were split by 3.35 eV. The fitting of the Pt $4f_{7/2}$ signal gave two peaks located at 71.6 eV and 72.7 eV, corresponding to metallic platinum and $Pt(OH)_2$ species, respectively. Analyzing the Bi 4f core level spectrum presented in Figure 5.6, two asymmetrical signals split by *ca.* 5.4 eV corresponded to the spin–orbits Bi $4f_{7/2}$ and Bi $4f_{5/2}$. The fitting of the Bi $4f_{7/2}$ signal presented two contribution peaks located at 158.6 eV and 160.4 eV. The first one could be assigned to a Bi_2O_3 species and the second one was tentatively assigned to $Bi(OH)_3$.[84,93] The oxidation states of platinum are the same on $Pd_{0.45}Pt_{0.45}Bi_{0.1}/C$ and $Pt_{0.9}Bi_{0.1}/C$ catalysts as for Pt/C material. Bismuth presented the same oxidation states as those observed in the case of the $Pd_{0.9}Bi_{0.1}/C$ catalyst, higher than those determined on $Pt_{0.9}Bi_{0.1}/C$. However, the ratio between the Bi_2O_3 and $Bi(OH)_3$ species was higher in the $Pd_{0.45}Pt_{0.45}Bi_{0.1}/C$ material than that found in $Pd_{0.9}Bi_{0.1}/C$. It was then proposed by the authors that the surface of the trimetallic $Pd_{0.45}Pt_{0.45}Bi_{0.1}$ nanoparticules was also composed of platinum and palladium interacting with bismuth adatoms or clusters.

5.5 Evaluation of the Catalytic Activity and Selectivity towards Sodium Borohydride Electro-oxidation

5.5.1 Electrochemical Methods

The BOR is limited by the diffusion of the borohydride ion from the bulk solution to the electrode.[26,32,78] To evaluate the kinetics and mechanism of the BOR on different catalysts, the Koutecky–Levich[94,95] equation is often used to draw some kinetic parameters from the polarization curves recorded from rotating disk electrode measurements at different rotation rates (Ω = 400, 900, 1600 and 2500 rpm). Considering that the catalytic layer is a thin film, the BOR mechanism can be written as follows:

(1) $BH_4^-{}_{\text{solution}} \rightarrow BH_4^-{}_{\text{electrode}}$ (diffusion from the bulk electrolyte to the electrode, j_l^{diff})

Table 5.4 XPS characterization data for the $Pd_{0.45}Pt_{0.45}Bi_{0.1}$ catalyst.

Pd 3d/*eV*	Bi 4f/*eV*	Pt 4f/*eV*	Pd/*at%*	Bi/*at%*	Pt/*at%*	O/*at%*	C/*at%*	Metal oxidation state	Bi/%	Pd/%	Pt/%
335.8	–	–	0.85	–	–	–	–	Pd	–	–	–
336.7	–	–	0.74	–	–	–	–	PdO	–	–	–
–	–	71.60	–	–	0.96	–	–	Pt	–	–	–
–	–	72.70	–	–	0.77	–	–	$Pt(OH)_2$	–	–	–
–	158.6	–	–	0.52	–	–	–	Bi_2O_3	–	–	–
–	160.4	–	–	0.14	–	–	–	$Bi(OH)_3$	–	–	–
Total/at%			1.59	0.66	1.73	8.93	87.09	–	17	40	43

(2) $BH_4^-{}_{electrode} \rightarrow BH_4^-{}_{surface}$ (diffusion inside the catalytic film, j_l^{film})
(3) $BH_4^-{}_{surface} \rightarrow BH_4^-{}_{adsorbed}$ (adsorption step, j_l^{ads})
(4) $BH_4^-{}_{adsorbed} + nOH^- \rightarrow Products + ne^-$ (electron transfer, rate-limiting step (j_0 and b)

The Koutecky–Levich eqn (5.16) can then be written as follows:

$$\frac{1}{j} = \frac{1}{j_l^{diff}} + \frac{1}{j_k^{apparent}} = \frac{1}{0.2nFD_{BH_4^-}{}^{2/3}C_{BH_4^-}\nu^{-1/6}\Omega^{1/2}} + \frac{1}{j_k^{apparent}} \tag{5.16}$$

with:

$$\frac{1}{j_k^{apparent}} = \frac{1}{j_l^{film}} + \frac{1}{j_l^{ads}} + \frac{1}{j_k} = \frac{1}{j_l} + \frac{1}{j_0 \frac{\theta}{\theta_{eq}} e^{\frac{\eta}{b}}} \tag{5.17}$$

and:

$$\frac{1}{j_l} = \frac{1}{j_l^{film}} + \frac{1}{j_l^{ads}} \tag{5.18}$$

where j is the current density, j_l^{diff} is the diffusion limiting current density, j_k is thc kinctic current density, j_l^{film} is the BH_4^- limiting current density inside the film formed by the catalytic layer, j_l^{ads} is the limiting current density associated with the BH_4^- adsorption on the catalyst, j_0 is the exchange current, b is the Tafel slope and $\eta = E - E_{eq}$ is the overvoltage, $D = 1.6 \times 10^{-5}$ cm^2 s^{-1} is the BH_4^- diffusion coefficient, $C_{BH_4^-}$ (mol L^{-1}) is the BH_4^- concentration in solution, $\nu = 1.19 \times 10^{-2}$ cm^2 s^{-1} is the solution kinematic viscosity, F = 96485 C mol^{-1} and Ω (rpm) is the electrode rotation rate. A discussion on the value of the BH_4^- diffusion coefficient and the solution kinematic viscosity has recently appeared in the literature. Recent measurements conducted by Chatenet *et al.*[96] led to different values for these parameters compared to those measured by Denuault *et al.*;[97] the following values were determined in 10^{-2} M $NaBH_4$ + 1 M NaOH electrolyte: $D_{BH_4^-} = 2.26 \times 10^{-5}$ cm^2 s^{-1} and $\nu = 1.33 \times 10^{-2}$ cm^2 s^{-1}.

The calculation made using the parameters determined recently by Chatenet *et al.* led to lower values of the number of exchanged electrons drawn from the the Koutecky–Levich equation. However, a deviation of only *ca.* 20% on the number of exchanged electrons can be expected, depending on the choice of the parameters.[84] Rotating disk experiments conducted in our laboratory for the BOR using a massive gold electrode gave results that were in agreement with the values measured by Denuault *et al.*, while no hydrogen evolution was observed during the experiment. Thus, those parameter values are used in this chapter for the determination of the BOR kinetic parameters. In eqn (5.17), θ and θ_{eq} represent the coverage degree of the catalytic surface by species coming from borohydride adsorption at the potentials E and E_{eq}, respectively.

Assuming that the electron transfer step is the rate-determining step, θ could be taken equal to θ_{eq} for the entire considered potential domain.

The limiting current density j_l results from a mixed control by the diffusion of borohydride inside the catalytic film, and by adsorption of borohydride at the catalytic particles. Conversely to the diffusion-limiting current density, j_l^{diff} of borohydride in the bulk electrolyte, which is controlled by the rotation rate Ω, the diffusion-limiting current density of borohydride inside the film, j_l^{film}, and the adsorption-limiting current density of borohydride on the catalytic surface, j_l^{ads}, are independent of Ω. Since these current densities are both independent of the rotation speed Ω and of the electrode potential E, it is not possible to separate their contributions, so that the analysis of data will only give j_l. The diffusion-limiting current density in the electrolyte is given by the Levich's law (eqn 5.19):

$$j_l^{diff} = 0.2nFD_{BH_4^-}{}^{2/3}C_{BH_4^-}\nu^{-1/6}\Omega^{1/2} \tag{5.19}$$

From this equation, the theoretical current density values can be calculated as a function of the number of exchanged electrons. The results are present in Table 5.5.

Plotting the curve $j^{-1} = f(\Omega^{-1/2})$ allows determination of the number of exchanged electrons from the slope (s_{KL}) of the obtained straight lines plotted from the polarization curves recorded at different electrode rotation rates according to eqn (5.20):

$$n = \frac{1}{Bs_{KL}} \tag{5.20}$$

where $B = 0.2FD_{BH_4^-}{}^{2/3}C_{BH_4^-}\nu^{-1/6}$ is constant.

The kinetic parameters j_0 and b for the BOR on the different catalysts were estimated using the modified Butler–Volmer eqn (5.21):

Table 5.5 Theoretical limiting diffusion current density values (j_l^{diff}/mA cm^{-2}), calculated from Levich's law (eqn 5.19) as a function of the number of exchanged electrons and of the rotation rate of the electrode.

n exchanged electrons	Ω/rpm 400	900	1600	2500
8	41	62	82	103
7	36	54	72	90
6	31	46	62	77
5	26	38	51	64
4	21	31	41	51

Table 5.6 Kinetic parameters determined from eqns (5.16)–(5.21) for the BOR at Pd/C, Au/C and $Pd_{1-x}M_x$/C catalysts (where M = Au, Ni, Bi).

Catalyst	j_0/*mA cm*$^{-2}$	b/*mV dec*$^{-1}$	Potential range/*V* vs. *RHE*
Au	3.16×10^{-5}	177	0.5–0.7
Pd	3.26×10^{-2}	205	0.2–0.4
$Pd_{0.9}Au_{0.1}$	7.01×10^{-1}	352	0.2–0.4
$Pd_{0.7}Au_{0.3}$	1.31×10^{0}	449	0.15–0.35
$Pd_{0.5}Au_{0.5}$	5.13×10^{-1}	335	0.15–0.35
$Pd_{0.7}Au_{0.3}$	1.59×10^{0}	512	0.2–0.4
$Pd_{0.9}Au_{0.1}$	1.83×10^{-1}	318	0.4–0.6
$Pd_{0.5}Ni_{0.5}$	1.44×10^{-1}	233	0.1–0.2
$Pd_{0.9}Bi_{0.1}$	4.76×10^{-1}	344	0.2–0.4

$$\eta = E - E_{eq} = b\left(\log\frac{j_l}{j_0} + \log\frac{j_k}{j_l - j_k}\right) \quad (5.21)$$

Table 5.6 gives the kinetic parameters and the number of exchanged electrons for the BOR determined using this mathematical treatment of the Koutecky–Levich equation for different Pt- and Pd-based nanocatalysts.

5.5.2 Evaluation of the BOR on Monometallic Nanocatalysts

The polarization curves for the oxidation of 10^{-2} M $NaBH_4$ in 1.0 M NaOH were recorded with different rotation rates of the electrode at a scan rate v = 5 mV s^{-1} for Au/C, Pd/C and Pt/C catalysts.

The electro-oxidation of borohydride at gold surfaces was and is still extensively studied because it is proposed that the direct oxidation pathway is favoured on a massive gold electrode. However, the behaviour of nanostructured catalysts differs generally from that of massive materials. The cyclic voltammogram of an Au/C catalyst recorded at 50 mV s^{-1}, between 0.05 V and 1.2 V *vs.* RHE in an N_2-saturated 1 M NaOH electrolyte, is presented in Figure 5.7. At high potentials, the current in the zone A1, recorded during the positive potential scan, was attributed to the surface oxidation of gold according to eqn (5.22):[98]

$$Au{\cdot}H_2O + 3OH^- \rightarrow Au(OH)_3 + H_2O + 3e^- \quad (5.22)$$

The oxidized surface is reduced during the negative potential scan, originating the peak C1. At lower potentials, two redox peaks A0 and C0 are visible at *ca.* 0.4 V *vs.* RHE. The oxidation current in peak A0 is related to OH^- ions adsorption at the gold surface.[99,100]

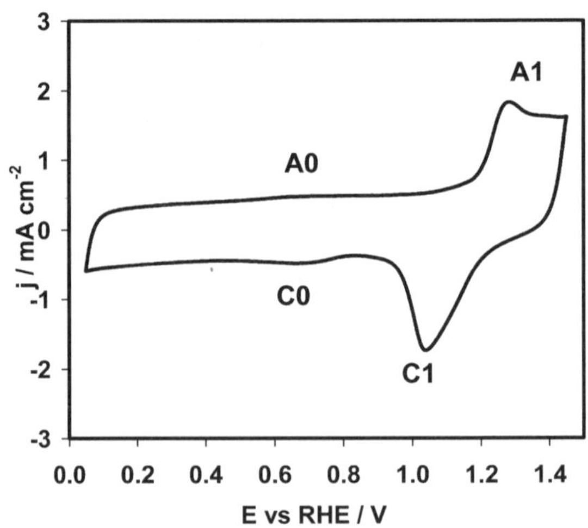

Figure 5.7 Cyclic voltammograms of Au/C catalysts recorded at 50 mV s^{-1} in an N_2-saturated 0.1 M NaOH electrolyte (T = 293 K).

Adžić and Avramoc-Ivić[101] proposed that the adsorption of hydroxyl ions with a partial charge transfer towards the gold surface, according to eqn (5.23), could be the determining step of ethylene glycol oxidation on Au(111), Au(110) and Au(100) surfaces:

$$\mathrm{Au} + \mathrm{OH}^- \rightarrow \mathrm{Au}_{\mathrm{ads}}^{\lambda-} + (1-\lambda)\mathrm{e}^- \tag{5.23}$$

The active surface area of the Au/C catalyst could be determined by measuring the electric charge involved during the surface oxide reduction in peak C1. According to different authors,[102,103] a charge of 400 μC cm^{-2} is involved in alkaline media for a polycrystalline smooth gold surface oxidation when the potential upper limit in the positive scan of the cyclic voltammograms is fixed at 1.45 V *vs*. RHE. The active surface area of the Au/C catalyst was estimated to be *ca*. 13 m^2 g^{-1}.

Figure 5.8(a) displays the polarization curves recorded at different rotation rates in a 10^{-2} M $NaBH_4$ + 1 M NaOH electrolyte. The onset potential of borohydride oxidation is *ca*. 0.2 V *vs*. RHE. The current densities recorded for Au/C for potentials higher than 0.7 V *vs*. RHE are twice higher with Ω = 1600 rpm than at Ω = 400 rpm, which indicates that the kinetics is really controlled by the BH_4^- ion diffusion in this potential range. With a nanostructured gold catalyst it appears that some hydrogen evolution occurs, leading to perturbations (oscillations) of the limiting diffusion current in the diffusion plateau.

As a typical example, the complete analysis for an Au/C catalyst according to the method developed in Section 5.1 is given below. Figure 5.8(b) shows Koutecky–Levich plots of j^{-1} *vs*. $\Omega^{-1/2}$ (with data taken from Figure 5.8a) at different electrode potentials. Straight lines are obtained, the slopes of which

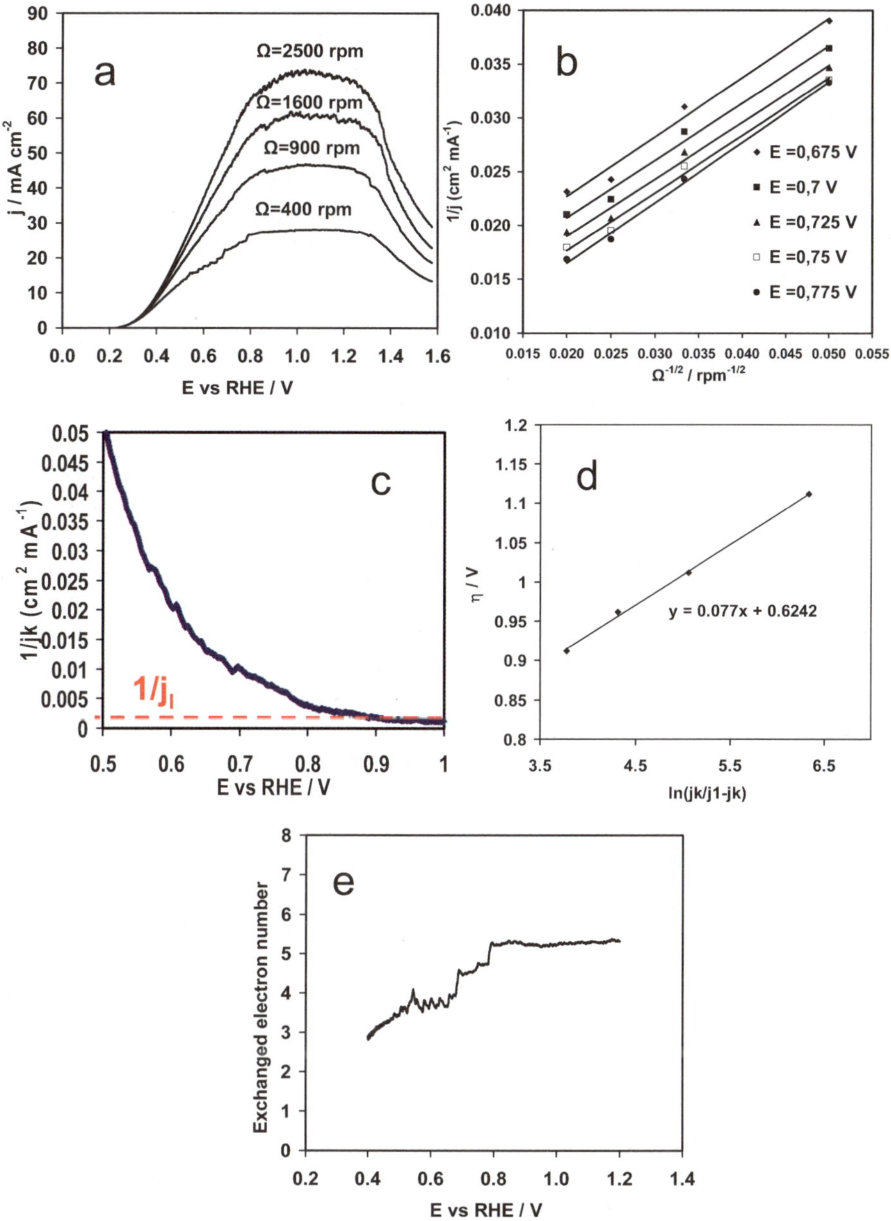

Figure 5.8 (a) Polarization curves of the $NaBH_4$ oxidation recorded on an Au/C catalyst with different electrode rotation rates Ω (400, 900, 1600, and 2500 rpm) at 5 mV s^{-1} in an N_2-saturated 10^{-2} M $NaBH_4$ + 1 M NaOH electrolyte (T = 293 K). (b) Related Koutecky–Levich plots $1/j$ *vs.* $1/\Omega^{-1/2}$ at different potentials. (c) Plot of $1/j_k$ *vs.* the electrode potential for the BOR. (d) Plot of the electrode potential E *vs.* $\ln[j_k/(j_l - j_k)]$ for the BOR on an Au/C electrode (Tafel plot). (e) Number of exchanged electrons as a function of the electrode potential.

yield n_t from 5 to 7 in the potential range from 0.5 to 1.3 V *vs.* RHE (Figure 5.8e), considering the parameter from Denuault *et al.* and the intercept of which at the origin gives j_k^{-1}. Then, plotting j_k^{-1} as a function of the electrode potential E (Figure 5.8c) allows determination of the limiting current density j_l. Knowing j_l it is possible to plot η *vs.* $\ln[j_k/(j_l - j_k)]$, which also gives a straight line (Figure 5.8d), from the slope of which the Tafel slope b is obtained, and from the intercept of which at the equilibrium potential, the exchange current density j_0 can be evaluated. The kinetic parameters, n_t, j_l, j_0 and b, obtained in this way, are given in Table 5.6, where the current densities are expressed in mA per geometric surface area.

Although electro-oxidation of $NaBH_4$ on platinum surfaces has been a relatively active research subject, few kinetic data are actually available compared with gold, for example. Recently, Finkelstein *et al.*[26] presented an extensive rotating disk investigation on Pt and Au bulk electrodes, where they proposed that $NaBH_4$ oxidation on platinum followed the direct pathway with *ca.* seven electrons involved per BH_4^- molecule at low potentials. This is one of the first attempts to understand the kinetics involved in BH_4^- oxidation on a platinum surface. As stated by the authors, the number of exchanged electrons per BH_4^- ion on a platinum surface is still controversial. Gyenge[56] considered a four-electron process with a Pt nanocatalyst. In Figure 5.9(a), Pt/C material exhibits typical voltammograms of clean catalytic nanoparticles supported on carbon and recorded in an alkaline medium. Considering a charge density of 210 μC cm^{-2} for a smooth polycrystalline platinum surface,[104,105] the determination of the charge involved in the peak C1 corresponding to hydrogen desorption from a platinum nanoparticle surface leads to an estimate for the active surface area of the Pt/C catalyst as *ca.* 23 m^2 g^{-1}. The polarization curves of borohydride oxidation at a nanostructured platinum catalyst recorded under the same experimental conditions as in Figure 5.8(a) are given for $\Omega = 0$ rpm in Figure 5.9(b) and for $\Omega \neq 0$ rpm in Figure 5.9(c).

The onset potential for the BOR is lower than that obtained with a Au/C catalyst, being close to 0.05V *vs.* RHE. Two zones can be observed: the first one (zone A) for potentials lower than 0.15 V *vs.* RHE is characterized by an intense oxidation peak at $\Omega = 0$ and by the superimposition of the polarization curves when different values of Ω are used. The mechanism in this potential region seems to involve hydrogen evolution *via* the hydrolysis of borohydride (bubbles have been observed at the electrode at the onset potential) and the oxidation of the formed hydrogen. The hydrogen accumulated at the platinum catalyst is not totally removed from the electrode surface at $\Omega = 0$ rpm, which blocks the reaction. At potentials higher than 0.15 V *vs.* RHE, the polarization curves recorded at different electrode rotation rates are separated: higher current densities are recorded for higher rotation rates, indicating that the reaction is diffusion limited, but no diffusion current plateau is reached. Moreover, the current densities recorded for $\Omega = 1600$ rpm are not twice those recorded at $\Omega = 400$ rpm. Therefore, at the Pt/C catalyst, mixed control (diffusion/charge transfer) of the reaction kinetics is proposed. The kinetics

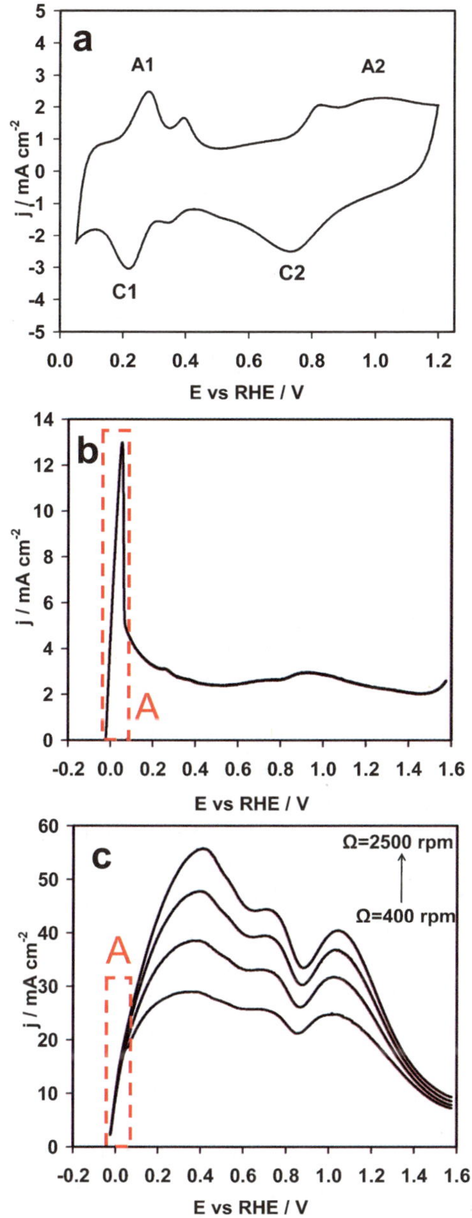

Figure 5.9 (a) Cyclic voltammograms of Pt/C catalysts recorded at 50 mV s^{-1} in an N_2-saturated 0.1 M NaOH electrolyte (T = 293 K). (b) Polarization curves of $NaBH_4$ oxidation recorded on a Pt/C catalyst without electrode rotation at 5 mV s^{-1} in an N_2-saturated 10^{-2} M $NaBH_4$ + 1 M NaOH electrolyte (T = 293 K). (c) Polarization curves of $NaBH_4$ oxidation recorded on a Pt/C catalyst with different electrode rotation rates Ω (400, 900, 1600 and 2500 rpm) at 5 mV s^{-1} in an N_2-saturated 10^{-2} M $NaBH_4$ + 1 M NaOH electrolyte (T = 293 K).

Table 5.7 Kinetic parameters determined from eqns (5.16)–(5.21) for the BOR using Pt/C, $Pt_{0.9}Bi_{0.1}$/C and $Pt_{0.45}Pd_{0.45}Bi_{0.1}$/C catalysts.

Catalyst	$j_0/mA\ cm^{-2}$	b/*mV dec*$^{-1}$	Potential range/*V* vs. *RHE*
Pt	3.26×10^{-2}	205	0.2–0.4
$Pt_{0.9}Bi_{0.1}$	7.01×10^{-1}	352	0.2–0.4
$Pd_{0.45}Pt_{0.45}Bi_{0.1}$	1.31×10^{0}	449	0.15–0.35

parameters were determined using the detailed method described earlier and are given in Table 5.7. The calculation of the number of exchanged electrons per BH_4^- molecule gave *ca.* five electrons for potentials higher than 0.15 V *vs.* RHE.

As stated earlier, few data are available in the literature concerning the BOR on nanostructured Pd/C catalysts. The typical cyclic voltammogram of Pd/C in alkaline media is given in Figure 5.10(a). Figure 5.10(b) shows the current density (referred to the geometric surface area) *vs.* potential curves for a platinum Pd/C catalyst. The currents are roughly proportional to the electrode rotation rates Ω. $NaBH_4$ oxidation starts at very negative potentials on the pure palladium nanocatalyst. The current densities recorded for the Pd/C catalyst in the potential range at from 0.5 to 0.8 V *vs.* RHE are twice as high at Ω = 1600 rpm than those recorded at Ω = 400 rpm. This indicates that the kinetics is controlled by the BH_4^- ion diffusion/convection in the referred potential ranges. By comparing the polarization curve at Ω = 0 rpm (Figure 5.10c) with those recorded with $\Omega \neq 0$ rpm (Figure 5.10b), three potential ranges could be separated. Previous work by Simoes *et al.*[25] showed that in the potential range from −0.2 V to 0.1 V *vs.* RHE, where the reaction is not diffusion limited, the hydrolysis reaction of BH_4^- into BH_3OH^- accompanied with hydrogen evolution was involved, followed by the oxidation of the latter species. In the potential range from 0.1 to 0.4 V, both reactions, oxidation of hydrogen formed on the surface and direct oxidation of $NaBH_4$, occurred competitively. For potentials higher than 0.4 V the direct borohydride oxidation was favoured with six electrons exchanged per oxidized BH_4^- ion. The kinetics parameters drawn from the Koutecky–Levich equation are given in Table 5.6.

Simoes *et al.*[106] verified that the decrease of the NaOH concentration led to an increase in the kinetics of the hydrogen evolution reaction on the Pd/C electrode. This fact has been explained in terms of the reduced stability of BH_4^- at lower pH. The voltammograms they recorded at different rotation rates in 10^{-1} M NaOH + 10^{-3} M $NaBH_4$ solution displayed the same current density at potentials lower than 0.2 V *vs.* RHE as those recorded in 1 M NaOH + 10^{-2} M $NaBH_4$ solution, whereas the diffusion current is about 10 times lower than that recorded in 1 M NaOH + 10^{-2} $NaBH_4$, in accordance with the 10 times decrease of the $NaBH_4$ concentration. Therefore, it can be proposed that this first oxidation wave at low potential is due to the oxidation of hydrogen produced on

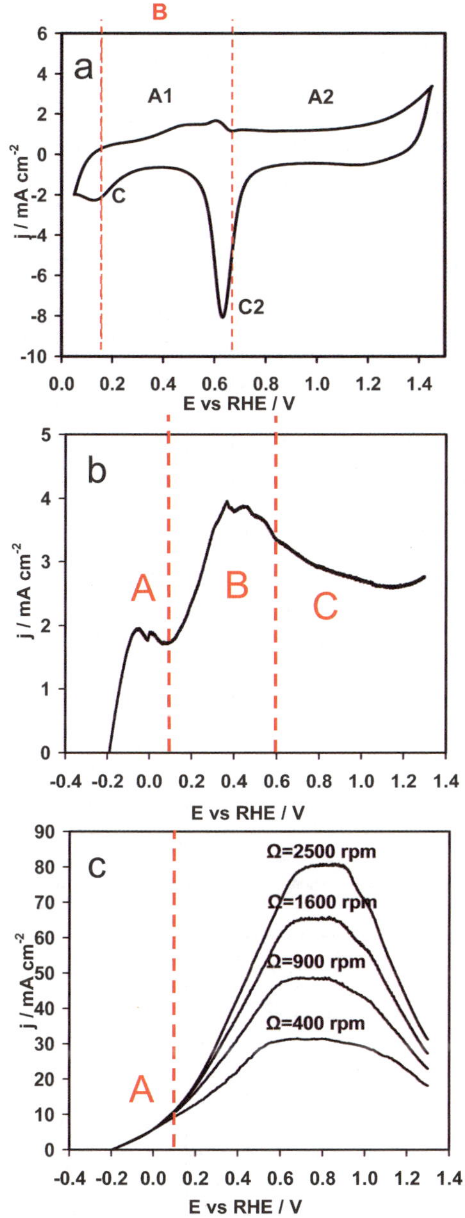

Figure 5.10 (a) Cyclic voltammograms of Pd/C catalysts recorded at 50 mV s^{-1} in an N_2-saturated 0.1 M NaOH electrolyte (T = 293 K). (b) Polarization curves of $NaBH_4$ oxidation recorded on a Pd/C catalyst with different electrode rotation rates Ω (400, 900, 1600, and 2500 rpm) at 5 mV s^{-1} in an N_2-saturated 10^{-2} M $NaBH_4$ + 1 M NaOH electrolyte (T = 293 K). (c) Polarization curves of $NaBH_4$ oxidation recorded on a Pd/C catalyst without electrode rotation at 5 mV s^{-1} in an N_2-saturated 10^{-2} M $NaBH_4$ + 1 M NaOH electrolyte (T = 293 K).

the electrode surface. By increasing the rotation rate, more BH_4^- species are transported on the electrode surface, which leads to a higher hydrogen evolution *via* the hydrolysis reaction. This can explain that the value of the number of exchanged electrons generally proposed is close to six, considering that the working potential range of a DBFC can correspond to the anode potential region where both direct and indirect reaction pathways are involved.

5.5.3 Evaluation of the BOR on Pd-Based Bimetallic Catalysts

Voltammograms using the Pd_xAu_{1-x}/C catalyst are presented in Figure 5.11.

The electrochemical method proposed by Rand and Woods[107] in the 1970s allowed determination of the surface composition of the catalysts, whereas a method developed by Òukaszewski and Czerwinski[108] allows determination of the active surface area of PdAu materials.[109] The Pd_xAu_{1-x}/C catalytic surface, reported in Table 5.8, was found palladium rich,[25] although the bulk

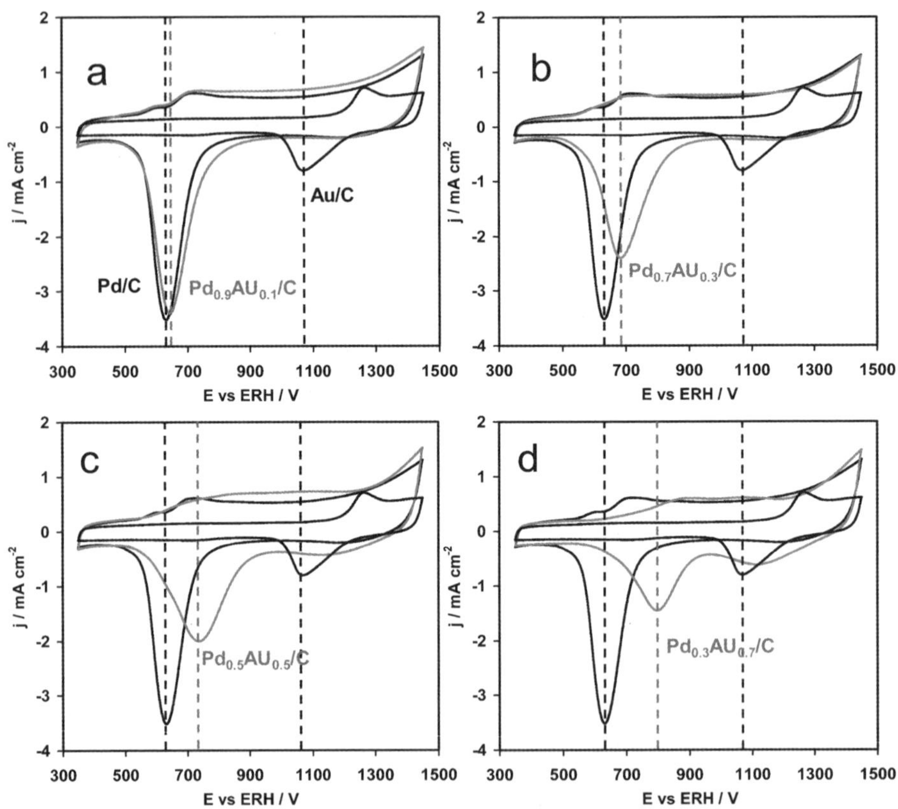

Figure 5.11 Cyclic voltammograms of Pd/C, Au/C and Pd_xAu_{1-x}/C catalysts recorded at v = 50 mV s^{-1} in an N_2-saturated 0.1 M NaOH electrolyte (T = 293 K, E_{max} = 1.45 V *vs.* RHE): (a) $Pd_{0.9}Au_{0.1}/C$; (b) $Pd_{0.7}Au_{0.3}/C$; (c) $Pd_{0.5}Au_{0.5}/C$; and (d) $Pd_{0.3}Au_{0.7}/C$.

Table 5.8 Electrochemical $Pd_{1-x}Au_x/C$ catalyst characterization by the Rand and Wood[107] method.

		Pd	$Pd_{0.9}Au_{0.1}$	$Pd_{0.7}Au_{0.3}$	$Pd_{0.5}Au_{0.5}$	$Pd_{0.3}Au_{0.7}$	$Pd_{0.1}Au_{0.9}$	Au
E_p^{alloy} *vs.* RHE/V		0.626	0.640	0.680	0.730	0.795	0.649, 0.726, 0.834	1.065
Surface Pd (at%)		100	97	88	76	61	95, 77, 53	0
ICP-OES								
Bulk comp. (at%)	Pd		90	69	50	31	11	–
	Au		10	31	50	69	89	–

compositions determined by ICP-OES were found close to the nominal ones and the crystallographic analysis indicated the formation of ordered PdAu alloys.

The activity of Pd_xAu_{1-x}/C catalysts for $x \geq 0.5$ was found to be the same as the Pd/C catalyst, as the polarization curves of borohydride oxidation are superimposed over the whole studied potential range (Figure 5.12a), whereas for lower values of x the activity tended to decrease down to that of Au/C, as shown in Figure 5.12(b).

The electrochemical method developed by Rand and Wood for the estimation of the surface composition was not applicable for PdNi material

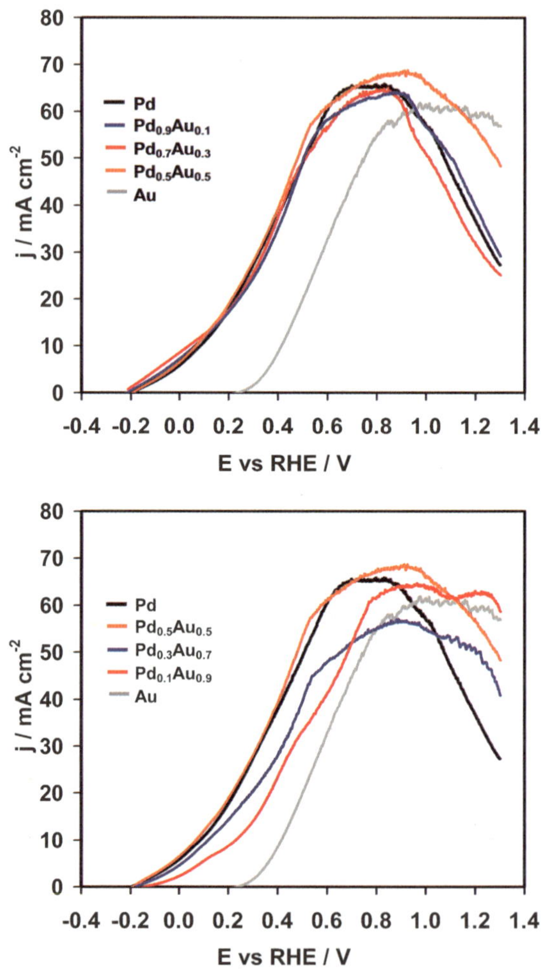

Figure 5.12 Polarization curves of $NaBH_4$ oxidation recorded on Pd/C, Au/C and Pd_xAu_{1-x}/C catalysts with an electrode rotation rate Ω = 1600 rpm, v = 5 mV s^{-1}, N_2 saturated, 10^{-2} M $NaBH_4$ + 1 M NaOH electrolyte (T = 293 K): (a) Pd-rich catalysts ($x \geq 0.5$); (b) Au-rich catalysts ($x < 0.5$).

as this catalyst exhibits separated potential regions for the reduction of surface oxides. In the case of the $Pd_{0.5}Ni_{0.5}/C$ catalyst, as an example, it appears clearly on the voltammogram in Figure 5.13(a) that palladium and nickel were present in the material, the reduction peak at high potential close to 1.3 V *vs.* RHE being related to the reduction of NiOOH to $Ni(OH)_2$, and the peak close to 0.65 V *vs.* RHE being due to the reduction of palladium surface oxides.[110,111] Figure 5.13(b) displays the polarization curves of borohydride oxidation recorded on Pd_xNi_{1-x}/C. As in the case of modification of palladium by gold, the replacing of half of the palladium atoms by nickel atoms did not affect the

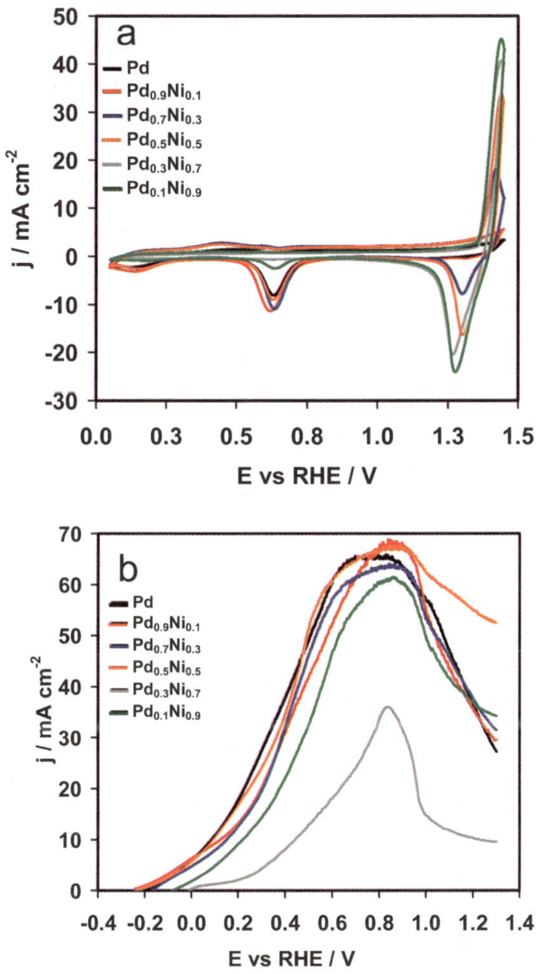

Figure 5.13 (a) Cyclic voltammograms of Pd/C and Pd_xNi_{1-x}/C (v = 50 mV s^{-1}, N_2-saturated 1.0 M NaOH electrolyte, T = 293 K). (b) Polarization curves of $NaBH_4$ oxidation recorded on Pd/C and Pd_xNi_{1-x}/C catalysts with an electrode rotation rate Ω = 1600 rpm, v = 5 mV s^{-1}, N_2-saturated 10^{-2} M $NaBH_4$ + 1 M NaOH electrolyte (T = 293 K).

catalytic activity of the catalyst. The activity was decreased only for a nickel atomic ratio higher than 50%.

Voltammograms of Pd/C and $Pd_{0.9}Bi_{0.1}$/C materials recorded in supporting electrolyte are presented in Figure 5.14. The presence of bismuth led to inhibition of the adsorption–absorption/reaction of hydrogen on palladium, since peaks in the potential region from 0.4 to 0.6 V *vs.* RHE recorded on the $Pd_{0.9}Bi_{0.1}$/C surface are smaller than those recorded on Pd/C. Those peaks are related to interactions between hydrogen and Pd(111) and Pd(100), respectively.[112] Casella and Contursi[93] verified a similar effect on palladium electrodes modified by bismuth adatoms. At higher potentials, an oxidation current starts to appear from *ca.* 0.7 V *vs.* RHE, which is related to the palladium surface oxidation.[113] Then, an intense oxidation peak located at *ca.* 0.9–0.95 V *vs.* RHE was assigned to surface oxidation processes of bismuth. During the negative-going scan, the reduction peak located at *ca.* 0.7 V on the $Pd_{0.9}Bi_{0.1}$ catalyst was assigned to the simultaneous reduction of Pd and Bi oxides,[93] which confirms the electronic interaction between both metals as suggested by XPS measurements.

The polarization curves obtained on the PdBi-based catalysts for the BOR at a rotation rate Ω = 1600 rpm are presented in Figure 5.15. The potential measured at open circuit is close to 0.1 V for $Pd_{0.9}Bi_{0.1}$/C and is displaced towards the higher values with the Bi atomic ratio in the catalyst, while the onset potential is *ca.* −0.2 V *vs.* RHE on Pd/C, $Pd_{0.5}Au_{0.5}$/C and $Pd_{0.5}Ni_{0.5}$/C catalysts. At potentials lower than 0.1 V *vs.* RHE the polarization curves recorded at different rotation rates were superimposed, indicating that the BOR was not limited by the diffusion of the BH_4^- species from the bulk electrolyte to the electrode in this potential range. For higher potentials, a

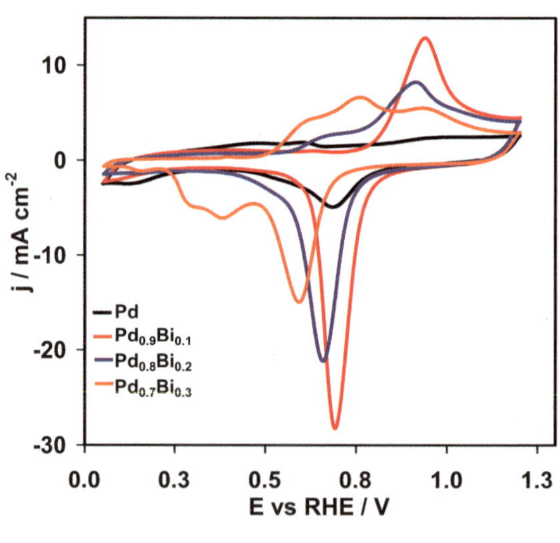

Figure 5.14 Cyclic voltammograms of Pd/C and $Pd_{1-x}Bi_x$/C catalysts (v = 50 mV s^{-1}, N_2-saturated 1 mol dm^{-3} NaOH electrolyte, T = 293 K).

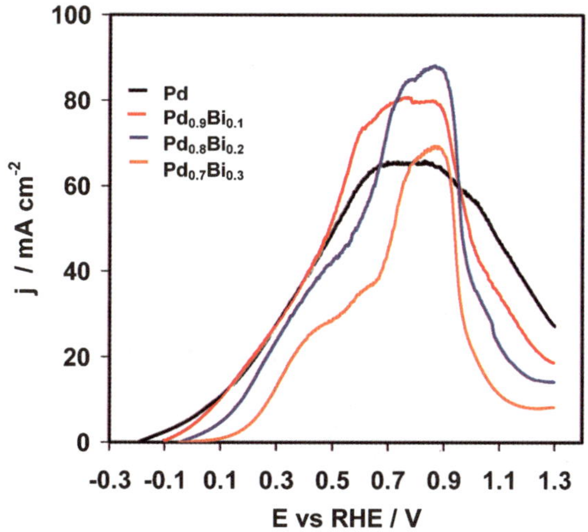

Figure 5.15 Polarization curves of $NaBH_4$ oxidation recorded on Pd/C and Pd_xBi_{1-x}/C catalysts with an electrode rotation rate Ω = 1600 rpm, v = 5 mV s^{-1}, N_2-saturated 10^{-2} M $NaBH_4$ + 1 M NaOH electrolyte (T = 293 K).

diffusion plateau is reached on this catalyst and the registered current density was doubled when Ω was increased from 400 rpm to 1600 rpm, indicating that the reaction was diffusion controlled; the limiting current density value indicated that eight electrons were exchanged.

Figure 5.16 compares the activity for the BOR obtained for the different most active palladium-based bimetallic catalysts. It can be seen that the activity of $Pd_{0.5}Au_{0.5}$/C and $Pd_{0.5}Ni_{0.5}$/C catalysts is identical to that of Pd/C. At very high potentials (higher that 0.8 V *vs.* RHE), when the Pd/C catalyst deactivates due to the surface oxidation, the current densities registered on $Pd_{0.5}Au_{0.5}$/C and $Pd_{0.5}Ni_{0.5}$/C catalysts are higher. This fact was interpreted in terms of interaction between palladium and the second foreign metal (Au or Ni). At lower potentials, the activity of these catalysts is comparable to that of monometallic Pd/C. The kinetic parameters for the BOR were determined for all the evaluated catalysts. The results are presented in Table 5.6. The values of the exchange current density (j_0) for the $Pd_{0.5}Au_{0.5}$/C and $Pd_{0.5}Ni_{0.5}$/C catalysts are higher than that evaluated with a Pd/C catalyst, indicating a higher intrinsic activity. However, the Tafel slope on the $Pd_{0.5}Au_{0.5}$/C catalyst is dramatically higher than that determined on Pd/C. The increased activity of the $Pd_{0.5}Ni_{0.5}$/C catalyst can be related to an improved activity towards BH_4^- hydrolysis and oxidation of the formed hydrogen. A careful analysis of the polarization curves obtained for the BOR on this catalyst in the potential region from −0.2 V to 0.1 V *vs.* RHE (where diffusion of the borohydride from the bulk electrolyte to the electrode does not limit the reaction rate) showed that the measured current density at Ω = 400 rpm is higher than that obtained at Ω = 1600 rpm. This fact indicates that the $Pd_{0.5}Ni_{0.5}$/C catalyst possesses a

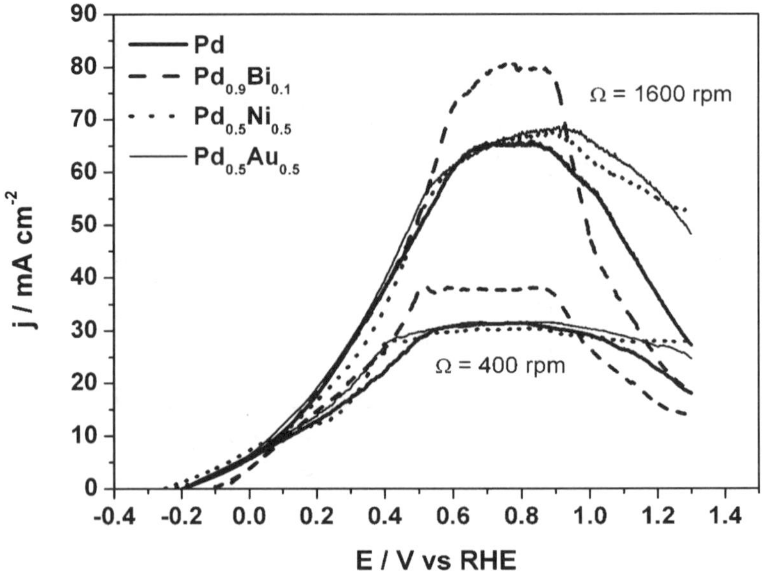

Figure 5.16 Polarization curves for $NaBH_4$ oxidation recorded at Ω = 400 rpm and 1600 rpm on Pd-based catalysts (v = 5 mV s^{-1}, N_2-saturated 10^{-2} mol dm^{-3} $NaBH_4$ + 1 mol dm^{-3} NaOH electrolyte, T = 293 K).

higher activity for borohydride hydrolysis and hydrogen oxidation at low potentials. The generated hydrogen is hardly removed from the electrode surface by the centrifugal force for lower rotation rates, which can perturb the BOR. The residence time of the BOR intermediate species seems to have an influence on the catalytic activity.

The $Pd_{0.9}Bi_{0.1}$/C catalyst is less active for the hydrolysis of borohydride. Even if the hydrolysis reaction can occur at low potentials, it is limited by the presence of bismuth adatoms or clusters at the surface of the palladium nanoparticles, which perturb the adsorption mode of the borohydride on palladium. A diffusion current plateau is reached at high potentials and the number of exchanged electrons reaches a value close to eight. The kinetic parameters calculated for the BOR on this catalyst show that its intrinsic activity is similar to that of the $Pd_{0.5}Au_{0.5}$/C catalyst. However, the faradic efficiency of the BOR seems to be improved on the $Pd_{0.9}Bi_{0.1}$/C catalyst. This is due to the low hydrogen evolution rate with $Pd_{0.9}Bi_{0.1}$/C, which made it difficult to measure, conversely to Pd/C, Au/C and bimetallic Pd_xAu_{1-x}/C catalysts which show relatively high activity for this reaction as shown in Figure 5.17. Chronoamperometries were carried out for each catalyst at open circuit potential and at 0, 0.2, 0.4 and 0.6 V *vs.* RHE for 1 h in 0.134 M $NaBH_4$ + 0.126 M NaOH electrolyte at different potentials.

Figure 5.17(a) shows that no borohydride hydrolysis is observed at the open circuit potential on the Au/C catalyst, but a small volume of H_2 was measured on Pd-based catalysts. As the potential is increased, the hydrogen generation

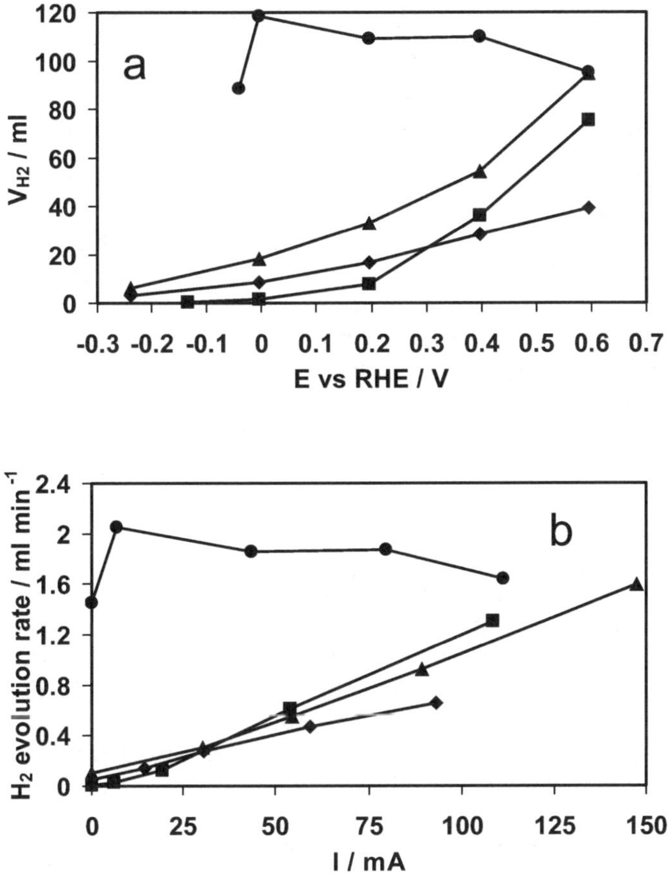

Figure 5.17 (a) Generated H_2 at different potentials and (b) chronoamperometric relationship with the H_2 generation rate for different potentials on (●) Pt/C, (◆) Pd/C, (▲) Au/C and (■) $Pd_{0.7}Au_{0.3}/C$ (0.124 M NaOH + 0.136 M $NaBH_4$, 50 mL solution, T = 293 K, metal loading = 0.4 mg cm^{-2}, electrode surface = 2.5 cm^2).

also increases. With the Au/C catalyst, an abrupt increase in H_2 production is observed for potentials higher than 0.2 V *vs.* RHE, which corresponds to the potential region where gold starts to be active for BH_4^- oxidation. With the Pd/C catalyst, the increase of the H_2 evolution flow with potential is lower than that recorded at the gold nanocatalyst. The alloyed $Pd_{0.7}Au_{0.3}/C$ catalyst presents a higher H_2 evolution flow than the Pd/C catalyst. This can be related to the presence of gold atoms on the surface. From the integration of the coulombic charge, H_2 production flow *versus* the oxidation current I can be plotted (Figure 5.17b). The presence of gold in the catalysts decreases the faradic efficiency since gold is not an active catalyst towards the hydrogen oxidation reaction. Therefore, the generated H_2 is not oxidized, and if it is not removed from the electrode the surface can be partially covered by gaseous H_2, reducing the available catalytic sites for BH_4^- oxidation.

5.5.4 Evaluation of the BOR on Pt-Based Multimetallic Catalysts

The voltammograms of $Pt_{0.9}Bi_{0.1}$/C material in alkaline media are given in Figure 5.18. The presence of bismuth also led to inhibition of the adsorption reaction of hydrogen on platinum as no H_{upd} feature is visible in the potential range from 0 to 0.4 V, in agreement with other studies performed on bulk electrodes.[84,114] The surface oxidation also starts at lower potential on $Pt_{0.9}Bi_{0.1}$/C than on Pt/C. Two quasi-reversible oxidation/reduction peaks are visible in the voltammogram of $Pt_{0.9}Bi_{0.1}$/C at *ca.* 0.75 V and 0.95 V *vs.* RHE. The first peak at 0.75 V was previously attributed to the oxidation of bismuth disseminated on the carbon surface, whereas the peak at higher potential was assigned to the oxidation of bismuth at the platinum surface.[84]

The polarization curves obtained on platinum for the BOR at $\Omega = 0$ rpm and $\Omega \neq 0$ rpm are presented in Figures 5.19(a) and 5.19(b), respectively. The $Pt_{0.9}Bi_{0.1}$/C catalyst presents the same onset potential for borohydride oxidation as the monometallic Pt/C catalyst. Regarding the polarization curves recorded at $\Omega = 0$ rpm (Figure 5.19a), the hydrogen oxidation peak due to BH_4^- hydrolysis is suppressed on the polarization curve recorded on the $Pt_{0.9}Bi_{0.1}$/C catalyst. Hydrogen bubbles were not detected at the electrode surface during the experiment in the whole studied potential range. This result indicated that $NaBH_4$ could be directly oxidized to $NaBO_2$ at very low potentials on the $Pt_{0.9}Bi_{0.1}$ catalyst, without hydrogen evolution through

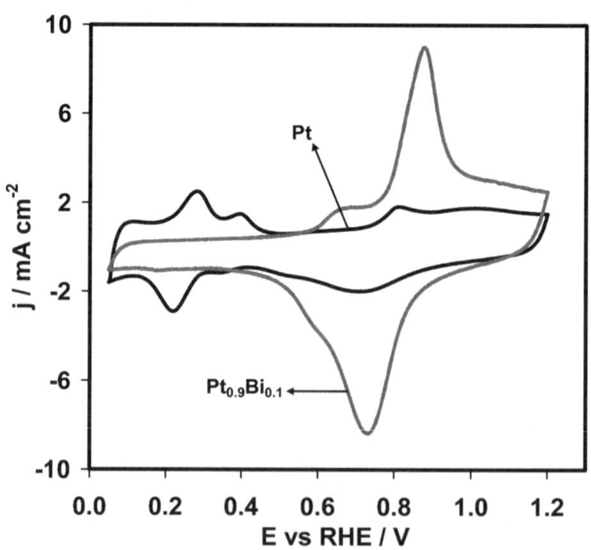

Figure 5.18 Cyclic voltammograms of $Pt_{0.9}Bi_{0.1}$/C catalysts (v = 50 mV s^{-1}, N_2-saturated 1 mol dm^{-3} NaOH electrolyte, T = 293 K).

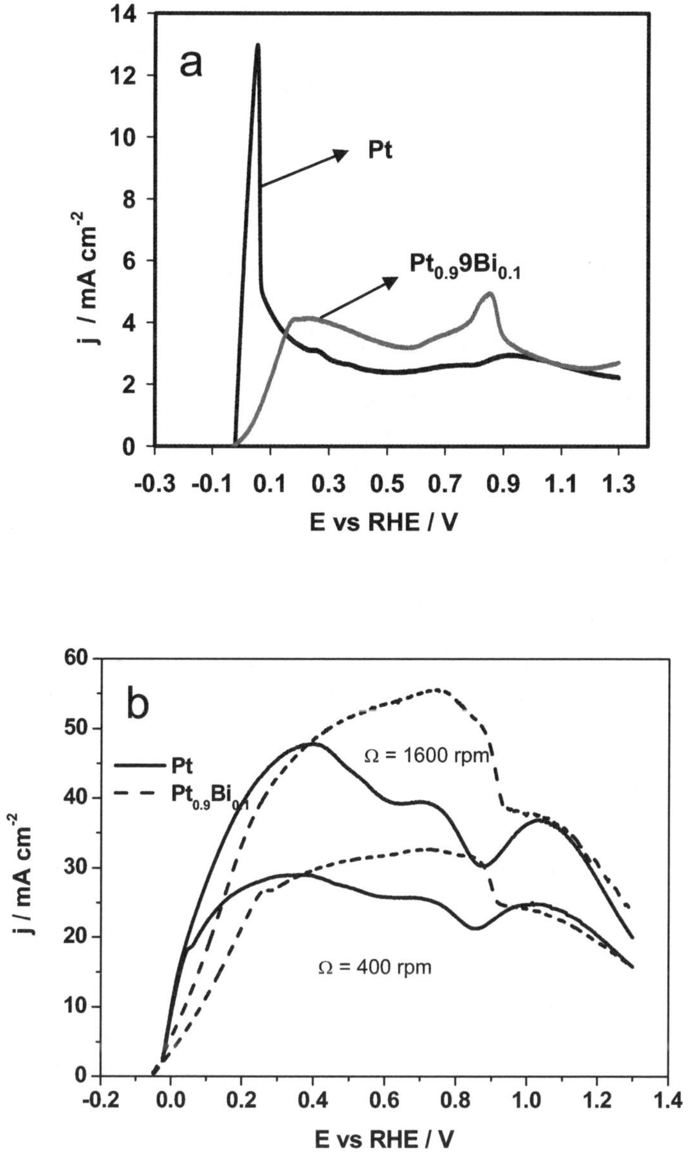

Figure 5.19 Polarization curves of the $NaBH_4$ oxidation recorded on Pt/C and $Pt_{0.9}Bi_{0.1}$/C catalysts at 5 mV s^{-1} in an N_2-saturated 10^{-2} M $NaBH_4$ + 1 M NaOH electrolyte (T = 293 K): (a) without electrode rotation; (b) with electrode rotation rates Ω = 400 and 1600 rpm

hydrolysis. The inhibition of borohydride hydrolysis was verified on the $Pt_{0.9}Bi_{0.1}$/C catalyst by adding 20 mg of Pt/C and of $Pt_{0.9}Bi_{0.1}$/C materials into two solutions containing 2 mol dm^{-3} $NaBH_4$ + 1 mol dm^{-3} NaOH. As soon as the Pt/C catalyst was added, bubbles of hydrogen were formed and significant hydrogen evolution occurred, whereas no or negligible hydrogen evolution was observed after addition of the $Pt_{0.9}Bi_{0.1}/C$ catalytic powder.

Regarding the polarization curves recorded at $\Omega \neq 0$ rpm (Figure 5.19b), at low potentials, between 0 and 0.2 V *vs.* RHE, the current densities measured for the different electrode rotation rates on the $Pt_{0.9}Bi_{0.1}/C$ catalyst are lower than those measured on Pt/C. Still, the polarization curves obtained with the $Pt_{0.9}Bi_{0.1}/C$ catalyst at Ω = 400 and 1600 rpm are separated over the whole potential range that was considered in the experiment, meaning that the BOR is strongly controlled by the diffusion of the borohydride ion, while on the Pt/C catalyst the BOR at low potentials evolves mainly through the oxidation of the hydrogen produced by the borohydride hydrolysis. Even if the reaction is mainly controlled by the BH_4^- diffusion, the current densities do not reach a diffusion plateau at high overpotentials on $Pt_{0.9}Bi_{0.1}/C$, as for Pt/C. This fact means that the reaction either is partially controlled by the kinetics or leads to the formation of reaction intermediates that can partially block the catalytic surface and that can be further oxidized or not. Furthermore, as already reported,[84] a very small hydrogen evolution was evidenced on the $Pt_{0.9}Bi_{0.1}/C$ catalyst during the experiments.

Voltammograms obtained for the $Pd_{0.5}Pt_{0.5}/C$ and $Pd_{0.45}Pt_{0.45}Bi_{0.1}/C$ catalysts are presented in Figure 5.20. The presence of bismuth has a clear effect on the hydrogen adsorption/absorption processes, indicating that bismuth atoms are disseminated on the surface. Two reversible bismuth-related redox processes are visible, depending on the potential applied to the electrode. The oxidation current peak located at 0.9 V *vs.* RHE has been attributed to the oxidation involving bismuth in interaction with Pd and Pt at

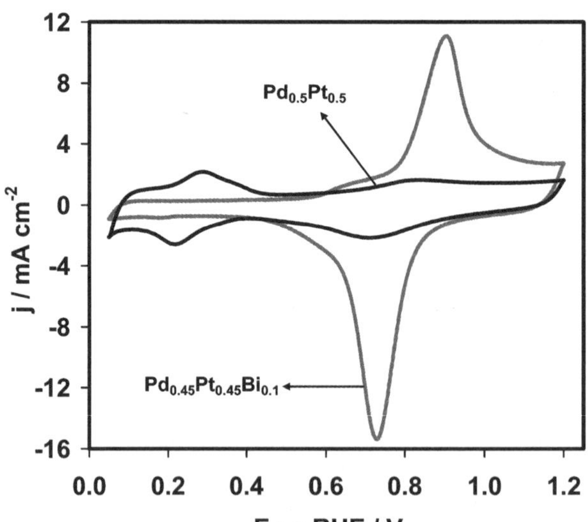

Figure 5.20 Cyclic voltammograms of $Pd_{0.5}Pt_{0.5}/C$ and $Pd_{0.45}Pt_{0.45}Bi_{0.1}/C$ catalysts (v = 50 mV s^{-1}, N_2-saturated 1 mol dm^{-3} NaOH electrolyte, T = 293 K).

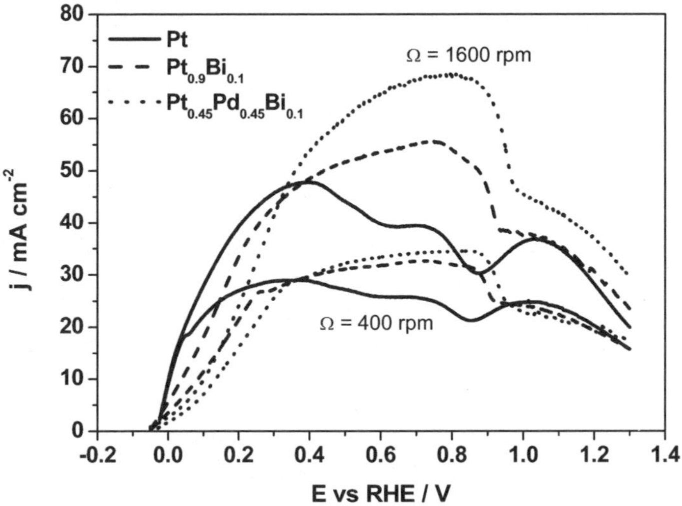

Figure 5.21 Polarization curves of the $NaBH_4$ oxidation recorded at Ω = 400 and 1600 rpm on Pt/C, $Pt_{0.9}Bi_{0.1}$/C and $Pd_{0.45}Pt_{0.45}Bi_{0.1}$/C ($v$ = 5 mV s^{-1}, N_2-saturated 10^{-2} M $NaBH_4$ + 1 M NaOH electrolyte, T = 293 K).

the surface of the nanoparticles.[84] At lower potentials, the oxidation current wave starting close to 0.5 V *vs.* RHE can be attributed to the bismuth agglomerates that are directly disseminated on the carbon Vulcan XC72 support.

Polarization curves recorded in presence of 10^{-2} M $NaBH_4$ are given in Figure 5.21 for the trimetallic $Pd_{0.45}Pt_{0.45}Bi_{0.1}$/C catalyst. The same onset potential for the oxidation of borohydride as that measured on the $Pt_{0.9}Bi_{0.1}$/C catalyst was recorded. Furthermore, the BOR was also controlled by the diffusion at low potentials. At higher potentials, a diffusion current plateau was achieved, with current densities in agreement with the theoretical ones calculated when the reaction is purely controlled by the diffusion when seven electrons are exchanged (Table 5.7). Considering the fact that a higher amount of hydrogen is produced during the experiment when compared with the $Pt_{0.9}Bi_{0.1}$/C catalyst, this result is in agreement with the exchanged number of electrons that is inferior to that found on the $Pt_{0.9}Bi_{0.1}$/C catalyst.

5.6 Summary

The open-circuit potential on a nanostructured Au/C catalyst is considerably higher when compared to that measured on palladium- and platinum-based catalysts. As was recently proposed, the BOR evolves on this catalyst with the production of hydrogen and less than eight electrons are exchanged on monometallic Au/C, Pt/C and Pd/C catalysts. Despite the fact that the Au/C catalyst has the lowest calculated Tafel slope for the BOR, the j_0 value is also the lowest, corresponding to poor intrinsic activity. In addition, this material displays the highest overvoltage for the BOR; it can therefore be concluded that it is not interesting for application as an anodic catalyst in a DBFC.

Based on the analysis of the BOR with the considered catalysts, a summary of the reaction mechanisms as a function of the electrode potential and of the catalyst composition is presented in Table 5.9. It can be seen that the BOR occurs on both $Pd_{0.5}Au_{0.5}/C$ and $Pd_{0.5}Ni_{0.5}/C$ catalysts by a mechanism which is identical to that determined on the Pd/C catalyst. Considering the kinetic parameters of the reaction, presented in Table 5.6, the bimetallic PdNi catalyst shows particular interest due to the fact that half of the palladium atoms comprising the catalyst are replaced by an element that is more abundant in nature and has a lower cost, while matching the monometallic Pd/C catalyst in terms of activity and selectivity. An identical analysis can be made for the $Pd_{0.45}Pt_{0.45}Bi_{0.1}/C$ catalyst. Despite the higher activity of this catalyst for borohydride hydrolysis than $Pt_{0.9}Bi_{0.1}/C$, it presents a similar activity and an identical overvoltage for the BOR compared to that for the $Pt_{0.9}Bi_{0.1}/C$ catalyst.

While the Pt/C catalyst is strongly active for $NaBH_4$ hydrolysis at the open-circuit voltage and at very low potentials, the presence of bismuth almost completely suppresses hydrogen evolution at low potentials. $Pt_{0.9}Bi_{0.1}/C$ is one of the first catalysts able to directly oxidize the BH_4^- ion at low potentials without significant hydrogen evolution, leading to *ca.* eight electrons exchanged. This is an important result, indicating that the surface modification of noble metals at the nanoscale could change the adsorption mechanism of the BH_4^- ion, orientating the oxidation reaction through the direct reaction pathway. Even if the $Pt_{0.9}Bi_{0.1}/C$ catalyst led to lower performance than Pt/C in terms of achieved current densities in the potential range up to 0.3 V, which is of interest for a direct borohydride fuel cell application, the suppression of the hydrogen evolution reaction can increase the autonomy (by increasing the fuel efficiency and further the available energy density) and safety (by limiting

Table 5.9 Mechanisms and number of exchanged electrons for the BOR as a function of the electrode potential for the different nanocatalysts.[a]

	Potential/*V* vs. *RHE*										
Catalyst	−0.2	−0.1	0	0.1	0.2	0.3	0.4	0.5	0.6	0.7	0.8
Au/C						Hyd + D		D → ~6 e^-			
$Pd_{0.5}Au_{0.5}/C$	Hyd + BH_3OH^- ox				H_2 ox + D			D → ~6 e^-			
Pd/C	Hyd + BH_3OH^- ox				H_2 ox + D			D → ~6 e^-			
$Pd_{0.5}Ni_{0.5}/C$	Hyd + BH_3OH^- ox			H_2 ox + D				D → ~6 e^-			
$Pd_{0.9}Bi_{0.1}/C$		Low (hyd + H_2 ox) + D						D → ~8 e^-			
$Pd_{0.45}Pt_{0.45}Bi_{0.1}/C$			D → ~7 e^- (low H_2 evolution)								
$Pt_{0.9}Bi_{0.1}/C$			D → ~8 e^- (mixed diffusion/kinetic control)								
Pt/C			Hyd + H_2 ox			(H_2 ox + D) → ~4–5 e^-					

[a]Hyd: hydrolysis reaction; D: direct oxidation reaction; H_2 ox: hydrogen oxidation reaction; BH_3OH^- ox: BH_3OH^- species oxidation reaction.

hydrogen linkage) of the system. Although further studies are needed to understand the reaction mechanism on noble metals and modified surfaces, the results presented in this chapter evidence that $NaBH_4$ can be directly oxidized at low potentials with high energy efficiency using nanostructured multi-metallic catalysts.

Acknowledgement

Part of works presented in this chapter was carried out under the framework of a contract (BODIPAC) from the French National Research Agency (ANR-PANH), funded by the French Ministry of Research.

References

1. M. Piana, M. Boccia, A. Filpi, E. Flammia, H. A. Miller, M. Orsini, F. Salusti, S. Santiccioli, F. Ciardelli and A. Pucci, *J. Power Sources*, 2010, **195**, 5875.
2. K. Matsuoka, Y. Iriyama, T. Abea, M. Matsuoka and Z. Ogumia, *J. Power Sources*, 2005, **150**, 27.
3. C. Coutanceau, L. Demarconnay, C. Lamy and J.-M. Léger, *J. Power Sources*, 2006, **156**, 14.
4. C. Ponce de Leon, F. C. Walsh, D. Pletcher, D. J. Browning and J. B. Lakeman, *J. Power Sources*, 2006, **155**, 172.
5. A. Ilie, M. Simões, S. Baranton, C. Coutanceau and S. Martemianov, *J. Power Sources*, 2011, **196**, 4965.
6. H. Hou, G. Sun, R. He, B. Sun, W. Jin, H. Liu and Q. Xin, *Int. J. Hydrogen Energy*, 2008, **33**, 7172.
7. I. Frenzel, H. Holdik, D. F. Stamatialis, G. Pourcelly and M. Wessling, *J. Membr. Sci.*, 2005, **261**, 49.
8. H. Yanagi and K. Fukuta, *ECS Trans.*, 2008, **16**, 257.
9. K. Fukuta, H. Inoue, Y. Chikashige and H. Yanagi, *ECS Trans.*, 2010, **28**, 221.
10. C.-C. Yang, *Int. J. Hydrogen Energy*, 2004, **29**, 135.
11. Y. Wang, L. Li, L. Hu, L. Zhuang, J. Lu and B. Xu, *Electrochem. Commun.*, 2003, **5**, 662.
12. P. K. Shen and C. Xu, *Electrochem. Commun.*, 2006, **8**, 184.
13. U. B. Demirci and P. Miele, *C. R. Chim.*, 2009, **12**, 943.
14. S. C. Amendola, P. Onnerud, M. T. Kelly, P. J. Petillo, S. L. Sharp-Goldman and M. Binder, *J. Power Sources*, 1999, **84**, 130.
15. H. I. Schlesinger, H. C. Brown and A. E. Finholt, *J. Am. Chem. Soc.*, 1953, **75**, 205.
16. F. Schubert, K. Lang, W. Schabacher and A. Burguer, *U.S. Pat.* 3 077 376, 1963.
17. Z. P. Li, N. Morigazaki, B. H. Liu and S. Suda, *J. Alloys Compd.*, 2003, **349**, 232.

18. Y. Kojima and T. Haga, *Int. J. Hydrogen Energy*, 2003, **28**, 989.
19. L. Kong, X. Cui, H. Jin, J. Wu, H. Du and T. Xiong, *Energy Fuels*, 2009, **23**, 5049.
20. Z. P. Li, B. H. Liu, J. K. Zhu, N. Morigazaki and S. Suda, *J. Alloys Compd.*, 2007, **437**, 311.
21. S. C. Amendola, S. L. Sharp-Goldman, M. S. Janjua, N. C. Spencer, M. T. Kelly, P. J. Petillo and M. Binder, *Int. J. Hydrogen Energy*, 2000, **25**, 969.
22. H. Senoh, Z. Siroma, N. Fujiwara and K. Yasuda, *J. Power Sources*, 2008, **185**, 1.
23. J. C. Walter, A. Zurawski, D. Montgomery, M. Thornburg and S. Revankar, *J. Power Sources*, 2008, **179**, 335.
24. S. Yamazaki, K. Kuratani, H. Senoh, Z. Siroma and K. Yasuda, *J. Power Sources*, 2010, **195**, 1107.
25. M. Simoes, S. Baranton and C. Coutanceau, *J. Phys. Chem. C*, 2009, **113**, 13369.
26. D. A. Finkelstein, N. Da Costa, J. L. Cohen and H. D. Abruna, *J. Phys. Chem. C*, 2009, **113**, 19700.
27. M. Chatenet, M. B. Molina-Concha and J.-P. Diard, *Electrochim. Acta*, 2009, **54**, 1687.
28. E. Gyenge, M. Atwan and D. Northwood, *J. Electrochem. Soc.*, 2006, **153**, A150.
29. G. Rostamikia and M. J. Janik, *Electrochim. Acta*, 2010, **55**, 1175.
30. J. I. Martins, M. C. Nunes, R. Koch, L. Martins and M. Bazzaoui, *Electrochim. Acta*, 2007, **52**, 6443.
31. B. H. Liu, Z. P. Li and S. Suda, *Electrochim. Acta*, 2004, **49**, 3097.
32. M. Chatenet, F. Micoud, I. Roche and E. Chainet, *Electrochim. Acta*, 2006, **51**, 5459.
33. M. H. Atwan, C. L. B. Macdonald, D. O. Northwood and E. L. Gyenge, *J. Power Sources*, 2006, **158**, 36.
34. M. E. Indig and R. N. Snyder, *J. Electrochem. Soc.*, 1962, **109**, 1104.
35. K. Wang, J. Lu and L. Zhuang, *J. Phys. Chem. C*, 2007, **111**, 7456.
36. M. H. Atwan, D. O. Northwood and, E. L. Gyenge, *Int. J. Hydrogen Energy*, 2007, **32**, 3116.
37. V. W. S. Lam and E. L. Gyenge, *J. Electrochem. Soc.*, 2008, **155**, B1155.
38. U. B. Demirci and F. Garin, *J. Alloys Compd.*, 2008, **463**, 107.
39. J. Delmas, L. Laversenne, I. Rougeaux, P. Capron, A. Garron, S. Bennici, D. Świerczyński and A. Auroux, *Int. J. Hydrogen Energy*, 2011, **36**, 2145.
40. J. Ma, J. Wang and Y. Liu, *J. Power Sources*, 2007, **172**, 220.
41. L. Wang, C. Ma, Y. Sun and S. Suda, *J. Alloys Compd.*, 2005, **391**, 318.
42. L. Wang, C. Ma, X. Mao, J. Sheng, F. Bai and F. Tang, *Electrochem. Commun.*, 2007, **7**, 1477.
43. Z. P. Li, B. H. Liu, K. Arai and S. Suda, *J. Electrochem. Soc.*, 2003, **150**, A868.
44. Z. P. Li, B. H. Liu, K. Arai, N. Morigasaki and S. Suda, *J. Alloys Compd.*, 2003, **356/357**, 469.

45. R. K. Raman, N. A. Choudhury and A. K. Shukla, *Electrochem. Solid-State Lett.*, 2004, **7**, A488.
46. N. A. Choudhury, R. K. Raman, S. Sampath and A. K. Shukla, *J. Power Sources*, 2005, **143**, 1.
47. Z. P. Li, B. H. Liu, K. Arai and S. Suda, *J. Alloys Compd.*, 2005, **404/406**, 648.
48. R. K. Raman, S. K. Prashant and A. K. Shukla, *J. Power Sources*, 2006, **162**, 1073.
49. Y.-G. Wang and Y.-Y. Xia, *Electrochem. Commun.*, 2006, **8**, 1775.
50. J. Ma, Y. Liu, P. Zhang and J. Wang, *Electrochem. Commun.*, 2008, **10**, 100.
51. G. Selvarani, S. K. Prashant, A. K. Sahu, P. Sridhar, S. Pitchumani and A. K. Shukla, *J. Power Sources*, 2008, **178**, 86.
52. C. Kim, K.-J. Kim and M. Y. Ha, *J. Power Sources*, 2008, **180**, 154.
53. C. Kim, K.-J. Kim and M. Y. Ha, *J. Power Sources*, 2008, **180**, 114.
54. R. K. Raman and A. K. Shukla, *J. Appl. Electrochem.*, 2005, **35**, 1157.
55. P. I. Iotov, S. V. Kalcheva and A. M. Bond, *Electrochim. Acta*, 2009, **54**, 7236.
56. E. Gyenge, *Electrochim. Acta*, 2004, **49**, 965.
57. P. Krishnan, T.-H. Yang, S. G. Advani and A. K. Prasad, *J. Power Sources*, 2008, **182**, 106.
58. B. Molina Concha, M. Chatenet, C. Coutanceau and F. Hahn, *Electrochem. Commun.*, 2009, **11**, 223.
59. M. Chatenet, F. H. B. Lima and E. A. Ticianelli, *J. Electrochem. Soc.*, 2010, **157**, 697.
60. F. A. Coowar, G. Vitins, G. O. Mepsted, S. C. Waring and J. A. Horsfall, *J. Power Sources*, 2008, **175**, 317.
61. B. Molina Concha and M. Chatenet, *Electrochim. Acta*, 2009, **54**, 6119.
62. B. Molina Concha and M. Chatenet, *Electrochim. Acta*, 2009, **54**, 6130.
63. M. H. Atwan, D. O. Northwood and E. L. Gyenge, *Int. J. Hydrogen Energy*, 2005, **30**, 1323.
64. B. H. Liu, Z. P. Li and S. Suda, *J. Electrochem. Soc.*, 2003, **150**, A398.
65. M. Kubokawa, M. Yamashita and K. Abe, *Denki Kagaku*, 1968, **36**, 788.
66. J. Q. Yang, B. H. Liu and S. Wu, *J. Power Sources*, 2009, **194**, 824.
67. Z. P. Li, B. H. Liu, J. K. Zhu and S. Suda, *J. Power Sources*, 2006, **163**, 555.
68. B. H. Liu, Z. P. Li and S. Suda, *J. Power Sources*, 2008, **175**, 226.
69. J.-H. Kim, H.-S. Kim, Y.-M. Kang, M.-S. Song, S. Rajendran, S.-C. Han, D.-H. Jung and J.-Y. Lee, *J. Electrochem. Soc.*, 2004, **151**, A1039.
70. K. T. Park, U. H. Jung, S. U. Jeong and S. H. Kim, *J. Power Sources*, 2006, **162**, 192.
71. A. Verma, A. K. Jha and S. Basu, *J. Power Sources*, 2005, **141**, 30.
72. R. Jamard, A. Latour, J. Salomon, P. Capron and A. Martinent-Beaumont, *J. Power Sources*, 2008, **172**, 287.
73. N. Duteanu, G. Vlachogiannopoulos, M. R. Shivhare, E. H. Yu and K. Scott, *J. Appl. Electrochem.*, 2007, **37**, 1085.
74. K. Deshmukh and K. S. V. Santhanam, *J. Power Sources*, 2006, **159**, 1084.
75. C. Ponce de León, F. C. Walsh, A. Rose, J. B. Lakeman, D. J. Browning and R. W. Reeve, *J. Power Sources*, 2007, **164**, 441.

76. H. Cheng and K. Scott, *J. Power Sources*, 2006, **160**, 407.
77. H. Cheng and K. Scott, *J. Appl. Electrochem.*, 2006, **36**, 1361.
78. H. Cheng and K. Scott, *J. Electroanal. Chem.*, 2006, **596**, 117.
79. B. H. Liu, Z. P. Li, K. Arai and S. Suda, *Electrochim. Acta*, 2005, **50**, 3719.
80. B. H. Liu and S. Suda, *J. Power Sources*, 2007, **164**, 100.
81. L. Gu, N. Luo and G. H. Miley, *J. Power Sources*, 2007, **173**, 77.
82. G. H. Miley, N. Luo, J. Mather, R. Burton, G. Hawkins, L. Gu, E. Byrd, R. Gimlin, P. J. Shrestha, G. Benavides, J. Laystrom and D. Carroll, *J. Power Sources*, 2007, **165**, 509.
83. M. Simões, S. Baranton and C. Coutanceau, *Appl. Catal. B*, 2010, **93**, 354.
84. M. Simões, S. Baranton and C. Coutanceau, *Electrochim. Acta*, 2010, **56**, 580.
85. Gloaguen, N. Andolfatto, R. Durand and P. Ozil, *J. Appl. Electrochem.*, 1994, **24**, 863.
86. H. Hiano, E. Iguchi, H. Uschida and M. Watanabe, *J. Phys. Chem. B*, 2006, **110**, 16544.
87. W. S. Rasband, ImageJ, US National Institutes of Health, Bethesda, MD, USA; http://rsbweb.nih.gov/ij/.
88. M. Simões, S. Baranton and C. Coutanceau, *Appl. Cata. B*, 2011, **110**, 40.
89. B. E. Warren, *X-Ray Diffraction*, Dover, New York, 1990.
90. P. Scherrer, *Nachr. Ges. Wiss. Göttingen, Math.-Phys.*, 1918, **Kl 26**, 98.
91. D. G. Narehood, S. Kishore, H. Goto, J. H. Adair, J. A. Nelson, H. R. Gutiérrez and P. C. Eklund, *Int. J Hydrogen Energy*, 2009, **34**, 952.
92. C. D. Wagner, W. M. Riggs, L. E. Davis, J. F. Moulder and G. E. Mouilenberg, *Handbook of X-Ray Photoelectron Spectroscopy*, Perkin-Elmer, Eden Prairie, MN, 1978.
93. I. G. Casella and M. Contursi, *Electrochim. Acta*, 2006, **52**, 649.
94. J. Koutecky and V. G. Levich, *Dokl. Akad. Nauk SSSR*, 1957, **117**, 441.
95. A. J. Bard and L. R. Faulkner, in *Electrochemical Methods: Fundamentals and Applications*, Wiley, Hoboken, NJ, 2nd edn., 2001, p. 337.
96. M. Chatenet, M. B. Molina-Concha, N. El-Kissi, G. Parrour and J.-P. Diard, *Electrochim. Acta*, 2009, **54**, 4426.
97. G. Denuault, M. V. Mirkin and A. J. Bard, *J. Electroanal. Chem.*, 1991, **308**, 27.
98. K. Juodkazis, J. Juodkazyt, V. Jasulaitien, A. Lukinskas and B. Šebeka, *Electrochem. Commun.*, 2000, **2**, 503.
99. B. Beden, A. Cetin, A. Kahyaoglu, D. Takky and C. Lamy, *J. Catal.*, 1983, **104**, 135.
100. A. Chen and C. Lipkowski, *J. Phys. Chem. B*, 1999, **103**, 682.
101. R. Adžić and M. Avramoc-Ivić, *J. Catal.*, 1986, **101**, 532.
102. R. Woods, in *Electroanalytical Chemistry*, ed. A. J. Bard, Dekker, New York, 1976, vol. 9, p. 2.
103. A. N. Kahyaoglu, PhD thesis, University of Poitiers, France, 1981.
104. A. Pozio, M. Francesco, A. Cemmi, F. Cardellini and L. Giorgi, *J. Power Sources*, 2002, **105**, 13.

105. C. Coutanceau, M. Croissant, T. Napporn and C. Lamy, *Electrochim. Acta*, 2000, **46**, 579.
106. M. Simões, S. Baranton. C. Coutanceau, C. Lamy and J.-M. Léger, *ECS Trans.*, 2009, **25**, 1413.
107. A. J. Rand and R. Woods, *J. Electroanal. Chem.*, 1972, **36**, 72.
108. M. Òukaszewski and A. Czerwinski, *Thin Solid Films*, 2010, **518**, 3680.
109. M. Mougenot, A. Caillard, M. Simoes, S. Baranton, C. Coutanceau and P. Brault, *Appl. Catal. B*, 2011, **107**, 372.
110. M. Grdén, K. Klimek and A. Czerwinski, *Electrochim. Acta*, 2006, **51**, 221.
111. M. Grdén, A. Czerwinski, J. Golimowski, E. Bulska, B. Krasnodebska-Ostrega, R. Marasi and S. Zamponi, *J. Electroanal. Chem.*, 1999, **460**, 30.
112. N. Hoshi, N. Nakamura, N. Maki, S. Yamaguchi and A. Kitajima, *J. Electroanal. Chem.*, 2008, **624**, 134.
113. M. Grdeń, M. Òukaszewski, G. Jerkiewicz and A. Czerwiński, *Electrochim. Acta*, 2008, **53**, 7583.
114. U. W. Hamm, D. Kramer, R. S. Zhai and D. M. Kolb, *Electrochim. Acta*, 1998, **43**, 2969.

CHAPTER 6

Bioelectrocatalysis in Direct Alcohol Fuel Cells

H. A. REEVE AND K. A. VINCENT*

Department of Chemistry, University of Oxford, Inorganic Chemistry Laboratory, South Parks Road, Oxford, OX1 3QR, UK
*E-mail: kylie.vincent@chem.ox.ac.uk

6.1 Introduction

Biological systems have adapted to extract energy from a wide variety of alcohols, ranging from methanol up to complex sugars. Complete fuel oxidation to CO_2 is possible, but is carried out by chains of enzymes which are typically each responsible for catalysis of a two-electron oxidation step. The term *bioelectrocatalysts* can encompass isolated enzymes or whole cells, and both systems have been demonstrated at the electrodes of fuel cells that extract energy from oxidation of alcohol functionalities in fuels. In this chapter we focus mainly on opportunities offered by isolated enzymes as efficient and selective catalysts for the oxidation of alcohol functionalities. For high-power generation from simple alcohols such as ethanol or methanol, enzymes are unlikely to compete with more conventional fuel cell catalysts due to their relatively large size (which means that active site density is low) and their limited stability over periods of weeks to months. Here we concentrate particularly on applications of biological catalysis for direct alcohol fuel cells designed to operate under specialized conditions or fed with more complex, bio-derived fuels or fuel mixtures, situations in which conventional catalysts are less effective.

RSC Energy and Environment Series No. 6
Catalysts for Alcohol-Fuelled Direct Oxidation Fuel Cells
Edited by Zhen-Xing Liang and Tim S. Zhao

Published by the Royal Society of Chemistry, www.rsc.org

Enzyme electrodes have been demonstrated for oxidation of methanol, ethanol, glycerol and a range of sugars, and one of the attractions of exploiting bioelectrocatalysts is the opportunities they raise for power generation from unusual fuels, such as fructose in fruit juices, lactose from milk, ethanol from fermented fruits or glucose in blood. Several electronics companies have shown interest in enzyme-based devices, including development at Sony of an enzyme-based bio-battery for powering portable music devices or toys on glucose.[1] A number of self-powered implantable devices using bioelectrocatalysts for blood sugar oxidation are under development, a market area where enzymes are already well established for glucose sensing. In most cases, fuel cells with bioelectrocatalysts have progressed little beyond proof-of-concept development stages, and here we will discuss some of the challenges and successes in "wiring" enzymes or whole cells to electrodes for alcohol oxidation, along with possible applications. We deal briefly with bioelectrocatalytic O_2 reduction since a bio-cathode can be applied in this capacity alongside a microbial- or enzyme-based alcohol-oxidizing anode.

Organisms have evolved to exploit many different routes for generating energy by oxidation of organic matter with electrons transferred to O_2 or other electron acceptors. From one viewpoint, biological respiratory circuits can be seen as miniature fuel cell systems where energy from coupled redox reactions is eventually stored or harnessed in chemical form, often as adenosine triphosphate (ATP), or in the form of transmembrane proton gradients.

To enhance the rates of these reactions, biological systems employ a suite of protein catalysts for fuel oxidation: enzymes tailored to operate selectively on specific fuel molecules. It is therefore unsurprising that biologists and chemists have sought to demonstrate the potential of biological catalysts operated *in vitro* in fuel cells and other devices. Rather than the precious metals usually associated with catalytic alcohol oxidation in traditional fuel cells, protein-associated organic cofactors play a major role in biological catalysis of alcohol oxidation. Biochemical procedures are well established for isolating enzymes in pure form, and fuel cells utilizing these isolated protein molecules as catalysts are commonly termed "enzyme fuel cells". Coupling whole biological cells to electrodes in "microbial fuel cells" can be more challenging from an electrical wiring perspective, as we discuss below, but has the advantage that the catalyst can be self-renewing and can operate on a wide range of fuels. Enzymes are capable of oxidizing a wide variety of alcohols, from methanol and ethanol up to complex sugars, and sophisticated strategies are emerging for engineering the substrate affinity of enzymes to further broaden the substrate range.[2,3] Alcohol-oxidizing oxidoreductase enzymes are of interest in other areas of biotechnology, including chemical sensors (enzyme-based glucose sensors are particularly well developed), and applications in pharmaceutical and fine chemical synthesis.[4] Scale-up of enzyme production and enzyme robustness and selectivity are therefore already under development for other applications.

6.1.1 Opportunities for Enzymes at the Anode of Fuel Cells

From an electrochemist's point of view, a key challenge in exploiting many of the fuel-oxidizing enzymes as catalysts outside their biological environment arises from the fact that alcohol oxidation is often accomplished by removal of a hydride (H^-) rather than release of a proton (H^+) and two electrons. For example, a major class of alcohol-oxidizing dehydrogenase enzymes (including the well-established liver alcohol dehydrogenase) depend upon the organic cofactor nicotinamide adenine dinucleotide (NAD^+; Figure 6.1A), which accepts hydride during fuel oxidation to form NADH (NAD-dependent dehydrogenase). A related class of dehydrogenases use the closely related cofactor nicotinamide adenine dinucleotide phosphate ($NADP^+$), generating

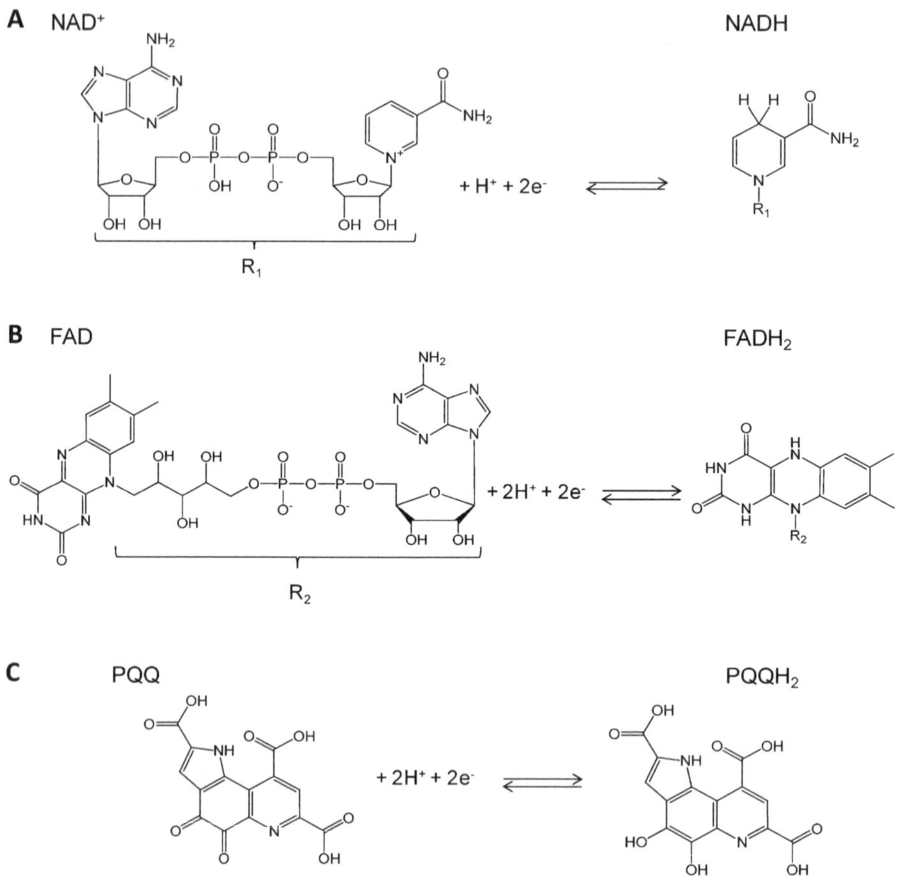

Figure 6.1 Reduction reactions associated with common redox cofactors required by many dehydrogenases: (A) nicotinamide adenine dinucleotide (NAD^+); (B) flavin adenine dinucleotide (FAD); and pyrroloquinoline quinone (PQQ).

NADPH. Other classes of fuel-oxidizing enzymes use oxidized cofactors such as flavin adenine dinucleotide (FAD) or pyrroloquinoline quinone (PQQ) in a similar way (Figures 6.1B and 6.1C). Widely exploited glucose oxidase from the black mould *Aspergillus niger* uses a deeply buried flavin centre to couple the oxidation of glucose (to gluconolactone) to the reduction of O_2 to H_2O_2, for example. Strategies for facilitating electron transfer between an electrode and the buried flavin centre of glucose oxidase have included the use of conductive linkers or gels[5–7] or networks of carbon nanotubes.[8] The vast literature on glucose oxidation at glucose oxidase electrodes is covered in a number of recent reviews.[7] Free cofactors such as NAD^+ cannot easily be cycled electrochemically because of high overpotential requirements at standard electrodes. A huge amount of work has gone into developing strategies for coupling oxidation reactions carried out by NAD^+-, FAD- or PQQ-dependent dehydrogenases to electron transfer at an electrode *via* soluble redox mediators, conducting "wires" or other cofactor-recycling enzymes. Some of the more successful strategies are summarized below.

A few fuel-oxidizing enzymes are designed for the transfer of electrons to integral or soluble redox cofactors or redox partner proteins, and are therefore much better suited to application on an electrode. For instance, the acetic acid bacterium *Gluconobacter oxydans* expresses a range of membrane-bound alcohol-oxidizing enzymes which are responsible for partial oxidation of ethanol and many sugars ("oxidative fermentation"), coupled to electron transfer to the carrier ubiquinone within the membrane, and eventually to O_2 (Figure 6.2).[9] A number of these membrane-associated *Gluconobacter* enzymes have been isolated and explored for electrocatalytic applications, either in sensors or fuel cells.[10]

The crystal structures of these membrane-bound dehydrogenases have not yet been solved; however, structures are available for several similar

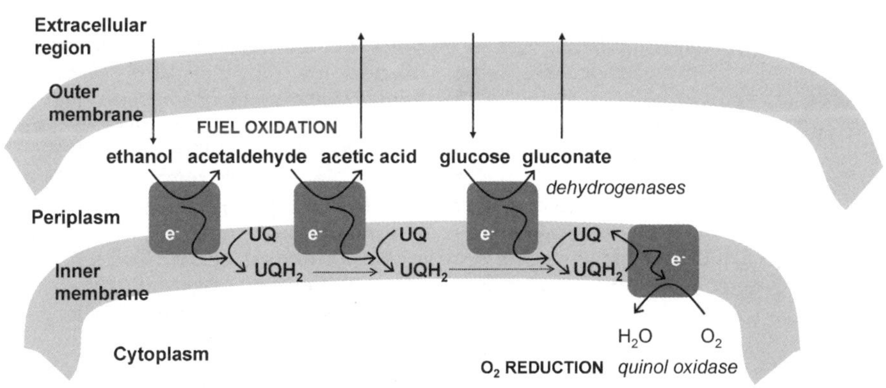

Figure 6.2 Part of the respiratory chain of the bacterium *Gluconobacter oxydans*. "Fuel oxidation" in the periplasmic compartment is coupled to O_2 reduction in the cytoplasm *via* the membrane-soluble electron carrier ubiquinone (UQ), which is reduced to ubiquinol (UQH_2).[9,10]

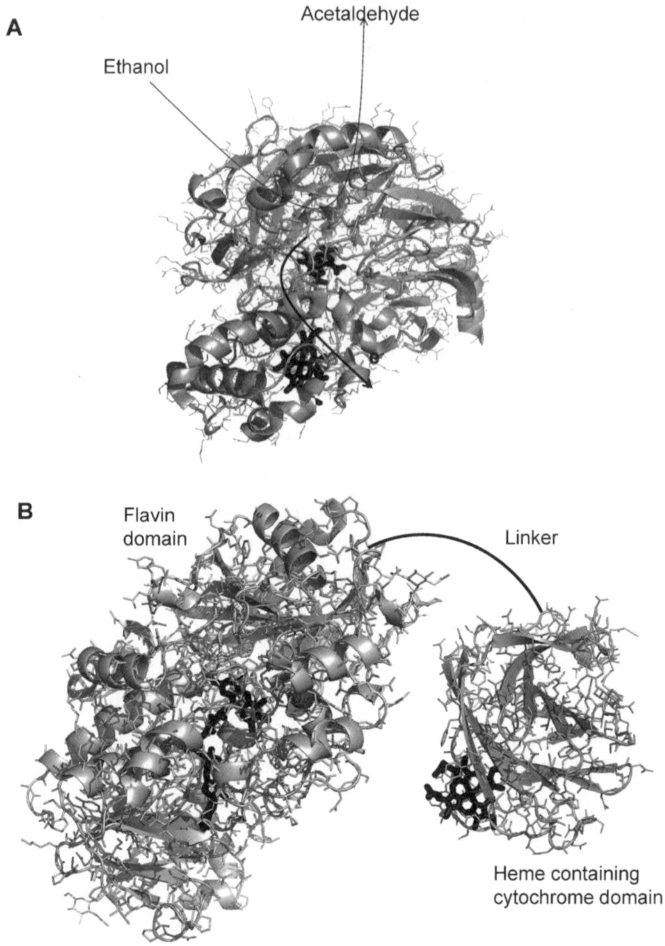

Figure 6.3 Examples of dehydrogenase enzymes equipped for direct electron transfer at an electrode. The redox cofactors are shown as *dark sticks*, the polypeptide backbone of the protein is shown in *ribbon form*, and the amino acid side chains are shown as *light sticks*. (A) Ethanol dehydrogenase from *Comamonas testosteroni* (PDB code 1KB0);[11] ethanol is oxidized at the PQQ active site, and electrons are transferred away *via* the heme relay centre. (B) Cellobiose dehydrogenase from *Phanerochaete chrysosporium*, showing the catalytic flavin domain (PDB 1KDG)[12] and the heme electron transfer domain (PDB 1D7C),[13] which are joined by a polypeptide linker in the native enzyme (the crystal structure of intact cellobiose dehydrogenase has not been solved).

dehydrogenases.[11–13] The crystal structure of an ethanol dehydrogenase from *Comamonas testosteroni* is shown in Figure 6.3A[11] and comprises a PQQ catalytic site where hydride is extracted from the alcohol, and a heme site which accepts electrons from the PQQ.

6.1.2 Opportunities for Enzymes at the Cathode of Fuel Cells

Respiratory O_2 reduction by cytochrome *c* oxidase or quinol oxidase (as shown in Figure 6.2) is poorly suited for fuel cell applications because the physiological reaction is coupled to proton pumping across a lipid bilayer membrane and there is a large overpotential for the reduction of O_2 to water. However, several copper-containing oxidase enzymes carry out this reaction close to the thermodynamic O_2/H_2O couple at mildly acidic or near-neutral pH values; indeed, bilirubin oxidase and laccase electrodes have been shown to exhibit a smaller overpotential than platinum electrodes for O_2 reduction to water.[14–16] These so-called "blue" copper oxidases are well suited for electron uptake from the electrode because they incorporate an electron transfer "blue" copper centre in addition to the active site (Figure 6.4).[17]

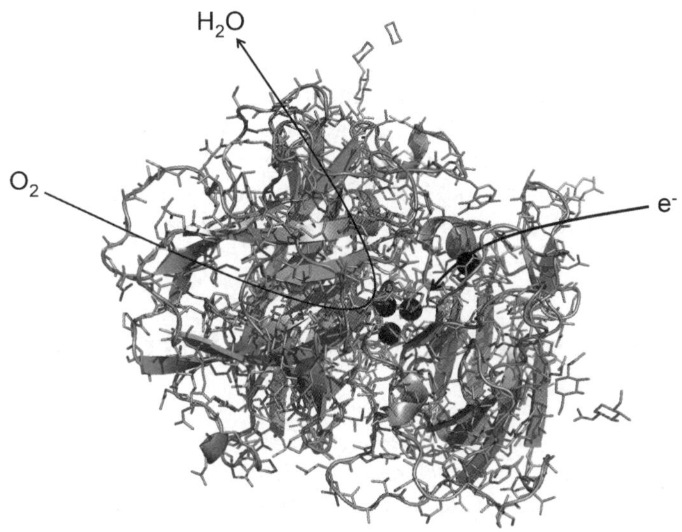

Figure 6.4 Crystal structure of laccase from *Trametes versicolor* (PDB 1KYA). The Cu atoms are shown as *spheres*: the catalytic active site consists of three Cu atoms (a "type II/type III" centre) and the fourth ("type I" or "blue" copper centre) acts as an electron relay site.[17]

6.1.3 Limitations in Assembling Energy Devices with Enzymes

Despite the enzyme engineering that is taking place for applications in other areas of biotechnology, the timescales and cost of purification of isolated enzymes may impede large-scale use in energy devices; strategies that are viable for biocatalyzed generation of high-value pharmaceuticals are often uneconomical for energy technologies. Secondly, although enzymes often exhibit high turnover frequencies per active site, their large overall size means that the volumetric active site density is low. This is even more of a problem for whole cell (microbial) catalysis. Although many groups are working to improve enzyme stability by immobilization or entrapment, enzyme and microbial fuel cells are likely to be limited to specialized conditions of operation for which high power output is not required or where conventional catalysts fail, for example obtaining energy from water treatment, implantable fuel cells or oxidation of specialized bio-derived fuels. Other limitations include enzyme activity being compromised by comparatively small changes in operating conditions, and temperature, pH and in some cases substrate or product concentrations. Critically there are difficulties in wiring enzymes to an electrode to enable fast and efficient electron transfer.

Most alcohol-oxidizing enzymes catalyze reactions in two-electron steps, and therefore complete oxidation of fuels requires several types of enzyme in a cascade (see, for example, the stepwise partial oxidation of ethanol by *Gluconobacter* shown in Figure 6.2).[18] Non-redox enzymes such as isomerases are also involved in the metabolism of sugars and other complex alcohols, and may be required as part of a cascade without contributing to the anodic current. In electrodes built from intact cells of micro-organisms, many individual enzymes make up the large and complex oxidation pathways. Owing to the presence of multiple redox pathways in biological cells, microbial fuel cells can use more than one fuel at a time and are often most useful in specialized applications such as energy generation coupled with water treatment.

6.2 Strategies for Wiring Enzyme Electrocatalysts to Electrodes

Obtaining efficient connectivity between biocatalysts and electrodes presents some key challenges. For efficient fuel cell design, it is often desirable to immobilize the biocatalyst to avoid limiting the anodic current by slow diffusion of enzyme or electron transfer mediator molecules and loss of catalyst with the fuel exhaust. The term direct alcohol fuel cell implies that alcohol oxidation is the primary electrochemical process at the anode, but in enzyme electrodes, fuel oxidation has often been achieved *via* a complicated series of electron transfer steps. For example, in a glucose dehydrogenase anode for glucose oxidation developed by Sony in conjunction with Japanese research institutions, the primary electrode process is oxidation of a soluble electron-transfer mediator, vitamin K_3 (2-methylnaphtho-1,4-quinone), which then takes up electrons from molecules of the reduced enzyme cofactor NADH that has been released from the enzyme following glucose (fuel) oxidation (Figure 6.5).[19] All components are trapped in a poly-ion matrix, and despite the complexity of the electron transfer system, the onset potential of *ca.* 0 V *vs.* SHE suggests relatively low overpotential loss at the anode, and the electrode exhibits glucose concentration-dependent sugar oxidation.

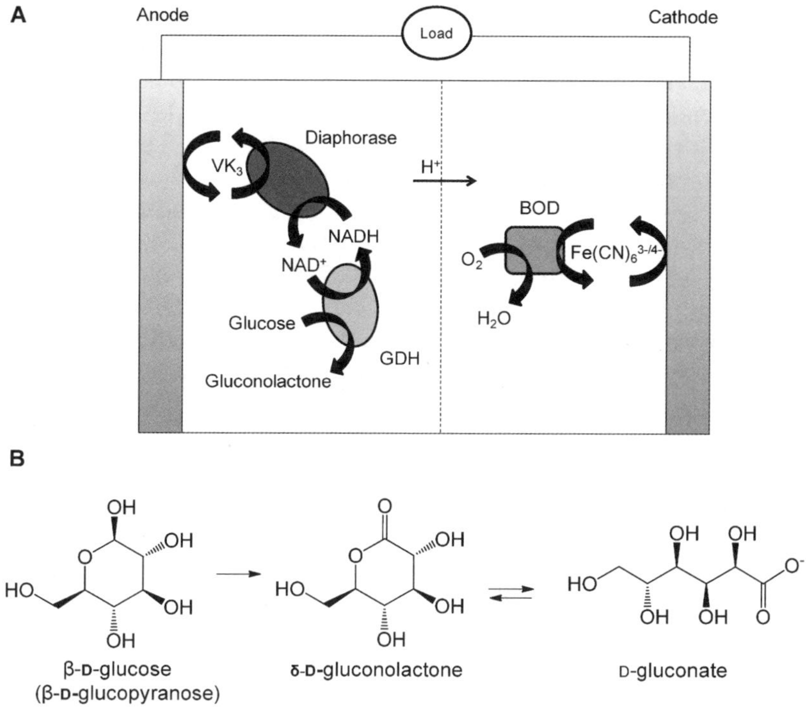

Figure 6.5 (A) Schematic representation of an enzyme fuel cell developed by scientists working with Sony.[19] At the anode, a cocktail of redox agents is required to couple glucose oxidation by glucose dehydrogenase (GDH) to vitamin K_3 (VK_3) oxidation at the electrode *via* a system of continuous cofactor regeneration by an NAD^+/NADH cycling (diaphorase) enzyme. At the cathode, O_2 reduction by bilirubin oxidase (BOD) is mediated by the ferricyanide/ferrocyanide couple. (B) The structure of glucose, the oxidation product (gluconolactone) and its hydrolyzed isomer (gluconate).

6.2.1 Cofactor Supply to $NAD(P)^+$-Dependent Dehydrogenases

Owing to the expense of NAD(P)H cofactors, they are typically supplied at catalytic levels. Interest in using NAD-dependent enzymes in biotechnology has led to much research into electrocatalytic regeneration of both $NAD(P)^+$ and NAD(P)H. Direct electrochemical reduction of NAD^+ at conventional graphite or platinum electrodes generally requires large overpotentials and often leads to enzyme-inactive side products and electrode fouling.[20] Some success in thermodynamically efficient NAD^+ reduction and NADH oxidation has been achieved at modified electrodes; for example, these reactions occur at negligible overpotential relative to $E(NAD^+/NADH)$ at a poly(neutral red) modified electrode,[21] and a poly(methylene green) coated electrode has been used with an enzyme cascade for methanol oxidation by a series of NAD^+-dependent dehydrogenases.[22] Efficient NAD^+/NADH cycling at electrodes modified with diaphorase enzyme moieties which comprise iron–sulfur electron relay clusters in addition to a flavin catalytic centre has been demonstrated, and these electrodes also have the potential to be used alongside cofactor-dependent dehydrogenases at fuel cell anodes.[23,24] Entrapment of the soluble cofactor is desirable so that it is not lost with the fuel exhaust. Miyake *et al.* used the carbodiimide coupling reaction to link NAD^+ to a high surface-area network of Ketjen black particles on which they immobilized a NAD^+-dependent glucose dehydrogenase; electrocatalytic oxidation of the reduced cofactor was reported with an onset potential of about +100 mV *vs.* SHE [a significant overpotential relative to the $E(NAD^+/NADH)$ potential, −320 mV], facilitating glucose oxidation at a similar potential.[25] There is thus scope for further optimization of tethered cofactor electrodes.

6.2.2 Redox Hydrogels

An approach pioneered by Heller and co-workers exploits partially mobile, tethered $Os^{III/II}$ complexes to mediate electron transfer between enzymes and electrodes and has now been used widely for wiring many different redox enzymes.[26] The osmium complex is incorporated into a water-soluble polymer which can be crosslinked to form a redox-active hydrogel around the enzyme. This arrangement offers the advantage of a high mediator concentration in the hydrated film, providing good electronic contact to a large number of trapped enzyme molecules, regardless of orientation. Altering the ligands on the Os centre allows the redox polymer to be tuned for the potentials of various proteins, and such systems have been used for both O_2-reducing bio-cathodes as well as sugar-alcohol oxidizing anode systems.[14,27–30] An example of such an electrode used for O_2 reduction by laccase is shown in Figure 6.6, together with a comparison with platinum; the laccase electrode shows a significantly smaller overpotential for O_2 reduction with respect to the potential of the O_2/H_2O redox couple corrected for the experimental conditions. The hydrogel approach also facilitates electron transfer to the buried active site of glucose oxidase,[31] and has been extended to high surface area carbon supports. Electrocatalytic currents greater than 20 mA cm^{-2} geometric area were obtained for glucose oxidation with an enzyme-loaded redox polymer coated onto multi-wall carbon nanotubes grown on carbon fibre paper (Figure 6.6).[32] The redox hydrogel approach readily leads to membrane-free "miniaturized" fuel cells for implantable self-powered devices and biosensors,[28,33] and a sensor relying upon glucose oxidase in an Os hydrogel electrode was utilized by Therasense in human clinical trials for patients with Type 1 diabetes in 2003.[34]

6.2.3 Tethers or Conductive Linkers

Many different approaches have been trialled for tethering proteins to electrodes *via* conducting molecular wires, including linkers that incorporate native cofactors such as PQQ or artificial mediators. One method for ensuring

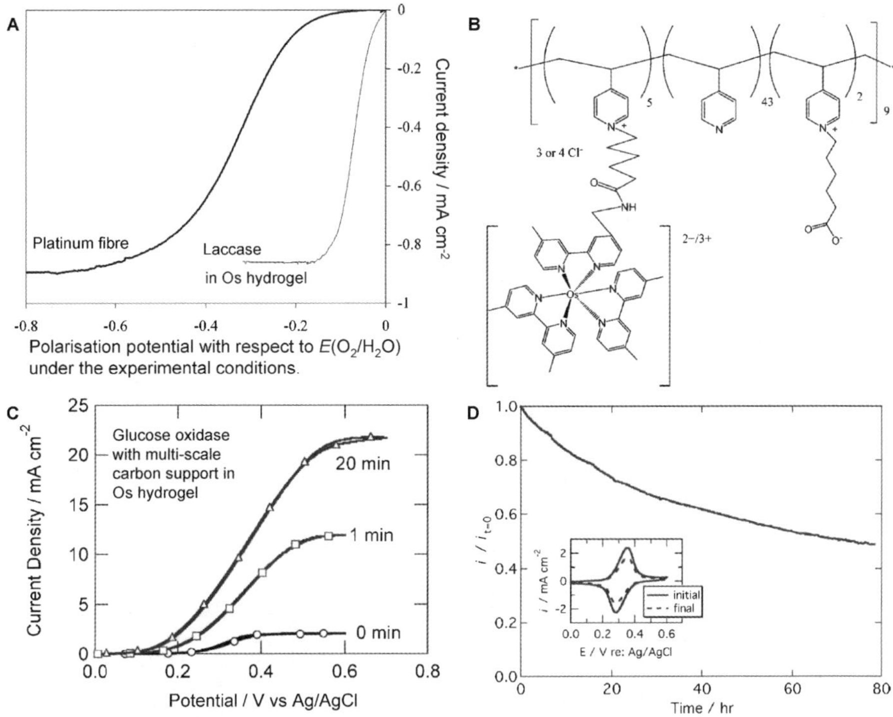

Figure 6.6 (A) Comparison of electrocatalytic O_2 reduction by a platinum fibre, and an electrode modified with the "blue" copper oxidase enzyme, laccase, in an Os-containing redox hydrogel as shown in (B). The scans in (A) were recorded at a scan rate of 1 mV s^{-1} at stationary electrodes in still solutions under air at 37 °C, with the platinum electrode in 0.5 M H_2SO_4 and the laccase electrode in 0.1 M pH 5 citrate buffer.[27] (C) Cyclic voltammograms recorded at 1 mV s^{-1} showing oxidation of glucose by *Aspergillus niger* glucose oxidase at a high surface area electrode comprising a redox hydrogel containing poly(vinylpyridine)[Os(bipyridine)$_2$Cl]$^{2+/+}$ supported on carbon paper, either bare (0 min) or exposed to 1 min or 20 min of carbon nanotube growth by chemical vapour deposition,[32] rotated at 4000 rpm and immersed in pH 7.1 phosphate buffer at 37.5 °C with 50 mM glucose. (D) Stability of the glucose oxidation current density at the nanotube-treated electrode poised at 0.5 V *vs.* Ag/AgCl and rotating at 1000 rpm in N_2-saturated buffer containing 50 mM glucose. *Inset* in D: non-catalytic electron transfer to the FAD cofactor of the enzyme in the absence of glucose at the start and end of the experiment.[32] (Panels A and B are reproduced from Mano *et al.*[27] by permission of the American Chemical Society; panels C and D are reproduced from Barton *et al.*[32] by permission of The Electrochemical Society.)

that enzymes are correctly orientated and for fast electron exchange with an electrode is to use a cofactor-free ("apo") enzyme in conjunction with a cofactor-modified electrode.[35] Katz and co-workers reported thermodynamically efficient glucose oxidation (at an onset potential of about −0.4 V *vs.*

SCE) by apo-glucose oxidase reconstituted with its FAD cofactor at a gold electrode modified with a linker bearing a rotaxane redox shuttle trapped by a FAD stopper head-group (Figure 6.7).[6] The same authors have also used PQQ-based electron-transfer linkers with a FAD head-group for electrocatalytic glucose oxidation by glucose oxidase, or with an NAD^+ head-group for lactate oxidation by lactate dehydrogenase.[36] They have also reported glucose oxidation (albeit at a moderate overpotential) by an apo PQQ-dependent glucose dehydrogenase reconstituted on a PQQ-capped nanostructured gold electrode.[37] Providing that redox groups in the linkers can be tuned appropriately for the potential of different biocatalyzed alcohol oxidations, these approaches hold promise for applications of cofactor-dependent enzymes in fuel cells, but may prove too complex for cost-effective electrode construction. A key advantage of immobilizing the cofactor is that the electrodes can be used in situations of fuel flow without loss of the costly cofactor. These electrode assemblies also ensure that the enzyme is orientated appropriately for electron transfer.

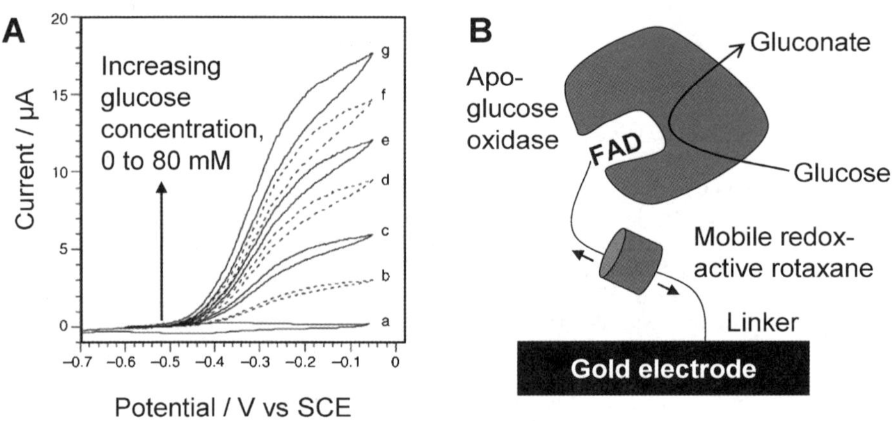

Figure 6.7 Electrocatalytic oxidation of glucose by apo-glucose oxidase reconstituted with its FAD cofactor on a cofactor-functionalized "conductive" tether on which a redox-active cyclophane rotaxane is able to shuttle electrons from the protein to the electrode.[6] The thermodynamic potential for the $FAD/FADH_2$ couple under these conditions is −0.51 V *vs.* SCE. Cyclic voltammograms shown in (A) were recorded in O_2-free 0.1 M phosphate buffer, pH 8.0. Panel B is a schematic representation of the electron transfer system. (Panel A is reproduced from Katz *et al.*[6] by permission of Wiley-VCH.)

6.2.4 Direct Electron Transfer

The electron transfer situation is vastly simplified if the redox enzyme is able to exchange electrons directly with the electrode, and this is often aided by the presence of internal redox centres in the protein.[38] There are now many reports of enzyme electrodes for alcohol oxidation involving direct electron transfer with no requirement for natural or artificial electron transfer mediators. Some of the PQQ- and flavin-dependent dehydrogenases incorporate a heme-containing domain of similar structure to a cytochrome (small heme electron transfer protein), which acts as an internal electron relay system (examples are shown in Figure 6.3). One such enzyme is the membrane-bound fructose dehydrogenase from *Gluconobacter* spp., which comprises a heme electron transfer centre in addition to the flavin catalytic site[39] and has been shown to oxidize fructose with a high current density when adsorbed on various carbon particles on a carbon paper electrode (Figure 6.8A).[40–42] Fusion of a heme-containing subunit onto a soluble PQQ-dependent glucose dehydrogenase was also shown to give rise to direct electron transfer electrocatalysis of glucose oxidation.[43] The two-domain construct of a cellobiose dehydrogenase similar to that shown in Figure 6.3B engages in efficient oxidation of lactose when mixed with single-walled carbon nanotubes and spotted onto a pyrolytic graphite electrode (Figure 6.8B).[44]

Modification of a planar electrode support with carbon particles or nanotubes is helpful in increasing the surface area, but may also assist with wiring of enzymes having buried catalytic centres, presumably because regions of the electrode are then able to penetrate into the protein shell. For example, oxidation of glucose by glucose oxidase which has a deeply buried flavin active site was achieved at a mild overpotential by pressing the enzyme into a disk with carbon nanotubes.[8] A nanostructured carbon cryogel was also used to

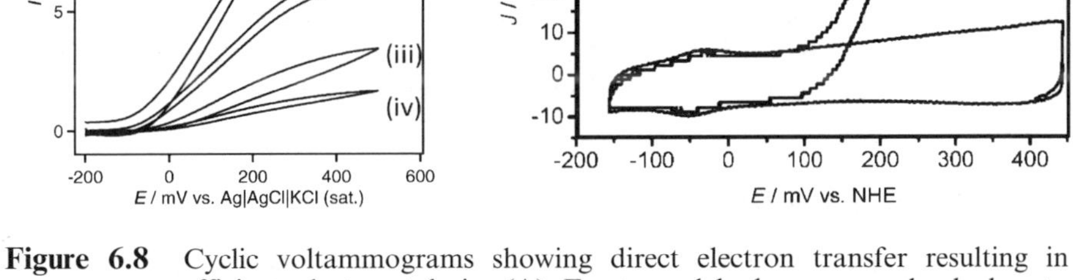

Figure 6.8 Cyclic voltammograms showing direct electron transfer resulting in efficient electrocatalysis. (A) Fructose dehydrogenase adsorbed onto various particles immobilized on a carbon paper electrode in pH 5.0 McIlvaine buffer containing 200 mM fructose, recorded at 20 mV s^{-1}; the particles were (i) Ketjen black; (ii) Vulcan XC-72R; (iii) Tokai Carbon carbon nanospheres; and (iv) Lamp black 101.[41] (B) Cellobiose dehydrogenase from *Phanerochaete sordida* adsorbed on single-walled carbon nanotubes in the presence and absence of 0.1 mM lactose.[44] (Panel A is reproduced from Tsujimura *et al.*[41] and panel B is reproduced from Tasca *et al.*,[44] both with permission of the American Chemical Society.)

"wire" *Acinetobacter calcoaceticus* PQQ-glucose dehydrogenase, which lacks a heme electron relay centre in addition to the PQQ active site.[45]

One of the challenges in immobilization of relatively large enzyme molecules onto an electrode is to obtain uniform orientation. Various electrode modifications have been used to increase the electrode surface area and to present chemical functionalities which promote enzyme adsorption in suitable orientations for effective electron transfer. One example is the modification of carbon electrodes with anthracene functionalities *via* diazonium coupling to promote adsorption of laccase *via* the organic-substrate binding pocket with the type I copper centre close to the electrode.[15] Orientation has rarely been well characterized, in part because the rough nature of many carbon-based electrode surfaces is unsuitable for atomic force microscopy, and spectroscopic methods lack sensitivity for probing low-coverage layers of immobilized enzyme. *In situ* spectroscopic methods are beginning to emerge for studying immobilized enzyme systems and to provide insight into the heterogeneity of immobilization.[46–49]

6.2.5 Wiring Whole Cells for Microbial Fuel Cells: Mediated and Direct Electron Transfer

There is also interest in exploiting electrodes modified with biofilms of whole, living microbial cells for electrocatalysis of alcohol fuel oxidation.[50,51] Here, electronic communication across the cellular membrane to harvest electrons at an electrode represents a significant challenge. One strategy relies upon small molecule redox mediators that are able to cross cellular membranes, and some micro-organisms produce their own mediators, for example phenazine derivatives released by *Pseudomonas aeruginosa*.[51] Pathways have recently been discovered by which certain bacterial cells are able to communicate electronically with the outside environment *via* conductive proteins bristling with heme redox centres; the crystallographic structure of one such protein has been solved and is represented in Figure 6.9.[52,53] These "electronically wired" cells open up possibilities for direct electronic communication between intracellular redox reactions and electron uptake at an anode.[54]

6.3 Examples of Enzyme Electrocatalysis

Open circuit potentials close to 1 V should be possible for enzyme-based alcohol fuel cells, given an onset potential for oxidation of the alcohol functionalities of biologically relevant fuels below *ca.* -0.15 V, together with efficient biocatalyzed O_2 reduction commencing at *ca.* +0.8 V at near-neutral pH. In practice, poor wiring between enzyme active sites and electrodes often means that these open-circuit potentials are not reached in fuel cells, but ongoing improvements in enzyme "wiring" should continue to improve this situation.

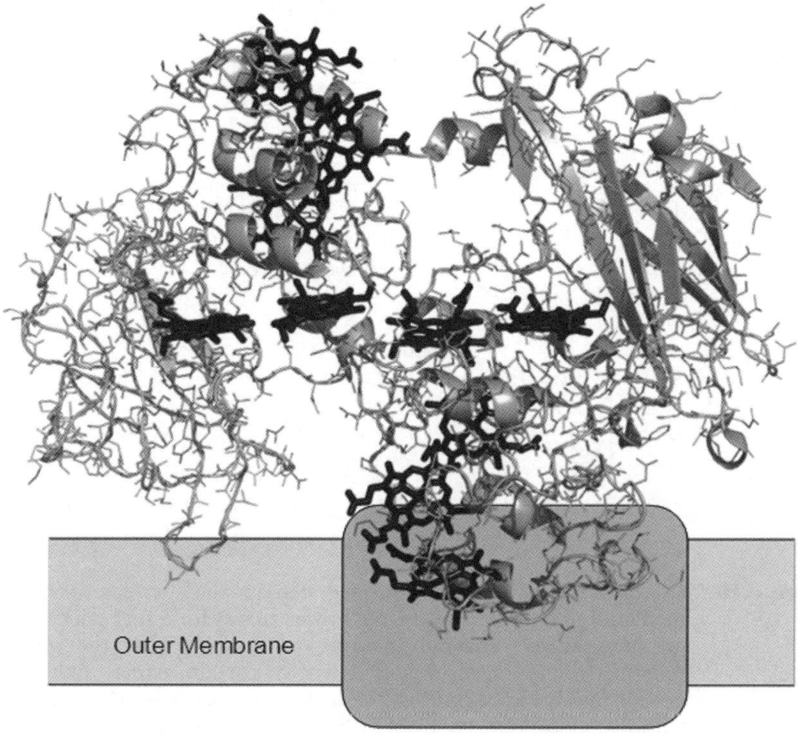

Figure 6.9 Representation of an extracellular membrane associated decaheme cytochrome from *Shewanella oneidensis* (PDB code 3PMQ).[52] Stick representations of the heme centers (*black*) are shown superimposed on the carton and line structure of the protein backbone and side chains. The membrane and channel for electron transfer are shown as *grey boxes*.

6.3.1 Methanol and Ethanol

No single enzyme is available for complete oxidation of methanol to CO_2, but the reaction can be achieved in a series of two-electron steps carried out by three NAD^+-dependent dehydrogenases, as shown in Figure 6.10A. This reaction sequence was demonstrated in an enzyme fuel cell by Palmore *et al.* in 1998, using mediated electron transfer to a diaphorase enzyme to regenerate NAD^+ from oxidation of the NADH released in each dehydrogenase (DH)-catalyzed step; in this study the dehydrogenases were contained in solution in the anode compartment.[18] A similar reaction sequence was reported by Addo *et al.* at an anode comprising poly(methylene green) polymerized on Toray carbon paper with the three dehydrogenases and cofactor NAD^+ trapped in an ion-exchanged Nafion derivative.[55] As expected, the power output was found to increase with each successive oxidation stage (Figure 6.10B). More complex NAD^+-dependent enzyme cascades have been demonstrated for ethanol oxidation.[56,57] Direct electron transfer driving electrocatalytic ethanol oxidation by a heme-PQQ enzyme has been demonstrated at metal electrodes,[58] and

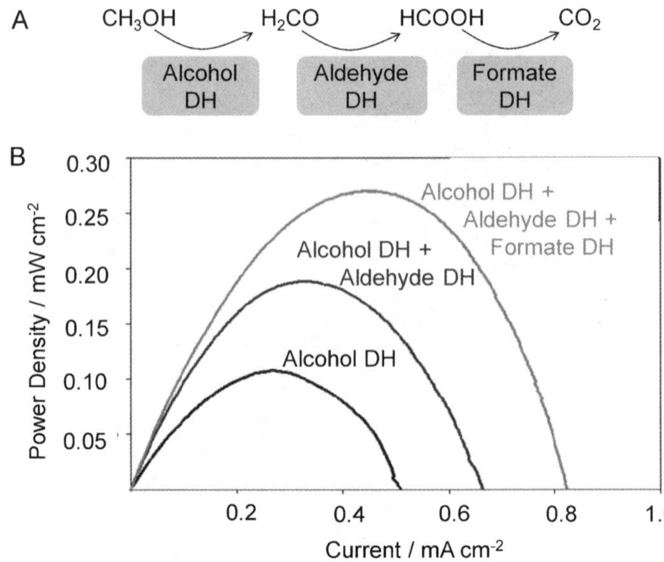

Figure 6.10 (A) Sequence of enzyme-catalyzed steps in the complete oxidation of methanol to carbon dioxide. (B) Power curves for a fuel cell comprising an immobilized enzyme cascade, as indicated, at the anode on poly(methylene green)-modified Toray carbon paper, linked *via* a Nafion NRE-212 membrane to a platinum cathode (Vulcan XC−72 with 20% platinum on PEMEAS E-TEK carbon cloth). The fuel cell was operated on a fuel/electrolyte solution containing 1.5 mM NAD^+, 6.0 M sodium nitrate, 100 mM methanol in 10.0 mM pH 7.4 phosphate buffer. (Panel B is reproduced from Addo *et al.*[55] with permission of Wiley-VCH.)

as further enzymes of this type are isolated and become available there is scope for avoiding the NAD^+-regeneration steps in biocatalyzed ethanol oxidation. These studies provide interesting demonstrations of the viability of exploiting a series of sequential enzyme steps, but enzyme anodes operated on simple alcohols such as methanol are unlikely to compete with more conventional metal-catalyzed systems in terms of power.

6.3.2 Sugars and Carbohydrates

There are many biological pathways for the selective oxidation of the alcohol functionalities of more complex carbohydrates, from simple monosaccharides such as glucose and fructose to polysaccharides such as cellulose. Enzyme fuel cells operating on glucose are the most advanced, since enzyme electrocatalysis is already extensively exploited in glucose sensors for diabetic patients. Biocatalyzed sugar oxidation opens up possibilities for fuel cells that produce electricity upon oxidation of biologically derived sugars which can readily be supplied as safe, non-toxic, renewable fuels, easily topped-up from liquid or solid sources. Lactose is oxidized by certain dehydrogenases, and this reaction could be used to harvest electrical energy from lactose removed from some milk sources. Demonstrations of soft drinks or fruit juices as fuels have captured the imagination and drawn attention to the area of biocatalysis. In most cases, incomplete oxidation of sugars is achieved, meaning that power extracted per gram of fuel is low, but nevertheless biocatalyzed sugar fuel cells have received attention for portable applications, as with the Sony fuel cell described above which is intended for glucose-powered toys or small electronics. Mano *et al.* demonstrated the operation of a glucose fuel cell in a grape, using as electrocatalysts glucose oxidase and bilirubin oxidase wired to the anode and cathode, respectively, *via* osmium-containing redox hydrogels.[28]

The other area where glucose is attractive as a fuel is for self-powered implantable devices where blood glucose and dissolved O_2 provide the fuel and oxidant, respectively. Implantable, miniaturized biofuel cells could eventually open up possibilities for powering pacemakers and artificial organs such as kidneys and livers. For intravenous devices, important considerations are clearly size, lifetime and confidence that toxic materials will not leak into the body. Implantable devices powered from glucose present in the body must operate at around pH 7, 37 °C and at physiological O_2 and glucose concentrations.

Glucose oxidase couples the conversion of glucose to gluconolactone with the reduction of O_2 to H_2O_2, and while it is desirable to harvest all electrons at an electrode, diversion of electrons to O_2 reduction is likely to occur to some extent. For this reason, Zebda *et al.* combined catalase with glucose oxidase at the anode of a fuel cell intended for applications in oxygenated glucose solutions (at concentrations mimicking biological fluids) to ensure rapid decomposition of H_2O_2.[8] Their fuel cell with a laccase carbon nanotube cathode, running at physiologically relevant glucose concentrations (5 μM) in air-equilibrated solution, produced a maximum power density of *ca.* 1 mW cm^{-2} at an operating cell potential of around 0.6 V. A decrease of only 4% power output was recorded during operation at 50 μM glucose after 1 month storage in glucose solution. Glucose oxidase and catalase have also been used together in solution, confined by a dialysis membrane, for an implantable fuel cell tested in rats.[59] For future developments, heme-containing PQQ-dependent dehydrogenases offer an advantage over glucose oxidase in that they channel electrons from glucose oxidation directly to an electrode *via* the heme centre without suffering from competing O_2 reduction and H_2O_2 generation.

Enzyme-catalyzed fructose oxidation is also promising for fuel cells, with high currents achieved at a Ketjen black electrode modified with a PQQ-containing fructose dehydrogenase,[42] and a maximum power output close to 1 mW cm^{-2} was achieved when the electrode was incorporated into a fuel cell.[40] The selectivity of dehydrogenases for sugars varies considerably, with some showing specificity for a single sugar, and others operating on a broad array of substrates.[45] Fungal extracellular cellobiose dehydrogenases (CDHs) also offer the potential for electrocatalytic oxidation of a range of sugars (mono-, di- and

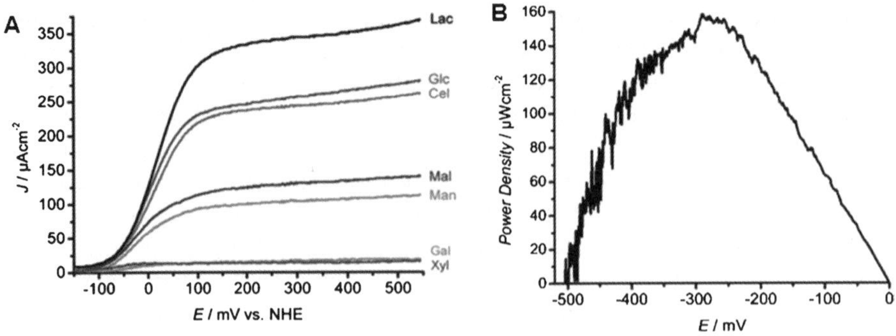

Figure 6.11 Cellobiose dehydrogenase from *Myriococcum thermophilum* with an Os polymer and SWCNTs on a PGE electrode: (A) acting on 0.1 M lactose, glucose, cellobiose, maltose, mannose, galactose and xylose; (B) as the anode in a fuel cell with a Pt black cathode, acting on 0.1 M glucose showing the dependence of power on the operating voltage. (Reproduced from Tasca *et al.*[30] with permission of the American Chemical Society.)

polysaccharides) under mediated or direct electron transfer conditions. Lactose oxidation by this group of enzymes has been studied most thoroughly, but electrocatalytic oxidation of a wider range of sugar "fuels" has been demonstrated at an electrode incorporating *Myriococcum thermophilum* cellobiose dehydrogenase in an osmium-containing redox hydrogel: lactose, glucose, cellobiose, maltose, mannose, galactose and xylose (Figure 6.11A). A similar electrode was demonstrated as the anode of a glucose fuel cell, as shown in Figure 6.11B.

6.3.3 Glycerol

Impure glycerol is a waste product from the transesterification process in biodiesel production from vegetable oil,[60] and this is a convenient fuel since it is water soluble, easily stored and non-toxic. Selective biocatalyzed oxidation of glycerol to glyceraldehyde or dihydroxyacetone mean that it may be possible both to extract energy and generate added-value products from mixed biomass waste. On the other hand, exploiting consecutive oxidation steps by an alcohol dehydrogenase, aldehyde dehydrogenase and oxalate oxidase, a 10-step complete oxidation of glycerol, can be achieved using just three enzymes.[61]

6.4 Microbial Fuel Cells

Microbial fuel cells are the subject of several recent reviews.[50,51,62] The use of whole organisms at the anode of fuel cells negates the need for enzyme isolation and potentially provides self-regenerating catalyst systems. Single species or mixed microbe consortia can be used, depending on the available "fuel" species, providing routes for partial or complete oxidation of multiple fuels and adaptation according to changing fuel availability. However, extracting the electrons released upon fuel oxidation is often more difficult at microbe electrodes, as discussed earlier, and microbial fuel cell systems are more likely to find application in specialized situations such as waste water treatment or cleanup of food waste rather than high-power generation. Some interesting examples of "fuels" for microbial fuel cells include canteen food scraps,[63] distillery water from bioethanol production[64] and nutrients in sea floor sediments.[65]

6.5 Summary

There are now many examples of efficient electron transfer between alcohol-oxidizing enzymes and electrodes, paving the way for application of these biocatalysts in energy devices. Isolated enzymes offer the possibility to selectively oxidize alcohol functionalities in a wide range of fuel molecules, from simple aliphatic alcohols up to complex sugars. Enzyme-catalyzed oxidation of fuels is usually incomplete, although complete oxidation to CO_2 can be achieved *via* multiple two-electron steps using enzyme cascades. In some instances, partial fuel oxidation may actually be an advantage, yielding higher-value chemicals alongside energy production from waste materials. The relatively large size of enzymes and the limited operating conditions available for these biomolecules (in terms of pH, temperature, pressure and solvent conditions) means that enzyme electrodes are probably unlikely to become widely used for high-power generation, but they may find roles in oxidation of niche fuels such as sugars in blood, fruit juice or milk. Increasing understanding of molecular pathways for direct electron transfer across the outer cell membranes of certain microbes such as *Shewanella* suggests that harnessing efficient microbial electrocatalysis should also be possible, allowing for self-regenerating electrode systems operating on a wider range of waste-derived fuels.

Acknowledgements

K.A.V. is grateful for Fellowship support from the Royal Society and the RCUK, and to the ERC for Starting Grant 258600. Protein structure images were prepared using the PyMOL Molecular Graphics System, version 1.3, Schrödinger LLC.

References

1. http://www.sony.net/SonyInfo/News/Press/200708/07-074E/index.html (last accessed April 2011).
2. F. G. Mutti, J. Sattler, K. Tauber and W. Kroutil, *ChemCatChem*, 2011, **3**, 109.

3. E. Hornung, M. Walther, H. Kuhn and I. Feussner, *Proc. Natl. Acad. Sci. U.S.A.*, 1999, **96**, 4192.
4. W. Kroutil, H. Mang, K. Edegger and K. Faber, *Adv. Synth. Catal.*, 2004, **346**, 125.
5. O. Yehezkeli, S. Raichlin, R. Tel-Vered, E. Kesselman, D. Danino and I. Willner, *J. Phys. Chem. Lett.*, 2010, **1**, 2816.
6. E. Katz, L. Sheeney-Haj-Ichia and I. Willner, *Angew. Chem. Int. Ed.*, 2004, **43**, 3292.
7. A. Heller and B. Feldman, *Chem. Rev.*, 2008, **108**, 2482.
8. A. Zebda, C. Gondran, A. Le Goff, M. Holzinger, P. Cinquin and S. Cosnier, *Nat. Commun.*, 2011, **2**, 370.
9. C. Prust, M. Hoffmeister, H. Liesegang, A. Wiezer, W. F. Fricke, A. Ehrenreich, G. Gottschalk and U. Deppenmeier, *Nat. Biotechnol.*, 2005, **23**, 195.
10. J. Tkac, J. Svitel, I. Vostiar, M. Navratil and P. Gemeiner, *Bioelectrochemistry*, 2009, **76**, 53.
11. A. Oubrie, H. J. Rozeboom, K. H. Kalk, E. G. Huizinga and B. W. Dijkstra, *J. Biol. Chem.*, 2002, **277**, 3727.
12. B. M. Hallberg, G. Henriksson, G. Pettersson and C. Divne, *J. Mol. Biol.*, 2002, **315**, 421.
13. B. M. Hallberg, T. Bergfors, K. Backbro, G. Pettersson, G. Henriksson and C. Divne, *Struct. Fold. Des.*, 2000, **8**, 79.
14. V. Soukharev, N. Mano and A. Heller, *J. Am. Chem. Soc.*, 2004, **126**, 8368.
15. C. F. Blanford, R. S. Heath and F. A. Armstrong, *Chem. Commun.*, 2007, 1710.
16. L. dos Santos, V. Climent, C. F. Blanford and F. A. Armstrong, *Phys. Chem. Chem. Phys.*, 2010, **12**, 13962.
17. T. Bertrand, C. Jolivalt, P. Briozzo, E. Caminade, N. Joly, C. Madzak and C. Mougin, *Biochemistry*, 2002, **41**, 7325.
18. G. T. R. Palmore, H. Bertschy, S. H. Bergens and G. M. Whitesides, *J. Electroanal. Chem.*, 1998, **443**, 155.
19. H. Sakai, T. Nakagawa, Y. Tokita, T. Hatazawa, T. Ikeda, S. Tsujimura and K. Kano, *Energy Environ. Sci.*, 2009, **2**, 133.
20. I. Katakis and E. Domínguez, *Mikrochim. Acta*, 1997, **126**, 11.
21. A. A. Karyakin, Y. N. Ivanova and E. E. Karyakina, *Electrochem. Commun.*, 2003, **5**, 677.
22. P. Kar, H. Wen, H. Li, S. D. Minteer and S. C. Barton, *J. Electrochem. Soc.*, 2011, **158**, B580.
23. C. D. Barker, T. Reda and J. Hirst, *Biochemistry*, 2007, **46**, 3454.
24. L. Lauterbach, Z. Idris, K. A. Vincent and O. Lenz, *PLoS ONE*, 2011, **6**, e25939.
25. T. Miyake, M. Oike, S. Yoshino, Y. Yatagawa, K. Haneda, H. Kaji and M. Nishizawa, *Chem. Phys. Lett.*, 2009, **480**, 123.
26. A. Heller, *Curr. Opin. Chem. Biol.*, 2006, **10**, 664.
27. N. Mano, V. Soukharev and A. Heller, *J. Phys. Chem. B*, 2006, **110**, 11180.

28. N. Mano, F. Mao and A. Heller, *J. Am. Chem. Soc.*, 2003, **125**, 6588.
29. S. Rengaraj, P. Kavanagh and D. Leech, *Biosens. Bioelectron.*, 2011, **30**, 294.
30. F. Tasca, L. Gorton, W. Harreither, D. Haltrich, R. Ludwig and G. Nöll, *J. Phys. Chem. C*, 2008, **112**, 13668.
31. N. Mano, F. Mao and A. Heller, *Chem. Commun.*, 2004, 2116..
32. S. C. Barton, Y. H. Sun, B. Chandra, S. White and J. Hone, *Electrochem. Solid State Lett.*, 2007, **10**, B96.
33. A. Heller, *Anal. Bioanal. Chem.*, 2006, **385**, 469.
34. B. Feldman, R. Brazg, S. Schwartz and R. Weinstein, *Diabetes Technol. Ther.*, 2003, **5**, 769.
35. I. Willner, Y. M. Yan, B. Willner and R. Tel-Vered, *Fuel Cells*, 2009, **9**, 7.
36. E. Katz, O. Lioubashevski and I. Willner, *J. Am. Chem. Soc.*, 2005, **127**, 3979.
37. M. Zayats, E. Katz, R. Baron and I. Willner, *J. Am. Chem. Soc.*, 2005, **127**, 12400.
38. A. Ramanavicius and A. Ramanaviciene, *Fuel Cells*, 2009, **9**, 25.
39. M. Ameyama, E. Shinagawa, K. Matsushita and O. Adachi, *J. Bacteriol.*, 1981, **145**, 814.
40. Y. Kamitaka, S. Tsujimura, N. Setoyama, T. Kajino and K. Kano, *Phys. Chem. Chem. Phys.*, 2007, **9**, 1793.
41. S. Tsujimura, A. Nishina, Y. Kamitaka and K. Kano, *Anal. Chem.*, 2009, **81**, 9383.
42. Y. Kamitaka, S. Tsujimura and K. Kano, *Chem. Lett.*, 2007, **36**, 218.
43. J. Okuda and K. Sode, *Biochem. Biophys. Res. Commun.*, 2004, **314**, 793.
44. F. Tasca, L. Gorton, W. Harreither, D. Haltrich, R. Ludwig and G. Noell, *Anal. Chem.*, 2009, **81**, 2791.
45. V. Flexer, F. Durand, S. Tsujimura and N. Mano, *Anal. Chem.*, 2011, **83**, 5721.
46. P. A. Ash and K. A. Vincent, *Chem. Commun.*, 2012, **48**, 1400.
47. Ł. Krzemiński, L. Ndamba, G. W. Canters, T. J. Aartsma, S. D. Evans and L. J. C. Jeuken, *J. Am. Chem. Soc.*, 2011, **133**, 15085.
48. S. J. Marritt, G. L. Kemp, L. Xiaoe, J. R. Durrant, M. R. Cheesman and J. N. Butt, *J. Am. Chem. Soc.*, 2008, **130**, 8588.
49. D. Millo, P. Hildebrandt, M.-E. Pandelia, W. Lubitz and I. Zebger, *Angew. Chem. Int. Ed.*, 2011, **50**, 2632
50. M. H. Osman, A. A. Shah and F. C. Walsh, *Biosens. Bioelectron.*, 2010, **26**, 953.
51. K. Rabaey and W. Verstraete, *Trends Biotechnol.*, 2005, **23**, 291.
52. T. A. Clarke, M. J. Edwards, A. J. Gates, A. Hall, G. F. White, J. Bradley, C. L. Reardon, L. Shi, A. S. Beliaev, M. J. Marshall, Z. Wang, N. J. Watmough, J. K. Fredrickson, J. M. Zachara, J. N. Butt and D. J. Richardson, *Proc. Natl. Acad. Sci. U.S.A.*, 2011, **108**, 9384.
53. R. S. Hartshorne, B. N. Jepson, T. A. Clarke, S. J. Field, J. Fredrickson, J. Zachara, L. Shi, J. N. Butt and D. J. Richardson, *J. Biol. Inorg. Chem.*, 2007, **12**, 1083.

54. A. Okamoto, R. Nakamura and K. Hashimoto, *Electrochim. Acta*, 2011, **56**, 5526.
55. P. K. Addo, R. L. Arechederra and S. D. Minteer, *Electroanalysis*, 2010, **22**, 807.
56. D. Sokic-Lazic and S. D. Minteer, *Biosens. Bioelectron.*, 2008, **24**, 939.
57. S. Aquino Neto, J. C. Forti, V. Zucolotto, P. Ciancaglini and A. R. de Andrade, *Biosens. Bioelectron.*, 2011, **26**, 2922.
58. T. Ikeda, D. Kobayashi, F. Matsushita, T. Sagara and K. Niki, *J. Electroanal. Chem.*, 1993, **361**, 221.
59. P. Cinquin, C. Gondran, F. Giroud, S. Mazabrard, A. Pellissier, F. Boucher, J.-P. Alcaraz, K. Gorgy, F. Lenouvel, S. Mathé, P. Porcu and S. Cosnier, *PLoS ONE*, 2010, **5**, e10476.
60. M. A. Dasari, P.-P. Kiatsimkul, W. R. Sutterlin and G. J. Suppes, *Appl. Catal., A*, 2005, **281**, 225.
61. R. L. Arechederra and S. D. Minteer, *Fuel Cells*, 2009, **9**, 63.
62. M. Zhou, M. Chi, J. Luo, H. He and T. Jin, *J. Power Sources*, 2011, **196**, 4427.
63. R. Kannaiah Goud and S. Venkata Mohan, *Int. J. Hydrogen Energy*, **36**, 13753.
64. P. T. Ha, T. K. Lee, B. E. Rittmann, J. Park and I. S. Chang, *Environ. Sci. Technol.*, 2012, **46**, 3022.
65. K. Scott, I. Cotlarciuc, I. Head, K. P. Katuri, D. Hall, J. B. Lakeman and D. Browning, *J. Chem. Technol. Biotechnol.*, 2008, **83**, 1244.

CHAPTER 7

Challenges and Perspectives of Nanocatalysts in Alcohol-Fuelled Direct Oxidation Fuel Cells

E. H. YU*[a], X. WANG[a], X. T. LIU[a] AND L. LI[b]

[a] School of Chemical Engineering and Advanced Materials, Newcastle University, Merz Court, Newcastle upon Tyne, NE1 7RU, UK; [b] School of Chemistry and Chemical Technology, Jiao Tong University, Shanghai, 200240, China
*E-mail: eileen.yu@ncl.ac.uk

7.1 Challenges with Current Direct Oxidation Fuel Cell Catalysts

In fuel cells, Pt is still the most commonly used catalyst for both anode oxidation and cathode oxygen reduction reactions. For the anode oxidation reaction, the poisoning intermediate species (*e.g.* CO) from incomplete oxidation of alcohols still poses a great challenge for Pt-based catalysts. At the cathode, poor kinetics and selectivity of Pt-based catalysts cause significant performance loss of the fuel cell. Furthermore, catalysts play a major role in fuel cell durability. Carbon corrosion, Pt dissolution and growth, Ru dissolution and de-alloying of binary and ternary catalysts can result in shortening the life of fuel cells.

RSC Energy and Environment Series No. 6
Catalysts for Alcohol-Fuelled Direct Oxidation Fuel Cells
Edited by Zhen-Xing Liang and Tim S. Zhao

Published by the Royal Society of Chemistry, www.rsc.org

7.1.1 CO Poisoning

Taking methanol as example, the electro-oxidation of methanol to carbon dioxide requires the transfer of six electrons, but it is highly unlikely that these electrons will transfer simultaneously. Figure 7.1 shows the steps and the different pathways for the electro-oxidation of methanol.[1]

At the top left is methanol and at the bottom right is carbon dioxide. Any route through the compounds from top left to bottom right is possible, and all have the same result: three right steps and three down steps, producing carbon dioxide and six proton–electron pairs.[1] The basic mechanism for methanol oxidation can be summarized as:

- Electrosorption (and dehydrogenation) of methanol onto the Pt substrate to form an adsorbed carbon-containing intermediate.
- Addition of oxygen to the adsorbed carbon-containing intermediates to generate CO_2.

Only a few materials are able to adsorb and dehydrogenate methanol. In acidic electrolytes, only platinum-based electrocatalysts have the required activity and chemical stability. The dehydrogenation and oxidation of methanol is believed to take place through a sequence of steps:

Adsorption and dehydrogenation:

$$CH_3OH + Pt(s) \rightarrow PtCH_2OH + H^+ + e^- \tag{7.1}$$

$$PtCH_2OH + Pt(s) \rightarrow Pt_2CHOH + H^+ + e^- \tag{7.2}$$

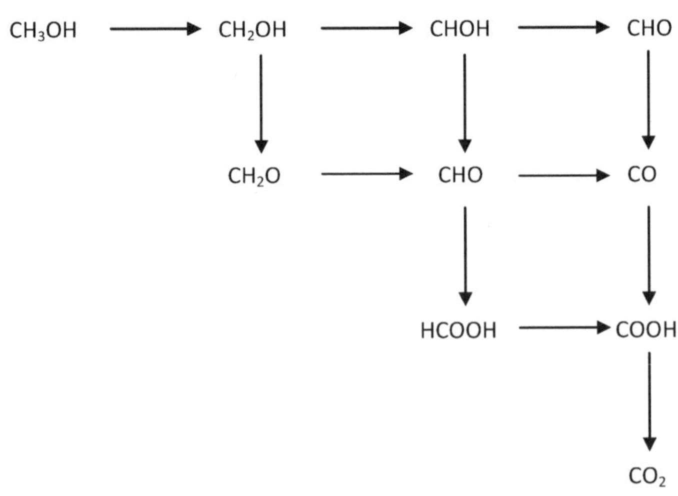

Figure 7.1 Pathways in the oxidation of methanol at a direct methanol fuel cell (DMFC) anode.

$$Pt_2CHOH + Pt(s) \rightarrow Pt_3COH + H^+ + e^- \tag{7.3}$$

$$Pt_3COH \rightarrow PtCO + 2Pt(s) + H^+ + e^- \tag{7.4}$$

Oxidation:

$$Pt(s) + H_2O \rightarrow PtOH + H^+ + e^- \tag{7.5}$$

$$PtOH + PtCO \rightarrow PtCOOH \tag{7.6}$$

or:

$$PtCO + H_2O \rightarrow PtCOOH + H^+ + e^- \tag{7.7}$$

$$PtCOOH \rightarrow Pt(s) + CO_2H^+ + e^- \tag{7.8}$$

Additional reactions that have also been suggested include:

$$PtCH_2OH \rightarrow Pt(s) + HCHO + H^+ + e^- \tag{7.9}$$

$$Pt_2CHOH + PtOH \rightarrow 3Pt(s) + HCOOH + H^+ + e^- \tag{7.10}$$

$$Pt_2CHOH + H_2O \rightarrow 2Pt(s) + HCOOH + 2H^+ + e^- \tag{7.11}$$

$$Pt_3COH + PtOH \rightarrow 3Pt(s) + PtCOOH + H^+ + e^- \tag{7.12}$$

$$Pt_3COH + H_2O \rightarrow 2Pt(s) + PtCOOH + 2H^+ + 2e^- \tag{7.13}$$

The series of reactions in eqns (7.7)–(7.10) can be demonstrated in detail, as shown in Figure 7.2.[2]

At potentials below around 450 mV *vs.* RHE (reversible hydrogen electrode) the surface of platinum is poisoned by a layer of strongly bonded CO_{ads} species and further chemisorption of methanol cannot proceed until the surface-bound CO_{ads} is oxidized from the platinum surface. At potentials below 450 mV, this process occurs at a slow rate and hence the platinum surface remains poisoned throughout its useful potential range. In alkaline media, the CO poisoning effect is less severe since the binding of CO_{ads} is not as strong as in acidic media

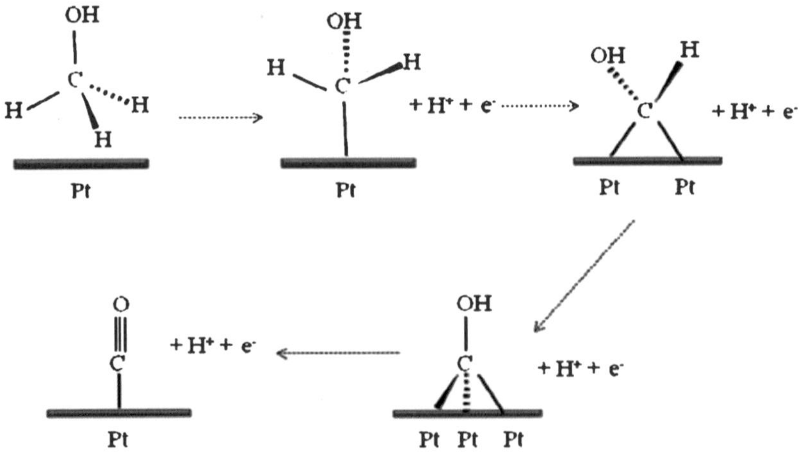

Figure 7.2 Electrosorption of methanol in a sulfuric acid electrolyte.[2]

due to the adsorption of OH.[3] However, the CO poisoning on the Pt catalyst still exists in an alkaline environment. This has resulted in an intensive search for alternative materials which can oxidize methanol at lower potentials, and in particular for additional secondary materials that might combine with platinum to promote the oxidation process.[2]

7.1.2 Oxygen Reduction Catalysts

For direct oxidation fuel cells (DOFCs), the oxygen reduction reaction (ORR) on the cathode is still the main reaction for receiving electrons. With the cathode in DOFCs, the fundamental challenges are the sluggish ORR kinetics and fuel crossover. The sluggish ORR kinetics require high loading of Pt in the cathode for performance with acceptable power output. Fuel crossover from the anode can be oxidized due to high potential on the cathode. This not only decreases fuel efficiency, but also causes a mixed potential which largely reduces the cathode performance. Therefore, the development of alcohol-tolerant cathode catalysts with high selectivity towards the ORR and unaffected by fuel crossover, and that promote a four-electron transfer process to avoid the H_2O_2 producing a two-electron pathway for the ORR, is highly desired for the implementation in DOFCs.

7.1.3 Carbon Corrosion

Carbon support corrosion is a major cause of the degradation of catalysts. In general, carbon corrosion is slow and negligible at potentials lower than 1.1 V *vs.* RHE. Nevertheless, the presence of catalysts such as Pt is believed to reduce the potential of carbon oxidation and accelerate carbon corrosion. At elevated temperatures (50–90 °C) and low pH (<1), carbon atoms are able to react with oxygen atoms and/or water to generate gaseous products such as CO and CO_2, which leave the cell with oxygen and the vapour stream.[4] Furthermore, during the start-up and shutdown of a fuel cell, local cathode potentials can reach as high as 1.5 V, which significantly accelerate the carbon corrosion.[4] Cai *et al.*

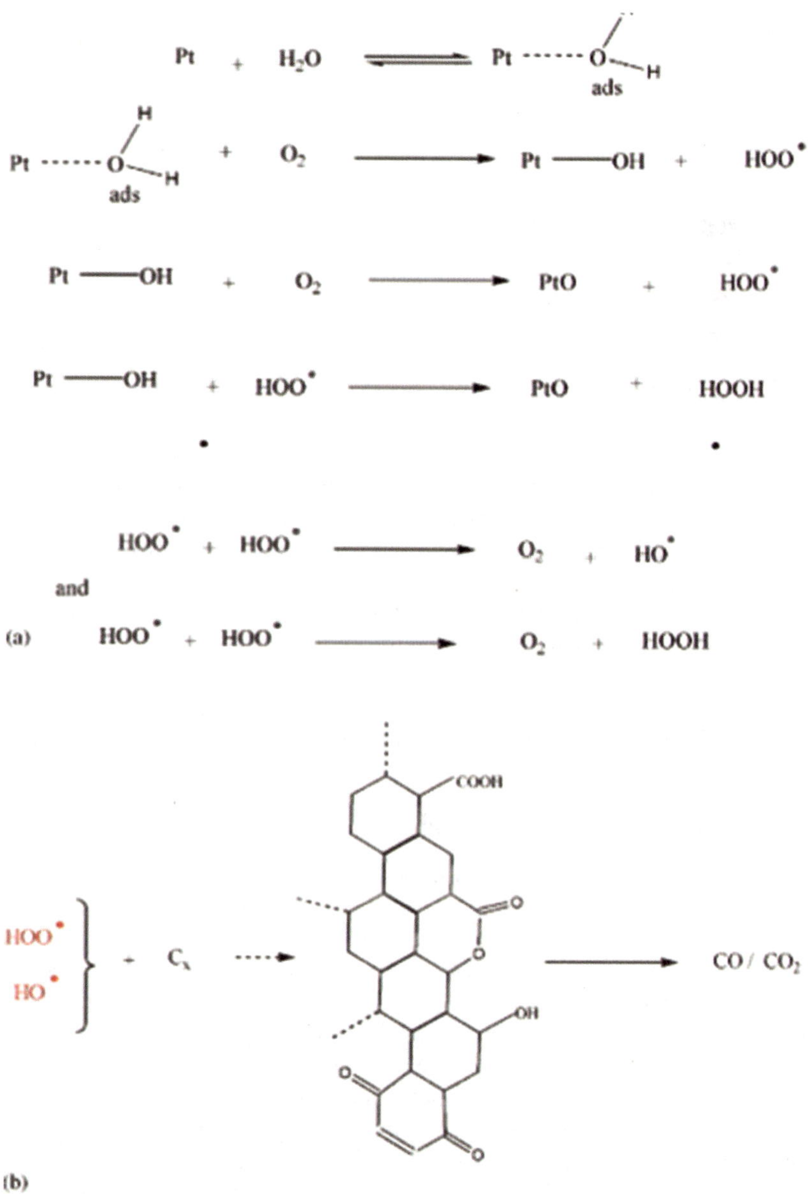

Figure 7.3 (a) Formation of radicals by reaction of Pt, O_2 and H_2O. (b) Carbon corrosion in the presence of Pt, O_2 and H_2O. (Reproduced from Cai *et al.*[5] with permission from Elsevier.)

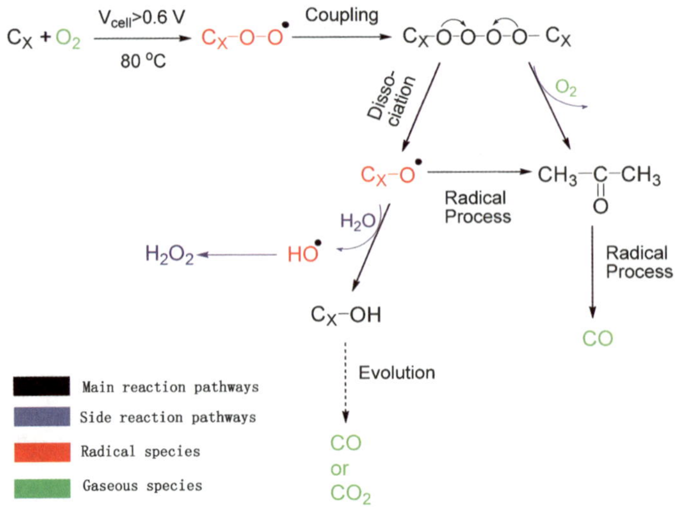

Figure 7.4 Schematic diagram of carbon corrosion in DOFCs in the absence of Pt.[5]

studied the effect of oxygen and water concentration on two commercial catalysts dispersed on carbon supports with different Brunauer–Emmett–Teller (BET) areas.[5] They suggested different carbon corrosion mechanisms with and without Pt. In the presence of Pt, Figure 7.3(a) shows that Pt can react with oxygen and water to form highly reactive OH and OOH radicals. These radicals can attack carbon, as shown in Figure 7.3(b), to form surface oxygen groups, which can then decompose at high temperatures to form gaseous CO and CO_2.

In the absence of Pt, because carbon has unpaired electrons, oxygen can react with carbon through the formation of radicals, as shown in Figure 7.4.[5]

7.1.4 Platinum Dissolution and Growth

Most commercial catalysts are carbon-supported nanoparticles of Pt and/or Pt alloys. They are widely used for both anodes and cathodes in DOFCs. Catalysts with particle size in the nanometre range possess a large electrochemical surface area (ECSA) and also have thermodynamic instability due to a high specific surface free energy. This could result in the growth of Pt particles. The possible mechanisms for Pt nanoparticle growth include local coalescence of agglomerated particles, which is the agglomeration of nonadjacent crystallites *via* Pt particle migration and subsequent ripening, and dissolution of the catalyst and subsequent reprecipitation of platinum.[6] Consequently, interrelated Pt particle dissolution and growth are observed and lead to the loss of ECSA. The dissolved Pt can then either deposit on adjacent platinum particles to form larger particles or alternatively diffuse into electrochemically inaccessible portions of the membrane–electrode assembly (*i.e.* sites not fulfilling the requirements of three-phase boundary). The particle growth rates and mechanisms may change at different operating conditions (potential, cell voltage cycling conditions, current density, state of hydration of the membrane, temperature). Ferreira *et al* classified coarsened Pt particles into two groups: spherical particles still in contact with the carbon support and nonspherical particles removed from the carbon support.[7] The former are from electrochemical Ostwald ripening, and the latter result from deposition in the ionomer by dissolved hydrogen. Both processes require preceding dissolution and growth of platinum particles.

7.1.5 Ruthenium Dissolution

Platinum–ruthenium catalysts are widely used as anode materials in proton exchange membrane fuel cells (PEMFCs) and DOFCs. Ruthenium dissolution from the PtRu anode catalyst at potentials higher than 0.5 V *vs.* RHE, followed by migration and deposition to the Pt cathode, can cause a decrease of activities of both anode and cathode catalysts, leading to deteriorated fuel cell performance. As reported by Holstein and Rosenfeld,[8] Ru loss from DOFC anodes can occur if the fuel cell is operated in conditions that result in ruthenium oxidation and reduction. High anode potentials might also occur when the fuel cell is shut down, owing to slow oxygen diffusion from the cathode under open circuit conditions. A return to normal operating conditions with a high methanol flow causes the catalyst to return to the reduced state. Both deliberate cycling of the anode to high potentials and repeated cycling between on and off states could eventually lead to significant ruthenium loss from the anode and a slow degradation of fuel cell performance.

7.2 Approaches to Catalyst Performance Enhancement

As the price of gold and platinum hits a record high on the market, catalyst development for DOFCs will be more and more focused on low or non-noble metal catalysts. Although promising fuel cell performances have been achieved from non-Pt catalysts, particularly Pd-based catalysts, further improvement in catalytic activity, catalyst stability and catalytic selectivity (especially for cathode catalyst) still remain as the major challenges.

7.2.1 Development of Composite Catalysts with Noble Metals

Even though the cost of noble metal catalysts is high and the effort in finding alternative nonprecious metal catalysts is underway, it is difficult to find any other materials to match the activity of the noble metals. Composite catalysts containing more than one catalyst composition can offer a way of tailoring catalyst properties for the desired applications. It provides solutions for the challenges discussed in previous sections. Further study on developing new catalysts and understanding the interactions between each composite on the catalytic activity is essential.

Multi-metal composite catalysts usually show improved catalytic activity compared to single metal catalysts. Alloys are normally formed in these catalysts, and the local structures of the alloyed catalyst nanoparticles are found to be critical for the catalyst properties, which are not well understood.[9] Another new method attracting much interest is to form a new core–shell catalyst structure.

7.2.1.1 Composite Catalyst Alloys with Multi-metal Compositions and their Functions

Compared to Pt alone for alcohol oxidation, the high catalytic activity of these binary and ternary catalysts was attributed to the bifunctional effect (promoted mechanism)[10] and to the electronic interaction between Pt and the alloyed metals (intrinsic mechanism).[10,11]

As mentioned in previous sections, a PtRu catalyst is the most common and effective composite catalyst used in DOFCs, particularly for methanol and organic fuels. Ru dissolved in a Pt matrix reduces the degree of anode CO poisoning because of the activity of Ru on water dehydrogenation. $Ru–OH_{ads}$ formed during water dehydrogenation reacts with $Pt–CO_{ads}$ formed from fuel oxidation, releases CO_2 and sets Pt free from CO adsorption.[12,13] As a result, the CO poisoning effect on the Pt catalyst is reduced. Tin has also shown a similar function to ruthenium.

Although PtRu and PtSn catalysts exhibit high activity, the majority of the oxidation products are still species containing a C–C bond, which will have an obviously negative effect on the fuel cell performance. Ternary catalysts made by adding a third element to modify the PtRu and PtSn catalysts present a higher specific activity of dehydrogenation and C–O and C–C bond cleavage during the alcohol oxidation process. Some ternary PtRuM (M = W, Mo, Ni) and PtSnM′ (M′ = Ni, Ru) catalysts with higher activity are reported in the literature.[14,15] In general, there are several methods to improve the stability of PtRu catalysts, including highly stable PtRu structures, stabilized PtRu by support effects or by using an additive. The different ways to improve the stability of PtRu catalysts are summarized in Table 7.1.

For alcohol oxidation, the combination of (alloying) elements such as Ru, Mo, Sn, Re, Os, Rh, Pb and Bi with Pt gives tolerance to the poisoning species compared to Pt alone.[16] Based on this bimetallic catalyst concept, further development on composite catalysts with binary, ternary and even quaternary compositions have been studied for DOFCs.[17,18] An example of the addition of Bi and Pd to Pt is shown in the study by Demarconnay *et al.*[19] The addition of Bi to Pt caused a positive shift of the onset potential for ethylene glycol (EG) oxidation by 70 mV and an increase of current density in alkaline media. The ternary catalyst PtPdBi/C did not change the EG oxidation onset potential compared to PtBi/C, but a higher current density was obtained.[19] In fuel cell

Table 7.1 Factors improving the stability of PtRu catalysts

Parameter	Type	Stability	Ref.
PtRu structure	Fcc PtRu alloys	Fcc PtRu alloy > Pt–RuO_xH_y	16
	Annealed RuO_xH_y	Annealed RuO_xH_y > untreated RuO_xH_y	17
	$HRuO_{ns}$–Pt and RuO_{2ns}–Pt/C	High stability of HRuO and RuO_2 nanosheets. RuO_{2ns}–Pt/C > Pt/C	18
Support	Carbon	PtRu/C > PtRu	19
	1-AP-MWCNTs[a]	PtRu/AP-MWCNTs > PtRu/MWCNTs ≈ PtRu/C	20
	CMM[b]	High stability	21
Additive	SiO_2	PtRu–SiO_2/C > PtRu/C	22
	TiO_x	PtRu–TiO_x/C > PtRu/C	23
	$FePO_4$	$FePO_4$ prevents the diffusion of dissolved Ru species into the electrolyte	24
	Au	Au/PtRu > PtRu	25

[a]1-AP-MWCNTs: 1-aminopyrene-functionalized multi-walled carbon nanotubes. [b]CMM: carbon mesoporous materials.

tests, the open circuit voltage (OCV) increased from 0.66 to 0.83 and 0.81 V and the maximum power density was raised from 19 to 22 and 28 mW cm^{-2} with Pt, PtBi and PtPdBi catalysts, respectively. Analyzed by *in situ* FTIR, a dilution effect was observed, *i.e.* the ability of Pt to break the C–C bond was decreased due to the addition of foreign atoms to Pt. However, PtPdBi alloy seemed to activate the oxidation of EG in the non-poisoning path to give oxalate.[20] Demarconnay *et al.* proposed that Bi favoured the adsorption of OH species and also affected the product distribution by changing the composition of the chemisorbed species, whereas Pd limits the poisoning of Pt sites by changing the composition of the chemisorbed species.[19]

Other PtM (M = Au, Pd, Cr, Co, Fe, Ni, V) alloys also have been investigated in order to improve the durability and activity of the catalysts. Zhang *et al.* reported that Pt nanoparticles modified by Au clusters showed negligible loss of activity after 30 000 potential cycles between 0.7 and 1.1 V *vs.* RHE.[21] The high stability for the AuPt/C catalysts was attributed to the stabilization of the Au clusters, which raised the Pt oxidation potential.

Fuel crossover to the cathode can cause "mixed potentials", which is a major problem hindering further improvement of fuel cell performance. The cathode ORR activity is reduced due to this effect. Apart from investigating novel membranes to decrease the permeability of fuels, investigating an ORR catalyst with improved activity and selectivity can be another way to deal with the challenge. Highly selective catalysts for the ORR have been prepared by several groups. Some alcohol-tolerant catalysts, including transition metal macrocycles, Ru-based chalcogenides, Pt-based alloys such as PtM/C (M = Fe, Ni, Cr, Co, Cu, Pd, Rh) and $PtMO_x$/C (MO_x = CuO, TiO_x) showed alcohol tolerance while retaining catalytic activity for oxygen reduction.[22–26] All factors such as the different Pt–M or Pt–MO_x composition, the degree of alloying, the morphology and the preparation method and particle size have important effects on the catalytic activity of these catalysts. Shukla *et al.* prepared PtPd/C catalysts by the wet-chemical method and found that the prepared PtPd/C catalysts exhibited much higher ORR activity than Pt/C in the presence of methanol.[27] Yu *et al.* synthesized a Pt/$CoSe_2$ nano-belt catalyst by *in situ* loading of Pt nanoparticles on a $CoSe_2$/diethylenetriamine (DETA) nano-belt through a polyol reduction approach.[28] This Pt/$CoSe_2$ catalyst displayed high ORR catalytic activity in acidic media and was highly resistant to methanol, even at concentrations up to 5 M. The improved activity seems related to the structural change because of the formation of alloys. For example, PtFe/C alloy crystallizing in an ordered face-centred tetragonal crystal structure with higher proportions of active platinum sites and a completely different nearest neighbour environment than Pt/C showed high oxygen-reduction activity in the presence of methanol.[23,29,38] PtNi is one of the most active catalysts for the ORR, due to a shift in the d-band centre of the surface Pt atoms and causing a weakened interaction between Pt and intermediate oxide species, thus freeing more active sites for O_2 adsorption.[30]

As a metal of the platinum group, Pd has similar chemical properties to Pt but only costs one-fourth of Pt. Pd has been studied as an alternative catalyst for use in fuel cells. It is an excellent electrocatalyst for the oxidation of primary alcohols in alkaline media and demonstrated good activity in several studies on propan-1-ol,[31] propan-2-ol,[31–33] ethanol,[34,35] ethylene glycol and glycerol[36] oxidation. However, the activity of Pd is still not comparable to Pt. Pd/C electrocatalysts alone have low activity and very poor stability for alcohol electro-oxidation.[32]

Pt- or Pd-based multi-metallic composite catalysts have shown enhanced catalytic activity compared to a single Pt or Pd catalyst. Pd binary catalysts with the addition of Ag, Au or Ru showed improved activity compared to a Pd catalyst alone. Pd/C promoted by Au, with the optimum ratio of Pd:Au = 4:1, demonstrated a significant increase of catalytic activity and stability for propan-2-ol oxidation.[37] Bunazawa and Yamazaki also prepared a PdAu/C catalyst for methanol fuel cells using ultrasonic synthesis and found that ethylene glycol was the optimum solvent, resulting in highly dispersed nanoparticles on a carbon support.[38] Lamy *et al.*[39] also used carbon-supported AuPd for a fuel cell with an anion exchange membrane and using glycerol as the fuel. The performance with AuPd was lower than using Au or Pd alone, and the highest power output was obtained with Pd/C catalysts. An enhancement was also observed from PdAg/C catalyst towards ethanol electro-oxidation in alkaline media.[40] The catalyst was prepared by a co-reduction method and PdAg alloy nanoparticles were formed. The PdAg/C catalyst exhibited high activity, enhanced adsorbed CO tolerance and stability for ethanol oxidation. Addition of Ru has the same enhancement on a Pd catalyst for ethanol oxidation.[41] With a PdRu catalyst, the ethanol oxidation current density was four times that from PtRu within the potential range 0.3–0.4 V.[41] Recently, Yi *et al.* studied ethanol oxidation with a PdRu binary catalyst supported on a carbonized TiO_2 nanotube in alkaline solution.[42] Introducing Ru to Pd enhanced the activity, and the $Pd_{87}Ru_{13}$ catalyst displayed the greatest activity. The onset potential for ethanol oxidation was −0.765 V *vs.* SCE with $Pd_{87}Ru_{13}$.[42] An *et al.* used PdNi as the anode catalyst for a direct EG fuel cell with a Tokuyama anion-exchange membrane and a non-Pt cathode catalyst.[43] A power density of 67 mW cm^{-2} was achieved, which is the highest reported for EG fuel cells.

Ternary and quaternary Pd catalysts were studied for ethanol oxidation in alkaline media by Bambagioni *et al.*[44] Pd was spontaneously deposited on Vulcan XC-72 supported NiZn and NiZnPd alloys using Pd^{IV} salts. The Pd(NiZn)/C and Pd(NiZnPd)/C catalysts exhibited promising activity for ethanol oxidation. The onset potential was −0.6 V *vs.* Ag/AgCl/KCl_{sat}, *i.e.* −0.4 V *vs.* NHE, and a specific current of 3600 A $(g\ Pd)^{-1}$ was obtained on both catalysts. The higher activity and stability of the catalysts were due to interaction with oxygen atoms from Ni–O moieties. Shen *et al.* reported that the ethanol oxidation activity of ternary PdIrNi/C is significantly higher than Pd/C and binary PdNi/C catalysts.[45] PdIrNi/C is more stable than binary PdNi/C due to the high efficiency of removing CO.

7.2.1.2 Composite Catalyst Alloys with Core–Shell Structures

Core–shell structure nanoparticle catalysts have become the prevalent research trend in the past few years, because they can significantly reduce the usage of the noble shell metal and retain similar or superior catalytic activity compared to bulk alloy catalysts. Alayoglu *et al.*[46] reported that the Pt-modified Ru core catalyst has significantly better catalytic activity when used as an anode catalyst for DOFCs than conventional PtRu nano-alloys, monometallic mixtures of nanoparticles or pure Pt particles. Crabb *et al.*[47] and Rose *et al.*[48] found Ru-modified Pt/C core–shell catalysts have a similar activity towards carbon monoxide oxidation to PtRu alloy catalysts. In addition, they proved that a bulk alloy phase is not an essential structure to improve CO tolerance of PtRu/C catalysts over that of Pt/C. This indicates that a core–shell structure catalyst has competitive catalytic activity when it is used as an anode electrode in hydrogen and direct alcohol fuel cell systems. Researchers have found that PtAu,[43,49] PtCu[50] and PtNi[51] core–shell structure catalysts exhibit good ORR activity as cathode electrode materials. A monolayer Pt on an Au core catalyst prepared Ma *et al.*[43] exhibited high activity towards the ORR; the total Pt mass-specific activity of the catalyst was 3.1–4.9 times that of a commercial Pt/C catalyst in electrochemical tests and 4.1 times that of a commercial Pt/C catalyst in single-cell test.

Core–shell structure particles can either have one or more monolayers of a second shell to cover the core metal completely to show their properties, or form a partial monolayer and create the same surface component as the bulk alloy does, and such core–shell structures can always form more intimate contact between the two metals.[48] Figure 7.5 shows a model of (a) a partial monolayer and (b) Pt on Au catalyst particles.

Hartl *et al.*[52] prepared a AuPt core–shell catalyst that only showed surface-area specific activity for the ORR equal to the polycrystalline Pt catalyst.

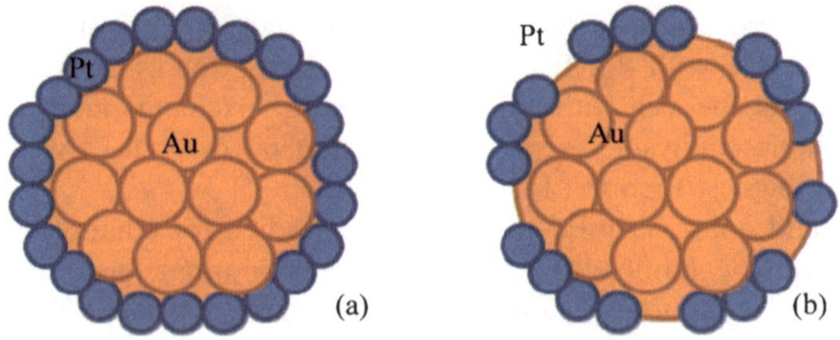

Figure 7.5 Pt on Au core–shell structure: (a) partial monolayer; (b) catalyst particles.

However, Ma *et al.*[43] indicated that cyclic voltammograms of their monolayer AuPt catalyst showed Au and Pt mixed features rather than complete Pt features. The clear evidence is that the properties of core–shell structure catalysts are also influenced by the core materials' catalytic and electronic properties, the so-called "ligand effect".[53–57] However, it is still not clear how the electronic properties of a Pt monolayer affect the whole catalyst system compared to pure Pt. One of the explanations might be the strain effect that alters the electronic structure of the metal and the strain effect gives a rise to changes in orbital overlap.[58–61] The d-orbital of each metal overlaps when a metal–metal bond forms and if there is a tensile strain force between the surface atoms, the d-orbital overlap is decreased leading to an increase in the d-band centre. This d-band centre shift affects the energy related to dissociative adsorption of adsorbates, particularly when the d-band centre increases; then surface atoms tend to strongly interact with the adsorbates, breaking the O–O bond.[59–63] Considering the bifunctional mechanism (dehydrogenation of methanol occurs on a Pt surface and a second metal, such as Ru and Sn, provides an –OH bond to weaken the CO–Pt bond, so reducing the overall methanol oxidation potential)[62] of anode Pt-based catalysts for direct alcohol fuel cells, researchers[47,48,63,64] have shown that core–shell nanoparticle bimetallic catalysts with less than a single monolayer shell metal coverage have the same surface structure and similar catalytic activities to the bulk alloy catalysts. EXAFS results[48] show such structures can create intimate contact between the core and shell materials, which benefits CO tolerance and alcohol dehydrogenation reactions.

The major drawback of the core–shell structure catalysts is the poor durability compared to the bulk alloy catalysts.[65,66] The shell metal dissolution and stripping off during the fuel cell operation particularly happens in anode catalysts of DOFCs. To overcome this problem, a carefully heat-treatment process is required to improve the catalyst stability but also retain high catalytic activities. In addition, it is hard to control the location of the shell metal deposition during the catalyst preparation of the core–shell catalysts. This will result in the accumulation of the shell material and leave more core metal surface exposed than is theoretical expected; this is called geometric effects. An example is that the order of activity for methanol oxidation was found to decrease in the order of Pt(100)Ru > Pt(110)Ru > Pt(111)Ru.[67,68] This indicates that the amount and control of the location of the shell metal will be the trend for core–shell structure catalyst research, as they dramatically influence the activity of the core–shell catalyst system.

Various preparation methods for core–shell structure catalysts have been published. Zhang *et al.*[57,69] and Vukmirovic *et al.*[70] reported a stable core–shell structure for the ORR consisting of Pt and another late transition metal deposited as a monolayer on the surface of Pt nanoparticles using an underpotential deposition (UPD) of the metal . Crabb *et al.*[47,63,64] used a novel controlled surface reaction to prepare Sn, Ru, Pd, Fe on Pt/C catalysts; they reported that the shell metals particles can be deposited on thee desired location of the Pt particles, and that the core:shell metal surface ratio can be carefully controlled. All catalysts show bimetallic properties and better catalytic activity for the use of PEMFCs. Recently, Pt on an Au/C catalyst was investigated and reported by several groups. Kristian *et al.* used a reduction strategy to synthesize a PtAu shell–core/C electrocatalyst that had a higher specific activity than the conventional Pt/C catalyst towards methanol oxidation.[71] However, Ma *et al.* showed their Pt on an Au catalyst prepared using a seed mediate method could not form a complete Pt monolayer on the Au particles. Use of suitable core and shell materials to obtain better activity with low cost, and novel preparation methods, are still being researched.

7.2.1.3 Improving Catalyst Activity with Metal Oxide Promoters

Apart from introducing multi-metals in the catalyst, the addition of metal oxides like CeO_2, NiO, Co_3O_4 and Mn_3O_4 also significantly promotes the activity and stability of Pt/C or Pd/C catalysts. Xu *et al.* investigated CeO_2-promoted Pt/C for alcohol oxidation in alkaline media[72] and also studied CeO_2 with Pt/C as the cathode ORR catalyst in PEM fuel cells.[73] Interestingly, the catalyst activities were both improved by adding CeO_2. For the oxidation, the poisoning resistance of the catalyst was improved due to the synergistic effect, while for the ORR reaction they believed that CeO_2 was first reduced in the cathode by an electrochemical reaction, and then was oxidized by the intermediate product H_2O_2 from the ORR to its original state. These processes promoted the reduced rate of oxygen and thus enhanced the performance of the fuel cell.[73] Later, they also studied the promoters for Pd/C catalysts for alcohol oxidation in alkaline media.[32,74] Pd–Co_3O_4 (2:1, w/w)/C showed the highest activity for the electro-oxidation of methanol, EG and glycerol, and the most active catalyst for the ethanol electro-oxidation was Pd–NiO (6:1, w/w)/C.[32]

Some metal oxides, in particular CeO_2, Mn_3O_4 and SnO_2, have multiple oxidation states and can store and release oxygen. It is believed that the formation of OH_{ads} species on metal oxides, due to the adsorption of OH^-, at lower potentials can transform CO^- or other carbonaceous species on the surface of Pt or Pd to CO_2, releasing active sites on Pt or Pd for further reaction.[75]

7.2.1.4 Non-precious Metal Composite Catalysts

Transition metal macrocycles, such as phthalocyanine and porphyrin, have been used as effective oxygen reduction catalysts since the work of Jasinski on metal phthalocyanines in the 1960s.[76] The active sites of these compounds are the four-nitrogen ring coupled with transition metals, commonly Fe or Co. These catalysts have shown highly selective catalytic activity for oxygen reduction. A novel preparation method of Fe-based non-noble catalysts showed comparable fuel cell performance to that using commercial Pt/C catalysts at a high cell voltage in PEMFCs.[77] Cobalt polypyrrole (PPy) catalysts, which are physically bonded to the carbon surface, were prepared and showed a four-electron transfer reaction towards the ORR in acidic media.[78] Bashyam *et al.* evaluated similar catalysts in PEMFCs and the short-term stability test (100 h) showed minimal degradation at 0.4 V and 80 °C.[79] Their high selectivity towards the ORR and low cost make these catalysts promising alternatives in DOFCs.

In order to improve the activity and stability of non-platinum catalysts, hybrid catalysts with cobalt phthalocyanine (CoPc) and fluorinated cobalt phthalocyanine (CoPcF16) on carbon-supported silver nanoparticles were prepared. Lower overpotential and higher kinetics were obtained compared to macrocycles or silver used alone. The activity of the ORR catalyst could be also tuned by adjusting the composition of the metal macrocycles.[80]

7.2.2 Novel Carbon Materials as Catalysts and Substrates

7.2.2.1 Novel Carbon Catalyst Substrates

Graphene has been attracting growing interest as a catalyst material due to its unique electronic properties, *e.g.* excellent two-dimensional electron conductivity. Seo *et al.* prepared Pt and Pd catalysts supported on graphene with a high metal loading (60 wt%).[81] The graphene-supported Pd (Pd/GNS) showed better catalytic activity than Pt/GNS. Co_3O_4 or graphene oxide alone is not catalytic for the ORR. However, Liang *et al.* found that the combination of the two materials showed significant activity towards the ORR and the oxygen evolution reaction (OER) in alkaline media.[82] They further explained that the unexpected activity is due to the chemical coupling effects between Co_3O_4 and graphene. Graphene is also used for the alcohol oxidation reaction as a catalyst support.[83] Further investigations on the mechanism effects of graphene as a catalyst support are desirable.

To address carbon corrosion, increasing efforts have been devoted to developing advanced support materials. Carbon materials with a high degree of graphitization, such as graphitized carbon black, carbon nanotubes and carbon nanofibres, and polypyrrole-modified carbon nanotubes,[84] as well as nanoscale graphite,[85] are believed to be more corrosion resistant than conventional carbon black.

7.2.2.2 N-Doped Carbon Nanotubes for the Oxygen Reduction Reaction

Carbon has activity for oxygen reduction, although the reaction is mainly a two-electron pathway from which hydrogen peroxide is produced. Recent reviews on N-doped carbon material showed the material had remarkable activity for the ORR in alkaline media.[86,97] The study from Gong *et al.* showed significantly enhanced ORR activity and stability from nitrogen-doped carbon nanotube (NCNT) arrays[87] in 0.1 M KOH. In their study, the ORR onset potential for NCNTs was comparable to a Pt/C catalyst; however, the activity of the NCNTs was not affected by the presence of 3 M methanol as well as H_2, glucose and formaldehyde, and the activity was not decayed after 100 000 cycles of continuous potentiodynamic sweeps. They attributed the high selectivity of the NCNTs on the ORR to a more positive ORR potential than the fuel oxidation potentials. These NCNTs even showed a four-electron pathway for the ORR. The effect of CO poisoning was also tested on NCNTs.[87] Unlike Pt, the non-metallic carbon catalyst was insensitive to 10% CO in oxygen.

Nitrogen heteroatoms in NCNTs play important roles in electrocatalytic activity for the ORR. There are mainly four types of nitrogen functional groups: pyridinic (N_1), pyrrolic (N_2), quaternary (N_3) and pyridine *N*-oxide (N_4).[88] The enhanced ORR activity may be from pyridinic or pyrrole/pyridine-type nitrogen. The importance of pyridinic nitrogen could lie in the availability of the extra lone pair of electrons on the nitrogen atom, which increases the electron density on the graphitic edge planes, commonly known as reactive sites.[89] The nitrogen groups can transform to more thermally stable structures from heat treatment.[90] However, a study by Nagaiah *et al.* showed that, after heat treatment, NCNTs contained the highest amount of quaternary N_3 groups, indicating the importance of these groups in enhancing activity.[88]

Nitrogen is also able to create defects on carbon, increasing the edge plane exposure and improving the catalytic activity.[91,92] It was demonstrated by Chen *et al.* that the structures and morphologies of NCNTs had a large influence on the ORR activity due to their impact on the surface defects. A higher degree of surface defect can enhance the activity. Their study showed that more defects formed on the NCNTs synthesized from ferrocene (Fc-NCNTs) than from iron phthalocyanine (FePc-NCNTs), which led to much higher ORR activity from the Fc-NCNTs.[93]

Various nitrogen groups and synthesis procedures have been used to produce N-doped nanocarbon materials. HNO_3,[94] NH_3 gas,[88,89,95] acetonitrile,[96] chemical vapour deposition[87,91] and carbonizing phthalocyanines coordinated to alkali metals or alkaline earth metals[97] have been used to introduce an N-group on carbon materials. Nitrogen-doped carbon materials have attracted lots of interest. However, the exact mechanisms of the reactivity of NCNTs are still not clear. There was a direct correlation observed between nitrogen content and the number of reducing sites on carbon.[89] The reductive character in high N-doped NCNTs resulted in a rise in pyridinic-type N-functionalities, which could enhance the chemical reactivity of the NCNTs.[86] Further research on such material to better understand the reaction mechanisms and improvement in catalytic activity are essential to make NCNTs an alternative cathode ORR catalyst for DOFCs.

7.2.3 Non-carbon-Based Catalyst Substrates

Catalyst substrates can significantly influence catalyst activity and stability. Novel carbon materials as catalyst substrates have been discussed in the last section. They have been extensively applied as catalyst substrates due to their favourable properties for PEM fuel cell electrodes. However, carbon corrosion, which occurs at potentials that are higher than 0.9 V *vs.* RHE,[98] has been found as a major issue concerning the catalyst durability, causing catalyst degradation and loss of activity.[99–104] This leads to a significant loss of the active catalyst area, thereby causing a decrease of the fuel cell performance.

Novel catalyst substrates ranging from tungsten carbide (WC),[105] TiO_2,[106–110] SnO_2,[101,111,112] WO_3,[113–116] GeO_2,[117–121] conductive ceramics of TiB_2,[122] TiN,[123] and nanocomposite substrates of WC and TiO_2[124] have all been investigated. The durability and high surface area of these substrates provide the catalysts and fuel cells with much improved stability and lifetime, and the catalyst loading can also be reduced.

7.2.3.1 Titanium Dioxide

TiO_2 has been reported to perform well as a catalyst substrate in PEMFC electrodes,[125] as it possesses good mechanical resistance and stability in acidic and oxidative environments.[126] Furthermore, TiO_2 may improve the Pt oxygen reduction activity through facilitating mechanisms such as reactant surface diffusion and oxygen spillover.[127,128] Huang *et al.*[108] reported a Pt/TiO_2 electrocatalyst showing comparable activity to that of a commercial Pt/C catalyst in fuel cell studies.

The TiO_2 catalyst showed nearly 10 times higher ORR activity (1.20 mA cm^{-2}) when compared to the Pt/C catalyst (0.13 mA cm^{-2}) in their accelerated durability test in half-cell conditions. The Pt/TiO_2 catalyst also showed only a small voltage loss (0.09 V at 0.8 A cm^{-2}), even after 4000 cycles. Meanwhile, the Pt/C catalyst showed no activity in fuel cell testing after 2000 potential cycles due to severe carbon corrosion, Pt dissolution and catalyst particle sintering. Pt accelerates carbon corrosion;[129,130] this results in a rapid degradation of the Pt catalyst and a significant reduction of the electrochemical surface area. This effect does not influence the TiO_2 as much as the carbon substrate. However, the relatively low conductivity of TiO_2 imposes a certain limitation on its catalytic performance.

7.2.3.2 Tungsten Carbide

Tungsten carbide (WC) is another widely studied catalyst substrate owing to its excellent properties of high hardness, high oxidation resistance, excellent thermal conductivity and electrical conductivity in many fields. Zhang *et al.*[131] found the Pt/WC catalyst possessed higher stability than the traditional Pt/C (Vulcan XC-72) catalyst during the oxidation tolerance test. Chhina *et al.*[132] measured the ORR activity of Pt supported on WC and carbon both before and after accelerated oxidation cycles between +0.6 V and +1.8 V *vs.* RHE. Tafel plots show that the ORR activity remained high even after accelerated oxidation tests for Pt/WC, while the ORR activity was extremely poor after accelerated oxidation tests for HiSpec 4000. Different to TiO_2, WC was found to have platinum-like behaviour for the chemisorption of hydrogen and oxygen;[133] it has been considered as an alternative electrocatalyst to Pt for hydrogen[134] or methanol fuel cells.[135] WC is a promising alternative electrocatalyst to replace noble metal catalysts for its low price and excellent capability of resistance to catalyst poisoning by hydrocarbons, H_2S or CO,[136] but its catalytic activity is much less than that of platinum metal.[137] A major drawback preventing WC from being widely implemented in the catalyst industry is the low surface area. Carbide preparation has been inherited from the metallurgical industry, where direct carburization of WO_3, ammonium paratungstate or ammonium metatungstate with graphitic carbon at high temperatures (1400–1600 °C) is performed, leading to low specific surface area carbides.[138,139]

7.2.3.3 Carbon, Titanium Dioxide and Tungsten Carbide Composites

Carbon, TiO_2 and WC composite materials have also been studied as catalyst substrates for their combination of advanced properties. Pt deposited on a C/TiO_2 composite showed an increased electrochemically active area of Pt deposited on carbon, and also showed an improvement of methanol tolerance.[110,116] Kraemer *et al.*[109] synthesized a TiO_2/C substrate which allowed Pt deposited on the TiO_2 rather than C. It showed high thermal stability, owing to reduced direct contact between Pt and C, and it performed comparable to conventional Pt/C materials when incorporated in a PEMFC cathode. A carbon on TiO_2 core–shell material made by Lee *et al.*[111] also showed improved catalytic activity and stability toward methanol electro-oxidation compared to Pt on carbon black (Vulcan XC-72R) catalysts.

Shao *et al.*[105] showed that a Pt on WC modified high surface area carbon substrate catalyst has slightly better stability and ORR activity than that of Pt/C. Meng and Shen[110,140–142] studied the effect of adding Pt onto tungsten carbides and observed a 10 times larger activity on Pt–W_2C/C than on Pt/C for the ORR in an alkali medium. Zhao *et al.*[143] synthesized Pt supported on WC-modified MWCNT catalysts, which showed better performance than Pt/C catalysts for methanol oxidation in terms of the onset potential and peak current density. WC and TiO_2 composite materials are rarely studied, and there are no clear advantages of such a substrate reported. As mentioned previously, WC has been considered as an alternative electrocatalyst to Pt for hydrogen[134] and methanol fuel cells, and WC-based composite materials will become more and more attractive in catalyst research for fuel cells.

7.3 Summary

Catalyst development is one of the key factors affecting further development of DOFCs. For PEMFCs, noble metal catalysts, especially Pt-based catalysts, will still be the main catalysts employed in systems despite the efforts of using non-noble metal catalysts because of the low activity from these catalysts. Composite catalysts with binary, ternary and quaternary components, by introducing other transition metals, would provide the solution for Pt-based catalysts, mostly reducing the CO poisoning effect for anode oxidation and improving the selectivity in cathode ORRs. Making alloys or newly developed core–shell structures will be the main approach to obtain the desired catalytic properties. Ru, Sn, Au and Pd, as well as metal oxides, could be the essential composition in these catalysts. To overcome the degradation of catalysts due to carbon corrosion from the catalyst substrates, novel substrates involving either novel carbon materials, TiO_2 or WC will become more and more important in catalyst preparation. In conclusion, low cost, high activity and durability are the keys for the further development of catalysts in DOFCs.

References

1. J. Larminie and A. Dicks, *Fuel Cell System Explained*, Wiley-Blackwell, Chichester, 2003.
2. M. P. Hogarth and G. A. Hards, *Platinum Met. Rev.*, 1996, **40**, 150.
3. E. H. Yu, U. Krewer and K. Scott, *Energies*, 2010, **3**, 1499.
4. X. Yu and S. Ye, *J. Power Sources*, 2007, **172**, 145.
5. M. Cai, M. S. Ruthkosky, B. Merzougui, S. Swathirajan, M. P. Balogh and S. H. Oh, *J. Power Sources*, 2006, **160**, 977.
6. R. Borup, J. Meyers, B. Pivovar, Y. S. Kim, R. Mukundan, N. Garland, D. Myers, M. Wilson, F. Garzon, D. Wood, P. Zelenay, K. More, K. Stroh, T. Zawodzinski, J. Boncella, J. E. McGrath, M. Inaba, K. Miyatake, M. Hori, K. Ota, Z. Ogumi, S. Miyata, A. Nishikata, Z. Siroma, Y. Uchimoto, K. Yasuda, K. I. Kimijima and N. Iwashita, *Chem. Rev.*, 2007, **107**, 3904.
7. P. Ferreira-Aparicio, M. A. Folgado and L. Daza, *J. Power Sources*, 2009, **192**, 57.
8. W. L. Holstein and H. D. Rosenfeld, *J. Phys. Chem. B*, 2005, **109**, 2176.
9. M. Garza Castanon, S. Velumani, O. V. Kharissova, M. A. Jimenez and A. M. Kannan, *Int. J. Energ Res.*, 2011, **35**, 594.

115. O. Savadogo and P. Beck, *J. Electrochem. Soc.*, 1996, **143**, 3842.
116. J. Shim, C. R. Lee, H. K. Lee, J. S. Lee and E. J. Cairns, *J. Power Sources*, 2001, **102**, 172.
117. C. L. Campos, C. Roldán, M. Aponte, Y. Ishikawa and C. R. Cabrera, *J. Electroanal. Chem.*, 2005, **581**, 206.
118. D. H. Lim, W. D. Lee, D. H. Choi, H. H. Kwon and H. I. Lee, *Electrochem. Commun.*, 2008, **10**, 592.
119. A. O. Neto, L. A. Farias, R. R. Dias, M. Brandalise, M. Linardi and E. V. Spinacé, *Electrochem. Commun.*, 2008, **10**, 1315.
120. J. Wang, X. Deng, J. Xi, L. Chen, W. Zhu and X. Qiu, *J. Power Sources*, 2007, **170**, 297.
121. C. Xu and P. K. Shen, *Chem. Commun.*, 2004, 2238..
122. S. Yin, S. Mu, M. Pan and Z. Fu, *J. Power Sources*, 2011, **196**, 7931.
123. K. Kakinuma, Y. Wakasugi, M. Uchida, T. Kamino, H. Uchida and M. Watanabe, *Electrochemistry*, 2011, **79**, 399.
124. Z. Yang, X. Zhou, H. Nie, Z. Yao and S. Huang, *ACS Appl. Mater. Interfaces*, 2011, **3**, 2601.
125. T. Ioroi, Z. Siroma, N. Fujiwara, S.-i. Yamazaki and K. Yasuda, *Electrochem. Commun.*, 2005, **7**, 183.
126. Z. Liu, J. Zhang, B. Han, J. Du, T. Mu, Y. Wang and Z. Sun, *Microporous Mesoporous Mater.*, 2005, **81**, 169.
127. A. Bauer, K. Lee, C. Song, Y. Xie, J. Zhang and R. Hui, *J. Power Sources*, 2010, **195**, 3105.
128. B. Hammer and J. K. Nørskov, in *Advances in Catalysis*, ed. H. K. Bruce and C. Gates, Academic, New York, 2000, vol. 45, p. 71.
129. P. J. Ferreira, G. J. la O', Y. Shao-Horn, D. Morgan, R. Makharia, S. Kocha and H. A. Gasteiger, *J. Electrochem. Soc.*, 2005, **152**, A2256.
130. Y. Shao, G. Yin and Y. Gao, *J. Power Sources*, 2007, **171**, 558.
131. S. Zhang, H. Zhu, H. Yu, J. Hou, B. Yi and P. Ming, *Chin. J. Catal.*, 2007, **28**, 109.
132. H. Chhina, S. Campbell and O. Kesler, *J. Power Sources*, 2008, **179**, 50.
133. H. Bohm, *Nature*, 1970, **227**, 483.
134. H. Böhm and F. A. Pohl, *Wiss. Ber. AEG-Telefunken*, 1968, **41**, 46.
135. D. Baresel, W. Gellert, J. Heidemeyer and P. Scharner, *Angew. Chem. Int. Ed. Engl.*, 1971, **10**, 194.
136. J. Lemaître, B. Vidick and B. Delmon, *J. Catal.*, 1986, **99**, 415.
137. R. B. Levy and M. Boudart, *Science*, 1973, **181**, 547.
138. R. Koc and S. K. Kodambaka, *J. Eur. Ceram. Soc.*, 2000, **20**, 1859.
139. C. Liang, F. Tian, Z. Li, Z. Feng, Z. Wei and C. Li, *Chem. Mater.*, 2003, **15**, 4846.
140. H. Meng and P. K. Shen, *Chem. Commun.*, 2005, 4408..
141. H. Meng and P. K. Shen, *J. Phys. Chem. B*, 2005, **109**, 22705.
142. H. Meng and P. K. Shen, *Electrochem. Commun.*, 2006, **8**, 588.
143. Z. Zhao, X. Fang, Y. Li, Y. Wang, P. K. Shen, F. Xie and X. Zhang, *Electrochem. Commun.*, 2009, **11**, 290.

Subject Index